Sustainable Finishing Techniques in Textiles

Santosh Biranje · Mohammad Shahid ·
Ravindra V. Adivarekar
Editors

Sustainable Finishing Techniques in Textiles

Springer

Editors
Santosh Biranje
Department of Fibres and Textile Processing Technology
Institute of Chemical Technology
Mumbai, Maharashtra, India

Ravindra V. Adivarekar
Department of Fibres and Textile Processing Technology
Institute of Chemical Technology
Mumbai, Maharashtra, India

Mohammad Shahid
Department of Chemistry
Koneru Lakshmaiah Education Foundation
Guntur, Andhra Pradesh, India

ISBN 978-981-96-4859-7 ISBN 978-981-96-4860-3 (eBook)
https://doi.org/10.1007/978-981-96-4860-3

This Springer imprint is published by the registered company Springer Nature Singapore Pte Ltd.
The registered company address is: 152 Beach Road, #21-01/04 Gateway East, Singapore 189721, Singapore

Preface

The textile industry is a major part of the global economy but has also faced increasing pressure to adopt more sustainable practices. One of the most resource-heavy stages in textile production is the finishing process, where chemicals are applied to improve the quality of fabrics. This book, *Sustainable Finishing Techniques in Textiles,* explores the latest methods in textile finishing that aim to reduce environmental impact while improving the performance and functionality of fabrics. As awareness of environmental issues grows, the textile industry is being asked to find more eco-friendly ways of working. Traditional textile finishing methods have been criticized for negatively affecting water resources, human health, and the environment. In response, the industry is shifting towards sustainable, biodegradable materials, bio-based chemicals, and new techniques supporting ecological goals and consumer demands. This book highlights these emerging techniques and offers practical solutions that can help make the industry more sustainable. The chapters in this book cover a variety of sustainable finishing methods. Chapter 1 introduces biodegradable and sustainable finishes, such as enzymatic treatments, nanotechnology, and plant-based materials, transforming textile production. Chapter 2 discusses the challenges of making finishes more biodegradable and sustainable, as well as the regulatory pressures pushing for alternative solutions. Other chapters explore bioactive finishes (Chap. 3), which give textiles extra features like antibacterial or UV protection, and wellness finishes (Chap. 4), which enhance health, hygiene, and comfort. The book also covers the use of biotechnology (Chaps. 5 and 6), such as microbial enzymes, to replace harmful chemicals while improving the efficiency of textile processing. Additionally, polymeric materials (Chap. 8) and plasma treatments (Chap. 9) are discussed in terms of offering ways to treat fabrics more environmentally friendly. The use of phase change materials (Chap. 10) in textiles helps regulate temperature, and Chap. 11 focuses on flexible electronics, which are revolutionizing wearable technology in textiles. Agro-textiles (Chap. 12) also play an important role in sustainable farming, showing how textiles can help improve agriculture. Finally, the book addresses methods for assessing sustainable finishes (Chap. 13) and provides an overview of the challenges and innovations shaping the future of textile finishing (Chap. 14). The focus is on reducing resource use, minimizing waste, and improving the life cycle of textile products, which are essential steps toward a more sustainable textile industry. This book provides an accessible overview of sustainable finishing techniques, highlighting the challenges and opportunities ahead. It is intended as a resource for researchers, industry professionals, and policymakers who want to learn about sustainable changes in textile production. By using green chemistry, biotechnology, and other innovative methods, the industry can create textiles that are better for the environment and meet consumers' growing needs. We hope this book inspires new ideas and encourages further collaboration to help make textile production more sustainable.

Mumbai, India — Dr. Santosh Biranje
Vaddeswaram, India — Dr. Mohammad Shahid
Mumbai, India — Prof. Ravindra V. Adivarekar

Contents

About the Editors

Dr. Santosh Biranje Ph.D., is an Assistant Professor in the Department of Fibres and Textile Processing Technology at the Institute of Chemical Technology (ICT), Mumbai, India. He completed his Ph.D. at ICT Mumbai, specializing in Fibres and Textile Processing Technology, with a focus on biopolymer extraction and modification. Following his doctoral studies, Dr. Biranje pursued postdoctoral research at Jiangsu University, China, under the mentorship of Professors Sun and Liu Jun. During this period, he pioneered the development of 3D-printed wound healing material scaffolds made from nanocellulose, contributing to medical materials and regenerative medicine. In April 2022, he joined an Assistant Professor position at ICT Mumbai, where his research focuses on extracting nanocellulose for sustainable textile applications. Dr. Biranje's academic and professional trajectory underscores his unwavering commitment to advancing research in fibrous materials and biopolymers.

Dr. Mohammad Shahid Ph.D., is an Assistant Professor in the Department of Chemistry, at Koneru Lakshmaiah Education Foundation, India. Before joining KLEF, he worked at various esteemed institutions, including Jamia Millia Islamia, New Delhi; Soochow University, China; University of Glasgow, UK; the Institute of Chemical Technology, Mumbai, Dr. K. N. Modi University, Tonk and Mewar University, Chittorgarh. He earned his Master's degree in Chemistry from Shibli National College, Azamgarh, in 2006 and his Ph.D. in Chemistry from Jamia Millia Islamia, New Delhi, in 2014. Dr. Shahid has received several prestigious fellowships to support his research pursuits, including grants from UGC, India, China Postdoctoral Fellowship, EU Marie Curie Fellowship, and Dr. D. S. Kothari Fellowship. He has contributed to over 50 scientific publications with ~4000 citations and h-index of 24. In addition, has also edited and authored several books with renowned international publishing houses.

Dr. Ravindra V. Adivarekar Ph.D., is a distinguished professor in Fibre Chemistry at the Institute of Chemical Technology (ICT), Mumbai, where he has dedicated over 25 years of service. He holds multiple degrees, including an honorary Fellowship of the Textile Association (FTA). Dr. Adivarekar is renowned for his expertise in textile processing and served as the Head of the Department of Fibres & Textile Processing Technology from 2008 to 2021. He currently serves as the Dean of Human Resource Development (HRD) at ICT Mumbai and as the Director of ICT-IOC Bhubaneswar. His extensive research interests span eco-friendly coloration, natural fiber composites, medical textiles, nanotechnology in textile processing, natural dyes, microbial dyes, and the application of enzymes. A committed mentor, Dr. Adivarekar has guided numerous students and managed research projects with total funding exceeding 18 crores. He has contributed significantly to the academic community with over 200 research papers published in national and international journals, more than 70 conference proceedings, and three patents. Beyond his academic achievements, Dr. Adivarekar is an esteemed member of several textile-related organizations and has received significant recognition for his contributions to the industry. He actively engages in advisory roles and government committees, addressing environmental issues in textile manufacturing.

Chapter 1
Biodegradable and Sustainable Finishes of Textiles

Aravin Prince Periyasamy and Melkie Getnet Tadesse

Abstract Traditional textile wet processing poses challenges to attaining eco-friendly and sustainable textile practices and products, chiefly because of its detrimental impacts on human health and the environment. The textile finishing industry, the greatest producer of wastewater, is currently reevaluating its technology and chemical utilization to comply with environmental and consumer requirements. There is a growing interest in the importance of maintaining sustainability, carbon neutrality, and water footprint, and meeting consumer demands for sustainability. The use of sustainable and biomaterial-based textile finishing techniques can improve manufacturing practices, leading to advancements in ecological, economic, and social aspects. The integration of biotechnology in the chemical processing of textiles is a significant application of emerging technologies that support environmental friendliness and sustainability. Enzymatic treatment, also known as bio-washing or bio-stoning, is a sustainable textile finishing method that adheres to the principles of green chemistry. This method reduces the use of harsh chemicals, conserves energy and water, and minimizes waste generation. Biomaterials, including biopolymers and extracts obtained from plants, can be used for sustainable and renewable functional finishing on textiles. The use of plant extracts or microorganisms for biosynthesis is gaining popularity owing to its cost-effectiveness, diminished chemical usage, simplicity of implementation, and eco-friendly characteristics. This chapter examines sustainable finishing procedures that incorporate nanotechnology, plasma treatment, enzymes, biopolymers, and natural agents. Furthermore, it includes international environmental rules and activities related to textile products and manufacturing methods.

A. P. Periyasamy (✉)
Textile and Nonwoven Materials, VTT Technical Research Centre of Finland Ltd, Espoo, Finland
e-mail: aravinprincep@gmail.com

Department of Material Engineering, Technical University of Liberec, Liberec, Czech Republic

M. G. Tadesse
Sustainable Engineering (STE), Albstadt-Sigmaringen University, Albstadt, Germany

Ethiopian Institute of Textile and Fashion Technology, Bahir Dar, Ethiopia

S. Biranje et al. (eds.), *Sustainable Finishing Techniques in Textiles*,
https://doi.org/10.1007/978-981-96-4860-3_1

Keywords Sustainability · Textile finishing · Biopolymers · Enzymes · Polysaccharides

1.1 Introduction

The textile processing business is encountering growing environmental issues caused by the usage of hazardous chemicals in the manufacturing process, posing risks to human health and the environment [1–3]. According to the Ellen-Mac-Arthur Foundation (2016), it is projected that by 2050, the amount of plastic in oceans would match the entire weight of aquatic fish, leading to damage to marine ecosystems and the accumulation of hazardous substances in animal tissues [4, 5].

Biobased processing of textile materials has the potential to reduce the negative environmental effects of the textile industry by utilizing enzymes and biological agents to produce products [6, 7]. The climatic effects of the linear economy, such disruptions in freshwater availability and biodiversity, have a greater impact on indigenous populations, people of color, the elderly, women, and children. Mitigating these climatic repercussions will facilitate the achievement of the United Nations Sustainable Development Goals (SDGs), including ensuring access to clean water and sanitation, promoting responsible consumption and production, and addressing climate change [8, 9].

Future textiles signify a significant evolution in the industry through the integration of advanced materials, technologies, and attributes to address contemporary requirements for sustainability, performance, intelligent functionalities, and versatility (Fig. 1.1). They go beyond traditional textiles by including smart functionalities like sensing properties, and responsive materials for many applications [10]. The EU has initiated measures to advance sustainability across all textiles, showcasing its dedication to innovation within the textile and garment sectors. Biobased materials for textile finishing have attracted the attention and interest of technocrats and industry experts.

Textile finishing is a crucial phase in textile manufacturing that requires a substantial amount of water and energy. Textile processing industry effluents result in substantial water and air pollution. Industries need to adopt sustainable manufacturing techniques to address the growing demand for aesthetic and functional features in textile goods [11–15]. The heightened focus on "Green Chemistry" has resulted in the utilization of biopolymers derived from natural resources and plant extracts in textile finishing, hence improving the functional properties of textile materials. This chapter discusses the utilization of biopolymers, biotechnology, and plant extracts in the finishing of textile materials.

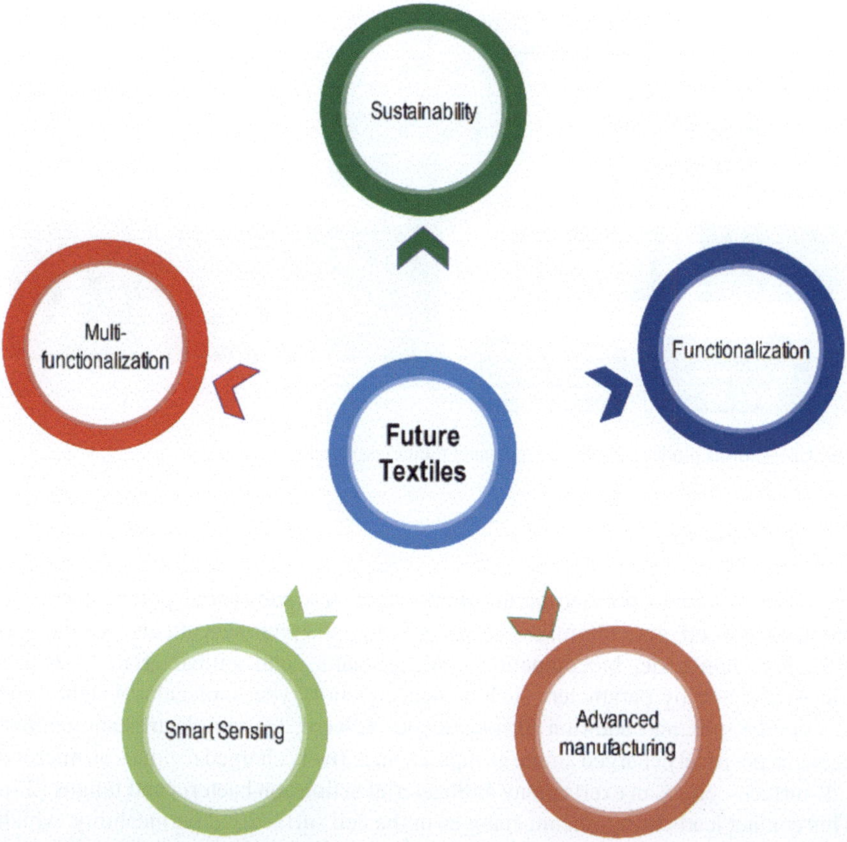

Fig. 1.1 Key aspect of future textiles

1.2 Biomaterials as a Source of Textile Finishing

1.2.1 Biopolymers

Polymers made from renewable sources such as maize and potatoes, known as bio-based polymers, are being recognized for their environmentally friendly and cost-efficient cellulosic textile products (Fig. 1.2). Researchers are investigating environmentally friendly compounds and creating sustainable cellulose-based textiles with desired practical characteristics [16]. Biopolymers are being increasingly utilized in enhancing the properties of cellulosic textiles as a safer and more environmentally friendly option compared to harmful textile chemicals. This section will analyze the roles of notable biopolymers in the functional finishing of cellulosic fabrics, classified by their synthesis methods. This method provides a long-lasting and environmentally beneficial option instead of harmful textile chemicals [17, 18].

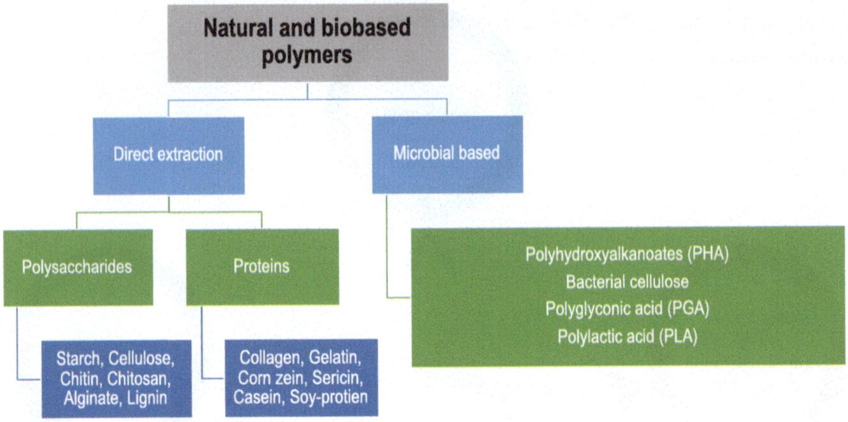

Fig. 1.2 Various biodegradable polymers and their classification

1.2.1.1 Chitosan

Chitosan, an amino polysaccharide biopolymer, has substantial potential uses in biomedical, food, agriculture, cosmetics, textiles, and pharmaceuticals (see Fig. 3a) [19]. It is non-toxic, biocompatible, biodegradable, and antibacterial. Its action can be affected by parameters such as deacetylation type, molecular weight, type of microbe, pH, and addition of nonaqueous solvent. Due to electrostatic contact between positively charged amine groups and negatively charged residues on microbe cell surfaces, chitosan exerts many antibacterial actions on bacteria and fungus [20]. This contact leads to significant changes in the cell surface and permeability, which impair the essential operations of microorganisms and result in cell death. Enhancing the positive-charge density of chitosan can boost its antibacterial effectiveness. Chitosan, whether modified or in nanocomposite form, can enhance qualities such as anti-crease and flame resistance. Yet, its limited capacity to bind to cellulosic materials is a significant disadvantage in the textile sector. Various methods have been used to enhance the adhesion of chitosan to cellulosic textiles and provide long-lasting and effective properties to treated materials. Various methods were used to enhance the attachment of chitosan or its derivatives to cellulosic textiles and to provide long-lasting and effective functionality to the treated materials, as shown in Table 1.1.

Chitosan applied to cotton, polyester, and wool fabric gave it antimicrobial properties [20]. Crosslinking compounds such as glutaric dialdehyde, formaldehyde-based, BTCA, and citric acid improved chitosan binding to textile substrates. Chitosan hydroxyl groups and polycarboxylic acid carboxylic groups crosslink to make finishing more durable [21].

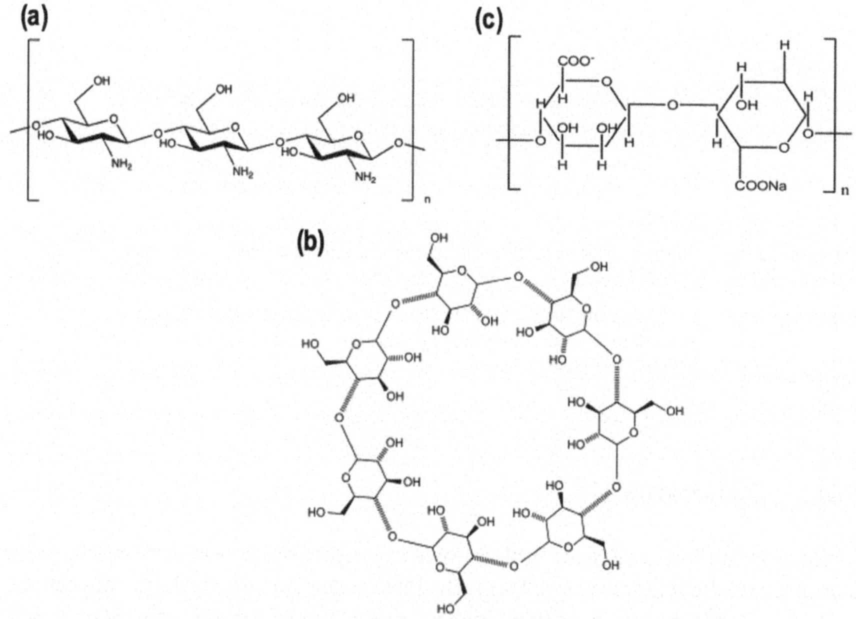

Fig. 1.3 Chemical structure of chitosan (**a**), cyclodextrin (**b**), and sodium alginate (**c**)

Table 1.1 Examples of applying chitosan for functional finishing of cellulosic materials

Type of additive	Description	Refs.
Antibacterial finish	The treated cotton materials improved crease recovery angle and antibacterial activity against G^+, G^-, and yeast	[22]
	The antibacterial effectiveness of all the treated cotton materials was outstanding	[23]
	BTCA increased the viscose fabric's amino group content and antibacterial activity against E. coli and Candida albicans by 100%	[24]
	Oxygen plasma-treated viscose fabric was more antimicrobial than air plasma-treated or untreated cloth	[25]
	Treated jute became more antibacterial as chitosan concentration increased. Chitosan boosts tensile strength and color retention	[26]
Aroma finishing	The resulting cotton fabric had fragrance therapeutic qualities from rosemary essential oil and chitosan	[27]
Flame-proof	Increasing the finishing formulation's chitosan concentration improved cotton fabric durability and flame retardancy	[28]
Multifunctional finish	Treatment made linen cloth durable, antibacterial, UV-protective, flame-proof, and anti-crease	[29]

Table 1.2 Instances of functional finishing of cellulosic textiles utilizing cyclodextrin

Type of additive	Comments	Refs.
Antibacterial finish	The cotton fabric with grafted β-cyclodextrin cavities harboring Ag^+ ions demonstrated antibacterial activity and regulated release	[32]
	Antibacterial activity and washability improved in the treated cotton fabrics. The treated sample is antibacterial because thymol kills germs	[35]
UV-protection finish	Improvement in dispersion dyeing and considerable UV properties of cotton fabrics	[36]
Insect repellent	Incorporating insect repellant in β-cyclodextrin cavities improves insecticide/irritate functionality over typical impregnation approaches	[37]
Multifunctional finish	–	[38]

1.2.1.2 Cyclodextrin

Potato, corn, and rice starch are broken down by enzymes to generate cyclic oligosaccharides called cyclodextrins (CDs) [30]. Pharmaceutical, food, flavor, cosmetics, and textile sectors benefit from these biocompatible, biodegradable, eco-friendly, and safe materials. CDs have three types: α-CD, β-Cyclodextrins, and γ-Cyclodextrins, each with 6–8 α-1,4-glycosidic linkages [30]. β-Cyclodextrin is commonly used to modify cellulosic materials but have the lowest solubility among Cyclodextrins (Fig. 3b) [31]. To address this issue, various chemical alterations were made to the –OH groups on the outer edges, such as hydroxypropyl derivatives, haphazardly methylated β-CD, and branched Cyclodextrins. Monochlorotriazine β-Cyclodextrin (MCT-β-Cyclodextrins) is a recognized derivative of β-Cyclodextrin. Cyclodextrins can encapsulate various compounds within their hydrophobic regions, creating host–guest complexes that enhance solubility, stability, volatility, sublimation, control of active ingredient release, and disorder. The efficacy of these complexes depends on various aspects, including guest molecule dimensions, van der Waals forces, water expulsion, hydrogen bonding, charge-transfer interactions, and hydrophobic interactions. β-Cyclodextrins have diverse applications including fragrance and antibacterial finishes, medical textiles, and UV-protection finishes [32]. They maintain the hydrophilic characteristics of cellulose and can form inclusion complexes with other molecules. The fixation of cyclodextrins on cotton does not impact the hydrophilic properties of cellulose. Table 1.2 displays different methods for β-Cyclodextrins utilization in cellulose-based textiles, including how they are attached, the substances they encapsulate, and the resulting properties [33, 34].

1.2.1.3 Sodium Alginate

Seaweeds produce sodium alginate, a polysaccharide made of polyguluronate (poly-G) and poly-mannuromate (poly-M) monomers and their copolymer [39]. The

Table 1.3 Some recent examples of functional finishing of cellulosic fabrics using biopolymers

Additive	Conditions	Descriptions	Refs.
Sericin	Pad-dry-cure	Showing better antimicrobial finish, by destroying G^+ and G^- microorganisms	[41]
Gelatin and pectin	Dip samples and dry at room temperature	Treated cotton fabric demonstrated ~83 and ~93% antibacterial effectiveness, respectively	[42]
Whey protein	Padding and drying	Treated cotton fabric enhances the flame-resistant qualities	[43]
Casein	Cotton treated with casein suspension	Improved flame and thermal resistance of treated cotton fabric	[44]

commercially available sodium alginate has molecular weights ranging from 32,000 to 400,000 g/mol. The chemical structures and physical qualities of different sodium alginate sources vary based on their compositions, arrangement of uronate residues, and molecular weights. Sodium alginate exhibits low solubility in cold water, resulting in the formation of a viscous solution. Viscosity increases with elevated molecular weight and decreases around pH 3–3.5. Sodium alginate demonstrates superior sol–gel formation when cross-linked with divalent cations (Cd^{2+}, Co^{2+}, Cu^{2+}, Mn^{2+}, Ni^{2+}, Pb^{2+}, and Zn^{2+}), whereas monovalent cations or Mg^{2+} are not recommended [40]. The industrial application of sodium alginate is associated with its uronic acid content. Potential uses include wound treatment, controlled release, encapsulation of biocides, biosorption of antimicrobial ions, and antimicrobial finishing. Moreover, it serves as a stabilizer in the creation of nanoparticles. Numerous research investigations have investigated the potential applications of various biopolymers in textile functional finishing. Sericin, pectin, whey protein, casein, and lignin are employed to impart antibacterial, flame-retardant, and UV-resistant properties to cellulosic fabrics. These biopolymers are expected to gradually replace conventional synthetic polymers soon owing to their cost-effectiveness, performance, and environmental advantages [40]. Table 1.3 outlines different biopolymer applications on textiles.

1.2.2 Bioactive Compounds Derived from Plants

Plant-based bioactive materials display a wide range of structures and chemical compositions. There has been significant research on bioactive materials and their impact on enhancing textile fabrics with antimicrobial properties, insect repellency, and UV protection. Plants are recognized for their significant contribution to the production of natural antimicrobials. Materials derived from various plant parts, including bark, leaves, roots, and flowers, encompass diverse chemicals such as pigments, phenolic compounds, quinones, flavonoids, tannins, polysaccharides, and essential oils, which exhibit potent antibacterial effects [45]. The application of biomaterials may encompass several strategies, including direct application, crosslinking, microencapsulation, β-cyclodextrin utilization, exhaustion method, and pre-surface modifications such as plasma treatment, UV radiation, and enzymatic treatment.

The study investigates the application of natural dyes in textile production. The textiles were treated with aloe vera, neem, neem extract, and their hybrid at pH 5, with a liquor ratio of 1:20 for 30 min by the exhaustion method [40]. All the samples that were treated showed antibacterial activity against E. coli and S. aureus, as well as antifungal activity against Aspergillus niger. The mixture exhibited superior and longer-lasting antibacterial effectiveness after washing as compared to using neem and aloe gel separately [46]. The extract of Tulsi leaves exhibited significant antibacterial action against the microorganisms tested. The cloth samples that were treated showed a 73% decrease in germs [47]. Eugenol cotton underwent microencapsulation and subsequent treatment with resin as a cross-linking agent, leading to a significant inhibitory zone and over 99% decrease in bacterial count against the studied pathogens. The prickly chaff flower extract exhibited varying antibacterial effects on S. aureus and E. coli [47]. Green tea extract and neem extract shown strong antibacterial effectiveness against S. aureus and E. coli using the zone of inhibition test [48].

Clove oil was used as a cross-linking agent, resulting in improved antibacterial activity against S. aureus and K. pneumoniae in the treated cotton fabric samples [45]. The antibacterial activity of natural dyes on cotton fibers was influenced by the technique of application. The colored cotton textiles exhibited outstanding antioxidant properties and antibacterial effects [45]. Indigo flavonoids underwent treatment with chitosan, resulting in the treated cotton fabric exhibiting outstanding antibacterial activity and UV-protection characteristics [49]. Viscose fabric samples were dyed with pad-dry-steam or cold pad-batch dyeing techniques, and the dye bath was repurposed to dye an additional pre-mordanted cloth. The results show that the discarded dyebath may be repurposed to produce colored linen fabric with satisfactory hues and practical properties, hence reducing the requirement for additional water and chemicals in the dyeing process [50].

1.2.2.1 Extracts from Plants

Cellulosic textiles that have been treated with natural antimicrobial agents such as aloe vera, Tulsi, prickly chaff flower, and green tea have shown potent antibacterial properties against a range of diseases [51]. Aloe vera leaves consist of around 200 active constituents, including nutrients, minerals, amino acids, and vitamins. These components have the potential to assist in the healing of wounds and burns, as well as in the decrease of inflammation. Aloe vera gel-treated cotton fabric exhibited considerable antibacterial efficacy against a range of organisms and fungal diseases [52]. Cotton textiles treated with aloe vera and neem demonstrated improved resistance to washing. Tulsi leaves exhibit properties that are antibacterial, insecticidal, antiprotozoal, diaphoretic, and expectorant in nature. The main components of these leaves include eugenol at 70%, methyl eugenol at 20%, and carvacrol at 3% [52].

The Eugenia caryophyllata tree yields clove, a spice that has antioxidant, insecticidal, antifungal, and antibacterial properties. Ginger, a member of the zingiberaceae family, is effective in alleviating cataracts, heart ailments, migraines, and kidney stones. Thespesia populnea has antibacterial, antiviral, wound healing, anticancer, and antisteroidogenic properties. The neem trees found in India possess abundant antibacterial properties, as seen by the green coloration of their leaves when treated with methanolic extracts. Neem extracts include approximately 300 medicinal compounds, including quercetin, nimbosterol, and lininoids. Aloe vera-based finishes are essential in cosmetic, pharmaceutical, and skin care products owing to their inherent ability to soothe the skin and retain moisture. Tulsi, an Indian substance, has a wide range of beneficial properties including antimicrobial, insecticidal, diaphoretic, antioxidant, antifungal, anti-allergic, anti-inflammatory, and antibacterial activities. Syzygium cumini, a culinary ingredient native to India, is rich in calcium, iron, and antibacterial compounds [53]. Ocimum sanctum, a widely used Ayurvedic herb, has antimicrobial, antifungal, and antibacterial properties. Aloe barbadensis has a pleasant fragrance and exhibits properties such as astringency, laxative effects, promotion of bowel movements, stimulation of menstruation, soothing of the skin, promotion of bile flow, increase in urine production, improvement of digestion, strengthening of the body, healing of wounds, counteracting of negative effects, prevention of cancer, and prevention of genetic mutations.

1.2.2.2 Essential Oils

Essential oils, which are a mixture of fragrant substances, provide protection against various microorganisms and have been used on textiles made of cellulose to create a long-lasting antimicrobial coating. In recent decades, there has been a resurgence in the use of essential oils for the purpose of promoting health and well-being. This trend may be attributed to the growing consumer demand for natural goods and alternative forms of medicine, particularly in the treatment of medical ailments such as pain, anxiety, depression, and sleeplessness. These oils are economically advantageous, exhibit less adverse effects in comparison to pharmaceutical treatments, and enhance the general health and well-being of patients. Aromatherapy is a fast-expanding kind of supplementary treatment that is gaining popularity globally. Studies conducted in the second decade of the twenty-first century have shown the notable medicinal benefits of aromatherapy in treating conditions such as dementia, anxiety, melancholy, pain management, dysmenorrhea, and sleep disturbances [27]. Nevertheless, several research investigations that purport to illustrate the impact of aromatic chemicals on human physiology and psychology encounter obstacles, and there is scant scientific data supporting the therapeutic advantages of essential oils on human health [54].

Neem oil is rich in Azadirachtin, a potent compound that effectively repels insects. It decreases the insects growth by blocking feeding, interferes with insect hormones, and hinders their development and ability to reproduce [41]. Azadirachtin exhibits resistance against nematodes and diminishes their feeding activity. Additional constituents of neem oil help eliminate insects by restricting their ability to feed themselves.

1.2.2.3 Natural Dyes

Natural colors have been well studied for their ability to provide several benefits to cellulosic textiles, including antibacterial, UV-protection, deodorizing, and insect-repellent characteristics. Natural dyes offer significant environmental advantages as they are derived from renewable sources and possess biodegradable properties [53]. These can be classified into four main groups: plants, minerals, animals, and microorganisms, with plants serving as the primary source. The various parts of the plant, such as roots, leaves, twigs, stems, heartwood, bark, and wood can be effectively utilized (Fig. 1.4). Natural dyeing has been a focus in research and development and the textile industry due to environmental issues, water contamination threats, and the sustainability of resources, processes, and products. Natural colors such as pomegranate, madder, Gromwell, indigo, Eucalyptus leaves, turmeric, hard, and henna have been effectively utilized for coloring and enhancing the functionality of cellulosic materials simultaneously. Nevertheless, natural dyes have limitations in terms of colorfastness and resistance to washing. To address these issues, other treatment methods have been utilized such as mordanting using metallic salt mordants, bio-mordants, pre-chemical alterations, enzymatic therapy, plasma pre-activation, UV-radiation, and sonication [53, 55].

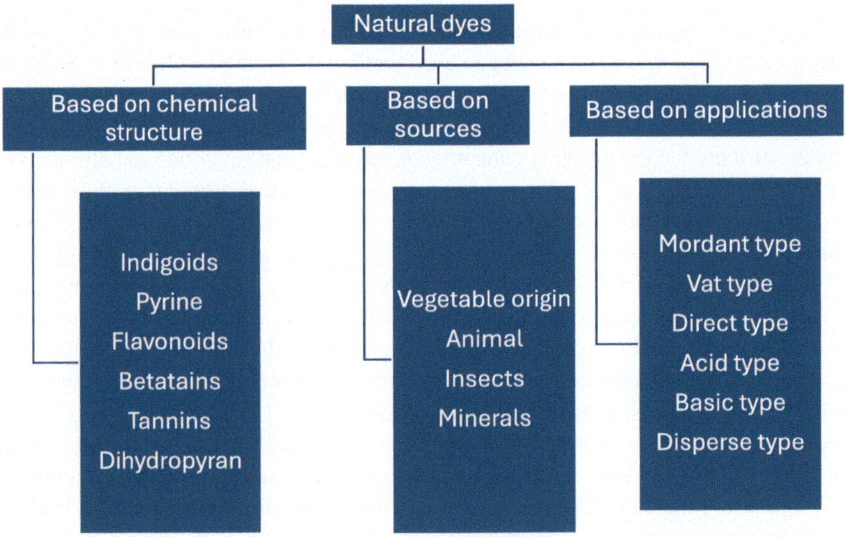

Fig. 1.4 Different types of natural dyes

1.2.3 Biotechnology in Textile Finishing

Enzymes are essential in various stages of textile processing, including desizing, scouring, bleaching, denim washing, and biopolishing. They operate swiftly and efficiently, utilizing minimal quantities while preserving essential resources like raw materials, energy, and water. Since the late 1980s, enzymes have been utilized, with a significant development taking place in 1987 for the stonewashing of denim. These are primarily utilized in the finishing processes of textiles, particularly for denim abrasion and finishing. In bio-stoning, cellulase enzymes work alongside pumice stones to effectively release indigo color from denim surfaces. This method reduces fabric harm, lessens the wear on equipment, and generates minimal dust [7, 40, 56, 57].

1.3 Finishes to Reduce the Microplastics

The electrofluidodynamic coating technique was used to apply a coating of biodegradable polymers, such as polylactic acid and poly(butylene succinate-co-butylene adipate) (PBSA), onto polyamide fabric [58]. This led to a remarkable reduction of 90% in the emission of microfibers when compared to the initial fabric. The uniformity of the coating technique enabled the production of nanoparticles at the nanoscale while preserving the fundamental characteristics of the fabric. The polyamide fabric, after being completed, released an average of 428 ± 92 and 456 ± 120 microfibers per gram of washed fabric, respectively. In comparison, the immaculate PA fabric released an average of 3966 ± 1425 microfibers per gram of fabric. The finishing procedure resulted in the formation of a consistent surface coating, which served to safeguard the fabric against any potential damage caused by mechanical or chemical reactions that may occur during washing. Nevertheless, the finishing process might diminish the mechanical qualities by 15–20%, and the endurance of the coating can be assessed by a maximum of 5 washing cycles. The efficacy of these coatings on polyester has not been evaluated [59]. Researchers [60] investigated the release of microfibers from PET fabric treated with chitosan. The pristine PET fabric released 3047 microfibers per gram, while the chitosan-finished PET fabric emitted 1726, 2497, and 2237 microfibers per gram at chitosan concentrations of 1%, 2%, and 3%, respectively. The findings reveal a reduction of 27% and 18% in microfiber content for chitosan concentrations of 3% and 2%, respectively. However, there is no consistent outcome when the data are replicated via repeated testing. Therefore, it is recognized as having lower efficacy when compared to other finishes such as pectin [14, 59] and PLA/PBSA. Figure 1.5 outlines the possible significant discoveries regarding the decrease of microfibers by different mechanical and chemical finishing methods. However, more research is needed to investigate the effects of enzymatic and plasma finishing [59].

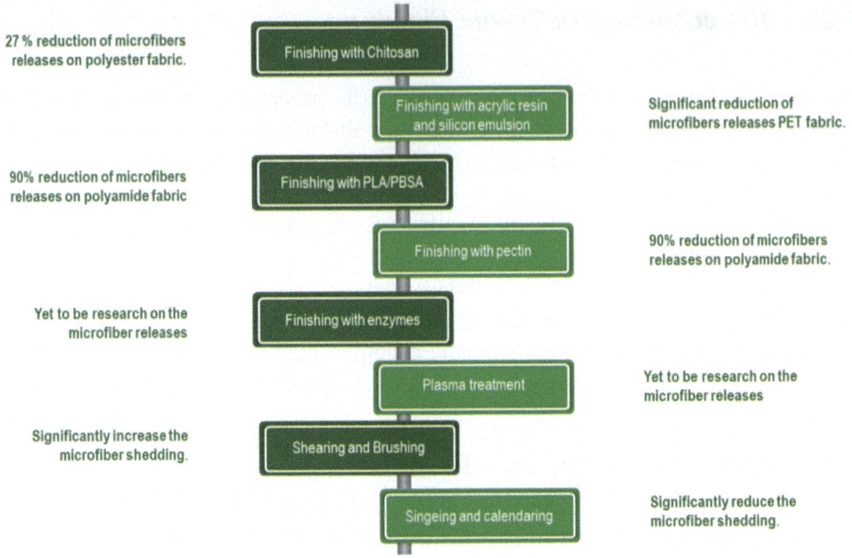

Fig. 1.5 Influence of mechanical and chemical finishing on microfiber releases

1.4 Conclusion

The incorporation of cutting-edge technology and eco-friendly materials into manufacturing practices is increasingly common in the textile sector. Diverse biopolymers and plant extracts are widely utilized in the textile industry, fulfilling various applications including medical textiles and aromatherapy, among others. The application of naturally sourced materials as finishing options holds the promise to revolutionize the industry, steering it toward a completely eco-friendly and sustainable future. The combination of contemporary chemistry and cutting-edge technologies will be crucial in reaching this objective. Biodegradable polymers offer a sustainable solution, being abundant, renewable, and biocompatible, while also possessing the ability to decompose naturally. This makes them a viable alternative to traditional harmful commercial chemicals and additives. Cellulosic textiles have been observed to incorporate or improve functional attributes, such as antimicrobial properties, UV protection, and flame resistance. Additional examination and evaluation are required to improve the output potential of these materials and successfully integrate them into textile finishing methods. Plant and herbal extracts, essential oils, and natural dyes are viewed as economically viable, non-hazardous, renewable, and environmentally sustainable resources, noted for their minimal ecological impact. Nonetheless, the longevity and resistance properties of these materials are constrained. Studies have been carried out to enhance their durability, yet the advancement of extraction methods and their application concerning economic and environmental issues continues to pose a challenge.

References

1. Periyasamy AP, Jiri M (2017) 10—Denim and consumers' phase of life cycle. In: Muthu SS (ed) Sustainability in Denim. Woodhead Publishing Limited, U.K., pp 257–282
2. Periyasamy AP, Militky J (2017) 7—Denim processing and health hazards. In: Muthu SS (ed) Sustainability in Denim. Woodhead Publishing Limited, U.K., pp 161–196
3. Periyasamy AP, Wiener J, Militky J (2017) 4—Life-cycle assessment of denim. In: Muthu SS (ed) Sustainability in Denim. Woodhead Publishing Limited, U.K., pp 83–110
4. Periyasamy AP (2023) Environmentally friendly approach to the reduction of microplastics during domestic washing: prospects for machine vision in microplastics reduction. Toxics 11:575. https://doi.org/10.3390/toxics11070575
5. Hedrich S, Janmark J, Langguth N et al (2022) Scaling textile recycling in Europe-turning waste into value. In: McKinsey. https://www.mckinsey.com/industries/retail/our-insights/scaling-textile-recycling-in-europe-turning-waste-into-value. Accessed 16 Jan 2023
6. Periyasamy AP (2023) Microfiber emissions from functionalized textiles: potential threat for human health and environmental risks. Toxics 11:406. https://doi.org/10.3390/toxics11050406
7. Vigneswaran C, Ananthasubramanian M, Kandhavadivu P et al (2014) 5—Enzymes in textile effluents. In: Vigneswaran C, Ananthasubramanian M, Kandhavadivu PBT-B of T (eds) Bioprocessing of textiles. Woodhead Publishing India, pp 251–298
8. Periyasamy AP (2024) Recent advances in the remediation of textile-dye-containing wastewater: prioritizing human health and sustainable wastewater treatment. Sustainability 16:495. https://doi.org/10.3390/su16020495
9. Periyasamy AP (2023) Nonwoven fabrics from agricultural and industrial waste for acoustic and thermal insulation applications. Textiles 3:182–200. https://doi.org/10.3390/textiles3020013
10. Bosowski P, Hoerr M, Mecnika V et al (2015) Design and manufacture of textile-based sensors. In: Dias T (ed) Electronic textiles: smart fabrics and wearable technology. Elsevier, Oxford, pp 75–107
11. Periyasamy AP (2021) Evaluation of microfiber release from jeans: the impact of different washing conditions. Environ Sci Pollut Res. https://doi.org/10.1007/s11356-021-14761-1
12. Periyasamy AP, Militky J (2020) Sustainability in textile dyeing: recent developments. In: sustainability in the textile and apparel industries, pp 37–79
13. Periyasamy AP, Militky J, Sachinandham A, Duraisamy G (2021) Nanotechnology in textile finishing: recent developments. Handbook of nanomaterials and nanocomposites for energy and environmental applications. Springer International Publishing, Cham, pp 1–31
14. Periyasamy AP, Periyasami S (2023) Critical review on sustainability in Denim: a step toward sustainable production and consumption of Denim. ACS Omega. https://doi.org/10.1021/acsomega.2c06374
15. Vik M, Periyasamy AP (2018) Chromic materials fundamentals, measurements, and applications, 1st edn. Apple Academic Press (CRC Press), New York, NY
16. Echavarri-Bravo V, Eggington I, Horsfall LE (2019) Synthetic biology for the development of bio-based binders for greener construction materials. MRS Commun 9:474–485. https://doi.org/10.1557/mrc.2019.39
17. Periyasamy AP, Venkatesan H (2019) Eco-materials in textile finishing. Handbook of ecomaterials. Springer International Publishing, Cham, pp 1461–1482
18. Venkatesan H, Periyasamy AP (2019) Eco-fibers in the Textile Industry. Handbook of ecomaterials. Springer International Publishing, Cham, pp 1413–1433
19. Sricharussin W, Sopajaree C, Maneerung T, Sangsuriya N (2009) Modification of cotton fabrics with β-cyclodextrin derivative for aroma finishing. J Text Inst 100:682–687
20. Lim S-H, Hudson SM (2003) Review of chitosan and its derivatives as antimicrobial agents and their uses as textile chemicals. J Macromol Sci, Part C: Polym Rev 43:223–269
21. Knaul JZ, Hudson SM, Creber KAM (1999) Crosslinking of chitosan fibers with dialdehydes: Proposal of a new reaction mechanism. J Polym Sci B Polym Phys 37:1079–1094

22. Hebeish A, Abdel-Mohdy FA, Fouda MMG et al (2011) Green synthesis of easy care and antimicrobial cotton fabrics. Carbohydr Polym 86:1684–1691. https://doi.org/10.1016/j.carbpol.2011.06.086
23. Ferrero F, Periolatto M (2012) Antimicrobial finish of textiles by chitosan UV-curing. J Nanosci Nanotechnol 12:4803–4810. https://doi.org/10.1166/jnn.2012.4902
24. Šauperl O, Volmajer-Valh J (2013) Viscose functionalisation with a combination of chitosan/BTCA using microwaves. Fibres Text East Eur 101:24–29
25. Karthikeyan N, Vijayalakshmi KA, Vignesh K (2016) Functionalisation of viscose fabric with chitosan particles using non-thermal oxygen plasma. Mater Technol 31:358–363. https://doi.org/10.1179/1753555715Y.0000000067
26. Bhuiyan MAR, Islam A, Islam S et al (2017) Improving dyeability and antibacterial activity of Lawsonia inermis L on jute fabrics by chitosan pretreatment. Text Cloth Sustain 3:1. https://doi.org/10.1186/s40689-016-0023-4
27. Muresan A, Muresan R (2009) Aromatherapeutic characteristics of cotton fabrics treated with rosemary essential oil. Cellul Chem Technol 43:435–442
28. El-Tahlawy K (2008) Chitosan phosphate: a new way for production of eco-friendly flame-retardant cotton textiles. J Text Inst 99:185–191. https://doi.org/10.1080/00405000701584311
29. Sheikh J, Bramhecha I (2018) Multifunctional modification of linen fabric using chitosan-based formulations. Int J Biol Macromol 118:896–902. https://doi.org/10.1016/j.ijbiomac.2018.06.150
30. Oliveira MF, Suarez D, Rocha JCB et al (2015) Electrospun nanofibers of polyCD/PMAA polymers and their potential application as drug delivery system. Mater Sci Eng, C 54:252–261
31. Gawish SM, Ramadan AM, Abo El-Ola SM, Abou El-Kheir AA (2009) Citric acid used as a cross-linking agent for grafting β-cyclodextrin onto wool fabric. Polym Plast Technol Eng 48:701–710
32. Bajpai M, Gupta P, Bajpai SK (2010) Silver(I) ions loaded cyclodextrin-grafted-cotton fabric with excellent antimicrobial property. Fibers Polym 11:8–13. https://doi.org/10.1007/s12221-010-0008-x
33. Buschmann HJ, Knittel D, Schollmeyer E (2001) New textile applications of cyclodextrins. J Incl Phenom 40:169–172. https://doi.org/10.1023/A:1011892600388
34. Haji A (2020) Functional finishing of textiles with β-cyclodextrin. In: Frontiers of textile materials. Wiley, pp 87–116
35. Rukmani A, Sundrarajan M (2012) Inclusion of antibacterial agent thymol on β-cyclodextrin-grafted organic cotton. J Ind Text 42:132–144. https://doi.org/10.1177/1528083711430244
36. Ibrahim NA, E-Zairy WR, Eid BM (2010) Novel approach for improving disperse dyeing and UV-protective function of cotton-containing fabrics using MCT-β-CD. Carbohydr Polym 79:839–846. https://doi.org/10.1016/j.carbpol.2009.10.008
37. Romi R, Lo NP, Bocci E et al (2005) Bioengineering of a cellulosic fabric for insecticide delivery via grafted cyclodextrin. Biotechnol Prog 21:1724–1730. https://doi.org/10.1021/bp050276g
38. Attarchi N, Montazer M, Toliyat T et al (2018) Novel cellulose fabric with multifunctional properties through diverse methods of Ag/TiO_2/β-cyclodextrin nanocomposites synthesis. Cellulose 25:1449–1462. https://doi.org/10.1007/s10570-017-1638-6
39. jian Shao Z, lian Huang X, Yang F et al (2018) Engineering sodium alginate-based cross-linked beads with high removal ability of toxic metal ions and cationic dyes. Carbohydr Polym 187:85–93. https://doi.org/10.1016/J.CARBPOL.2018.01.092
40. Eid BM, Ibrahim NA (2021) Recent developments in sustainable finishing of cellulosic textiles employing biotechnology. J Clean Prod 284:124701. https://doi.org/10.1016/j.jclepro.2020.124701
41. Rajendran R, Radhai R, Balakumar C (2012) Synthesis and characterization of neem chitosan nanocomposites for development of antimicrobial cotton textiles. J Eng Fiber Fabr 7:136–141
42. Tummalapalli M, Berthet M, Verrier B et al (2016) Composite wound dressings of pectin and gelatin with aloe vera and curcumin as bioactive agents. Int J Biol Macromol 82:104–113. https://doi.org/10.1016/j.ijbiomac.2015.10.087

43. Bosco F, Carletto RA, Alongi J et al (2013) Thermal stability and flame resistance of cotton fabrics treated with whey proteins. Carbohydr Polym 94:372–377. https://doi.org/10.1016/j.carbpol.2012.12.075
44. Xu F, Zhong L, Zhang C et al (2019) Novel high-efficiency casein-based P-N-containing flame retardants with multiple reactive groups for cotton fabrics. ACS Sustain Chem Eng 7:13999–14008. https://doi.org/10.1021/acssuschemeng.9b02474
45. Murugesh Babu K, Ravindra KB (2015) Bioactive antimicrobial agents for finishing of textiles for health care products. J Text Inst 106:706–717. https://doi.org/10.1080/00405000.2014.936670
46. Khurshid M, Ayyoob M, Asad M, Shah S (2015) Assessment of eco-friendly natural antimicrobial textile finish extracted from Aloe vera and neem plants. Fibres Text East Eur 23:120–124. https://doi.org/10.5604/12303666.1172176
47. Thilagavathi G, Rajendrakumar K, Rajendran R (2005) Development of ecofriendly antimicrobial textile finishes using herbs. PSG CT, Coimbatore, India
48. Garg H, Singhal N, Singh A et al (2023) Laccase-assisted colouration of wool fabric using green tea extract for imparting antioxidant, antibacterial, and UV protection activities. Environ Sci Pollut Res 30:84386–84396. https://doi.org/10.1007/s11356-023-28287-1
49. Jung JS (2017) Dyeing properties and functionality of cotton and silk dyed with fermented indigo and fresh indigo. In: Proceedings of academics world 72nd international conference, Seoul
50. Sharma A, Kadam S, Mathur P et al (2019) Re-using henna natural dyeing wastewater for coloration and multifunctional finishing of linen fabric. Sustain Chem Pharm 11:17–22. https://doi.org/10.1016/j.scp.2018.12.001
51. Ibrahim NA, El-Zairy EMR, Eid BM (2017) Eco-friendly modification and antibacterial functionalization of viscose fabric. J Text Inst 108:1406–1411. https://doi.org/10.1080/00405000.2016.1254583
52. Ali SW, Purwar R, Joshi M, Rajendran S (2014) Antibacterial properties of Aloe vera gel-finished cotton fabric. Cellulose 21:2063–2072. https://doi.org/10.1007/s10570-014-0175-9
53. Periyasamy AP (2022) Natural dyeing of cellulose fibers using syzygium cumini fruit extracts and a bio-mordant: a step toward sustainable dyeing. Sustain Mater Technol 33:e00472. https://doi.org/10.1016/j.susmat.2022.e00472
54. Negi A, Kesari KK (2022) Chitosan nanoparticle encapsulation of antibacterial essential oils. Micromachines (Basel) 13:1265. https://doi.org/10.3390/mi13081265
55. Brady PR (1992) Diffusion of dyes in natural fibres. Rev Prog Color Relat Top 22:58–78. https://doi.org/10.1111/j.1478-4408.1992.tb00090.x
56. Liu X, Kokare C (2016) Microbial enzymes of use in industry. In: Biotechnology of microbial enzymes: production, biocatalysis and industrial applications. Academic Press, pp 267–298
57. Sheikh J, Bramhecha I (2019) Enzymes for green chemical processing of cotton. In: Shahid-ul-Islam, Butola BSBT-TI, of Green Chemistry for Textile Technology P (eds) The impact and prospects of green chemistry for textile technology. Woodhead Publishing, pp 135–160
58. De Falco F, Cocca M, Guarino V et al (2019) Novel finishing treatments of polyamide fabrics by electrofluidodynamic process to reduce microplastic release during washings. Polym Degrad Stab 165:110–116. https://doi.org/10.1016/j.polymdegradstab.2019.05.001
59. De Falco F, Gentile G, Avolio R et al (2018) Pectin based finishing to mitigate the impact of microplastics released by polyamide fabrics. Carbohydr Polym 198:175–180. https://doi.org/10.1016/j.carbpol.2018.06.062
60. Piñol L, Rodriguez L, Brouta M et al (2014) Mapping and estimation of laundry effluent microplastics release

Chapter 2
Biodegradability and Sustainability in Textile Finishing—Current Challenges and Future Perspectives

Shriyasha Meghanath Tari and Ashok Ramchandra Athalye

Abstract Finishing, the last step of textile processing, is attributed to rendering crucial functional and commercial value addition to textiles. Depending on the customer's needs and end use, finishing can be categorized into two broad segments: perceptive and protective. Perceptive finishes enhance the sensory features of the fabric, like touch, smell, vision, sound, and taste. Fabrics treated with softeners, aroma, and wellness finishes stimulate the perceptive senses of the wearer, making the fabric in contact more comfortable and desirable. In contrast, protective finishes provide shielding to the fabric from external agents. Shielding properties like flame retardancy, water–oil repellency, UV protection, and antimicrobial and antistatic nature are enhanced by protective finishes. Investigating their sustainability and biodegradability becomes crucial as finishes are the last chemicals added and remain on the fabric. This chapter discusses the biodegradability and sustainability concerns of textile finishes and the limitations of existing finishing chemistries. The permissible limits and thresholds set for chemical finishes by various eco-regulations are also illustrated in this chapter. It further throws light on the environmental impact of some conventional textile finishes. The sustainable alternatives for conventional chemical finishes and recent product and process developments are also discussed.

Keywords Aquatic toxicity · Bioaccumulation · Functional effects · Persistent pollutants

---

S. M. Tari · A. R. Athalye (✉)
Department of Fibres and Textile Processing Technology, Institute of Chemical Technology, Mumbai, India
e-mail: ar.athalye@ictmumbai.edu.in

S. M. Tari
e-mail: tch22sm.tari@phd.ictmumbai.edu.in

S. Biranje et al. (eds.), *Sustainable Finishing Techniques in Textiles*,
https://doi.org/10.1007/978-981-96-4860-3_2

2.1 Introduction

Finishing plays a crucial role in determining the final properties of the fabric. Conventional textile finishing requires using chemicals that might harm the environment and humans [1]. Sustainable and environmentally friendly practices are the prime focus of textile processors. Textile finishing is a major source of organic and persistent chemicals [2]. This has raised concerns about the biodegradability of chemicals and sustainability in conventional finishing processes. Finishing of textiles can involve a wide range of chemical treatments or mechanical processes aimed at enhancing the appearance, texture, aesthetics, or functionality. Considering the biodegradability of textile finishes is important for a holistic view of sustainability. Finishes are expected to be durable and adhere to the fabric surface for a long time. Some of these chemicals remain on the fabric surface even after the end of the product life cycle. Effective textile waste management is still a challenge. The waste garments are either discarded in favor of landfilling or incinerated. Such anthropogenic textile waste disposal activity contributes to carbon emissions [3].

The fate of some applied finishing chemicals and their interaction with the ecosystem in the disposal of textile waste is still a question [4]. Also, the chemically finished textiles remain close to the human skin. Due to increasing environmental and ecological concerns, the sustainability aspects of textile finishing are greatly monitored by brands and are under constant scrutiny by eco-regulatory bodies. Studies have shown that finishing chemicals greatly influence the biodegradability of textiles. The biodegradation of textiles proceeds in three phases: (i) Bio-deterioration, (ii) Assimilation, and (iii) Mineralization.

Each of these three phases is carried out by a microorganism. Besides this, abiotic factors like sunlight, air, dust, moisture, and heat play a crucial role in the biodeterioration of textile polymers. Low molecular mass fragments of textile polymer can be absorbed by the microbial cell wall in the assimilation phase. However, chemical finishes and coatings on the textile materials can act as a barrier in the assimilation process. Some chemical finishes resist the microorganisms and obstruct the successive biodegradation phases [5]. Cotton fabric with different finishes like silicone softener, water repellent, and durable press was evaluated for its biodegradability rate. The results showed that fabric surfaces with a high order of crosslinking were more resistant to biodegradation. The observed rate of biodegradation decreased in the order of Softener > Water Repellent > No Finish > Durable Press [6]. Microfiber shedding and its bioaccumulation is a booming environmental concern. A relevant study investigated the influence of finishing agents like cationic silicone-based softeners and ammonium compound-based antimicrobial agents. Viscose fibers impregnated with finishing agents showed lower degradation (around 75%) than untreated fibers (reaching up to 85%).

The incorporation of antimicrobial agents reduced the biodegradation to 3–5%. These findings indicate that the applied textile finishes prolong the environmental half-lives of shed fibers [7]. In the chemical finishing segment, a wide range of perceptive and protective finishes are applied to textiles as per end use and customer demand.

In the subsequent sections of this chapter, the biodegradability and sustainability concerns of majorly consumed chemical finishes are discussed. The environmental impact and the new product developments to meet sustainable demands are also elaborated. This chapter has discussed existing challenges in the biodegradability and sustainability of textile finishing chemicals and processes with an overview of future perspectives, highlighting the need for innovative textile finishing technologies.

2.2 Persistence, Bioaccumulation, Toxicity, and Other Environmental Problems of Textile Finishing Chemicals

The energy-intensive and resource-consuming nature of the textile finishing industry is well known. Efforts are being made to make the textile value chain more sustainable by implementing water-less, resource and energy-saving technologies. However, considering the chemical aspects of raw materials and the post-production fate of these finishing chemicals is equally important. Certain finishes play a vital role in enhancing the functional properties of fabrics, such as water repellency, wrinkle resistance, flame retardancy, soil release, and antimicrobial characteristics. Fabrics are treated with a range of repellent finishes to resist the action of external agents like fire, water, oil, microbes, and dirt. Widely consumed repellent and protective textile finishes include durable water/ oil repellent finish, flame retardant, durable press, and antimicrobial finish. These finishes are in great demand due to their functional properties and are among the widely consumed finishes across the globe. Some of these protective finishing chemicals pose a serious threat to the ecosystem. They are synthesized using chemistries, which are not very sustainable. Some chemicals used to get these functional effects are non-biodegradable and can persist in the environment for unknown periods. Some of them are linked to bioaccumulation, eco-toxicity, and endocrine dysfunction. Such finishing chemicals are specifically discussed in the subsequent sections of this chapter, elaborating on their chemical nature, performance, and limitations.

2.2.1 Water and Oil-Repellent Finishes

These finishes provide waterproof characteristics to fabrics intended for outdoor clothing, tents, and umbrellas. These finishes form a hydrophobic barrier on the textile surface, causing water droplets to roll off the fabric rather than being absorbed. Using repellent finishing chemicals, functional applications like waterproofing, oil resistance, soil release, and chemical splash protection are incorporated on textile surfaces. The low surface energy of the finishing layer repels water molecules. This effect is generally facilitated by fluorinated compounds, which are known to have extremely low surface energies arising from fluorine atoms. Perfluorinated compounds (PFCs) are known for their excellent water and oil-repellent properties.

Traditionally, Perfluorooctanoic acid (PFOA), Perfluorooctanesulfonic acid (PFOS), Perfluorohexanesulfonic acid (PFHxS), Perfluorononanoic acid (PFNA) were used as repellent finishes. Fluorocarbons are lipid soluble and easily absorbed by the lungs [8]. Such per- and polyfluoroalkyl compounds were used to impart liquid repellency to textiles. These fluorine-based compounds are termed Per- and polyfluoroalkyl substances (PFAS). As defined by the Organisation for Economic Co-operation and Development (OECD), PFAS are any fluorinated substances that contain at least one perfluorinated methyl group (-CF3) or a perfluorinated methylene group (-CF2-). Due to reported cases of bioaccumulation and long-term persistence in the environment, these PFAS are banned by many environmental regulations. Perfluoroalkyl moieties in PFAS are highly resistant to metabolic degradation in the environment.

Therefore, PFAS are either non-degradable or eventually transform into stable transition products that are persistent in the environment. PFAS are classified as persistent compounds (vP) under the European Chemicals Regulation. As PFAS are either persistent or degrade into persistent products, they are popularly termed forever chemicals. Perfluorooctanoic acid (PFOA) and perfluorooctane sulfonate (PFOS) are the two most detected PFAS in various environmental settings. They are investigated greatly because of their proven persistence and endocrine dysfunctional capabilities. PFAS is also linked to infertility issues in women [9]. The bans and regulations on long-chain per- and polyfluoroalkyl substances (PFASs) have motivated the textile industry to find eco-friendly finishing substitutes for achieving waterproofing and oil-repelling effects. Four major chemistries of durable water-repellent (DWR) finishes prevail in the current textile industry. They consist of (i) Side-chain fluorinated polymers, (ii) Silicones, (iii) Hydrocarbons, (iv) Dendrimer, and inorganic nanoparticle chemistry.

Non-fluoridated DWR chemistries like silicones, hydrocarbons, dendrimers, and inorganic nanoparticles are not efficient in providing water-repellent characteristics compared to their fluorinated counterparts. Oil repellency cannot be achieved with non-fluorinated compounds especially. Amongst these four DWR finishes, hydrocarbon-based water-repellent finishes are believed to be the most environmentally benign compounds, followed by silicones and side-chain fluorinated polymers. The environmental aspects of dendrimers are not much known [10]. Following the ban on long carbon chain PFAS $C_nF_{2n} + 1$, $n \geq 7$, short-chain PFAS finishes, which were perceived to cause less bioaccumulation and toxicity, were being used as a replacement to long chain PFAS for many years (Brendel et al., However, the very recent studies have shown that short-chain PFASs have high mobility in soil and water and their degradation products are extremely persistent. This results in their fast distribution to water bodies and subsequently leads to contamination of drinking water resources [11]. Waste water treatments like adsorption, ion exchange, and advanced oxidation are ineffective in the remediation of short-chain PFAS [12].

A first-of-its-kind study reported the fate of PFAS released from durable water-repellent clothing during use. The impacts of aging, washing, and tumble drying on the concentration of extractable PFASs in (C-6) based sidechain fluoro polymers (SFPs) and perfluorooctane-based long-chain SFPs (C-8) were evaluated in this study. Perfluoroalkyl acid (PFAA) production and concentration in polyester and polyamide fabrics increased with the aging of the DWR-coated fabrics. This study proved that the aging and washing of fabrics release PFAS in the environment [13]. In a similar outdoor weathering experiment, the emission of (PFAS) from functional textiles was studied in Sydney, Australia. For six months, polyamide fabrics treated with SFPs-based water repellent were subjected to various natural stresses like heat, wind, precipitation, washing, abrasion, and direct sunlight on a rooftop. The results of this study revealed loss of both PFAS-containing textile fragments (microfibers) and loss of low molecular weight PFAS. Each SFP-textile finish emitted a distinct PFAA pattern under the influence of stress. Under some weathering conditions, the PFA concentrations exceeded regulatory limits for textiles [14].

Fluorocarbons with C8 chemistry were typically employed as textile finishes that were water and oil-repellent. These finishes release harmful perfluorooctanoic acid (PFOA) and perfluorooctane sulfonate (PFOS). PFOS and PFOA have also been linked to cancer, decreased fertility, and immune system impairment in children [15]. As a result, manufacturers of water-repellent finishes have consented to eradicate PFOS, PFOA, and their precursors from raw materials. Although PFOS and PFOA C8 chemistry is substituted by C6 and C_4 fluorocarbons, their oil and water repellency ratings and washing durability are significantly less than those of C8-based fluorocarbons. Recently, C4 fluorochemistry has also been explored for repellent finishes [16]. The commercially available popular oil and water repellents are shown in Fig. 2.1.

Finishing manufacturers have switched to non-fluoridated alternatives to overcome the environmental impact of fluorocarbons. To date, ongoing industrial research has focused on inventing non-fluoridated DWOR chemistries. Two eco-friendly oil and water repellents for cotton fabrics were synthesized using palmitic acid, stearic acid, formaldehyde-free bio-succinic, and maleic acid. The developed finishes provided comparable properties concerning the C6-based fluorocarbons [17]. Many such attempts have been made to replace C8 chemistry using natural waxes, chitosan coupled with layer-by-layer application techniques, etc. [18]. Silicone-based water repellents derived from siloxanes and silanes are widely explored for water-repellent finishes [19]. The major challenge remains in achieving enhanced water and oil repellence ratings without fluorochemicals. Non-fluorinated alternatives lack oil repellence, which limits their substitution in textile applications that require repellence toward non-polar liquids [20]. The relationship between repellence performance and the environmental impact of commercially available water-repelling chemistries is summarised in Table 2.1 [21].

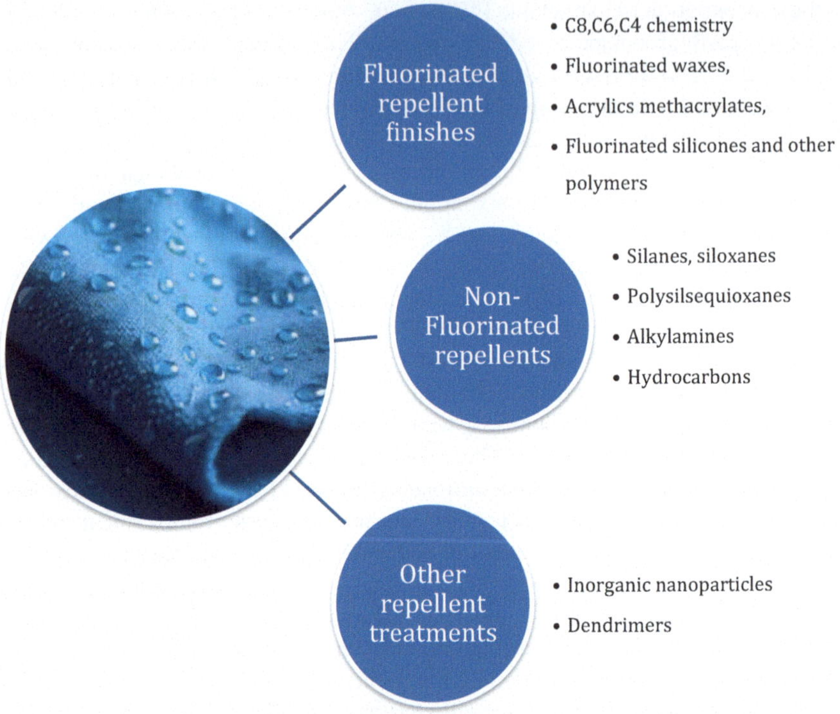

Fig. 2.1 Popular Chemistries of commercially available water and oil-repellent finishes

Table 2.1 Repellent chemistries and their environmental impact

Repellent chemistries	Water repellence	Oil repellence	Biodegradability	Bioaccumulation	Environmental Impact
C_8 -Fluorocarbons	+++++	+++++	Very persistent	Very High	+++++
C_6 -Fluorocarbons	++++	++++	Very persistent	Low	++++
C4 Fluorocarbons	+++	+++	Moderate to High persistence	Low	+++
Silicones	+++	–	Low to moderate persistence	Low	++
Hydrocarbons	+++	–	Low persistence	Low	++

2.2.2 *Flame Retardant Finishes*

Flame-retardant apparel is manufactured using intrinsic flame-retardant fibers or chemical fire-retardant coatings. Flame retardant (FR) finishes make textiles less flammable and protect them from ignition in the case of a fire. Flame retardant finishing treatments are essential to reduce the chance of fire-related mishaps and

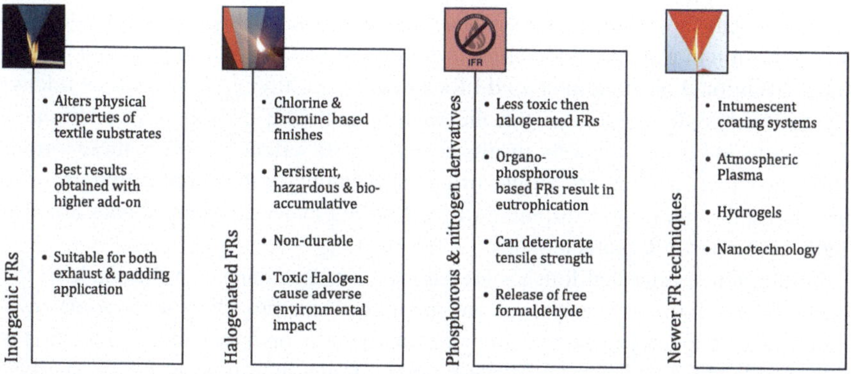

Fig. 2.2 Chronology of flame retardants finishing chemistries

restrict the spread of flames. These finishes are applied to various textiles, such as firefighters apparel, upholstery, curtains, bedding, and other home textiles. A consulting firm, Grand View Research, had reported that the textile flame retardant market was valued at USD 504.6 million in 2019 and is expected to grow at 3.9% CAGR from 2020 to 2027. Polyester provides the largest consumption share amongst textile flame retardants. Manufacturers seek to comply with strict safety requirements and regulations by adding flame retardant characteristics to fabrics in order to ensure the safety of both human life and property. The evolution of flame retardants with their application characteristics, environmental impact, and (fire retardant) FR performance is illustrated in Fig. 2.2. It also shows the recently launched innovative FR techniques.

Initially, hydroxides of Aluminium, Magnesium, and aluminum trioxide were used to shield thermoplastic fibers against fire. They create less smoke on combustion and can be applied through exhaust and padding techniques. Despite the ease of application, they are not preferred for rendering flame retardancy. A higher dosage of hydroxides is required to achieve optimum FR characteristics. The higher add-on of finishing chemicals negatively impacts the physical properties of fibers . Halogen-based FRs mainly constitute brominated and chlorinated compounds. For decades, brominated flame retardants in combination with antimony oxides were used to safeguard fabrics from fire. These are cost-effective means for achieving flame retardancy, but their toxicity profiles are critical. They emit more smoke and corrosive gases (halogen halides) during combustion. Many halogenated flame retardants like penta- and octa-bromodiphenyl ether and hexabromocyclododecane are banned. These flame retardants are listed in the Stockholm Convention because of their persistent characteristics. The UNEP defines halogenated FRs as persistent organic pollutants (POPs).

Currently, phosphorus and nitrogen-based flame retardants are being used to reduce the flammability of textiles. Phosphorous-based FRs are less toxic than their halogen-based counterparts. However, some studies have reported environmental issues like bio-concentration and biotransformation of organophosphorus-based flame retardants [22]. Organophosphorus molecules can leach into water streams during application and on washing of textiles. This can further lead to eutrophication in water bodies (*Indian Journal of Fiber and Textile Research*, 2023). The conventional phosphorus-based FRs contain N-methylol groups, which release free formaldehyde (carcinogen) in the environment [23]. Nitrogen, in addition to phosphorus in FR formulations, results in a synergistic effect. However, these formulations are costly and require a high curing temperature for their application. Besides this, they cause undesired stiffening of the finished material. Many silanes are also studied to impart FR characteristics.

Certain flame-retardant finishes are known to cause changes in the shade of the dyed material. Intumescent systems are an innovative approach to tackle some environmental problems associated with flame retardants. Condensed phase protection is used specifically in this kind of flame-retardant finish. Intumescent flame-retardant systems swell above a threshold temperature. As a result, the textile material's surface develops a foamed, charred layer that can effectively stop the material's combustion. With an increase in temperature, the foamy layer becomes less dense. As a result, it serves as an insulating layer, preventing heat transmission and oxygen penetration into the combustible material it covers [24]. Hydrogels are believed to be the next-generation flame retardants. The high-water content of hydrogel allows it to act as a fire-retardant layer. They can absorb huge amounts of heat energy and release the same when evaporation occurs. When combustible materials come in contact with the water vapor, they create an oxygen barrier that lowers the temperature, further ending combustion.

Conventional FRs based on halogens, phosphorus, nitrogen, and borax are effective in attributing flame retarding properties to fabrics but have certain limitations. Alteration of physical properties like tensile strength, stiffness, and shade change when finishing with FRs is undesirable. Achieving durability and FR performance with low chemical add-on is another challenge. Although intumescent systems, atmospheric plasma, hydrogels, and nanotechnology are promising solutions to overcome some of these challenges, they have their own challenges, such as cost, treatment parameters, and alteration of fiber properties on application.

2.2.3 Antimicrobial Finishes

Favorable conditions like large surface area, pH, humidity, and accumulated sweat provide ideal growing conditions for microbes and pathogens on textile materials. Anti-germ finishes protect the wearer from these microbes and provide shielding against dangerous diseases. The need for such antigerm coatings for textiles worn by frontline healthcare and hospitality personnel has been evident during the recent COVID-19 pandemic. Antimicrobial finishes are of great importance in applications like surgical wound dressings used in medical textiles, filters used in air and water purification systems, intimate apparel like undergarments, sportswear, military uniforms, and fabrics used in home furnishings.

Metal nanoparticles, especially silver nanoparticles (Ag-np), are sought-after choices for antimicrobial finishing. Silver ion (Ag+) is known to be a broad-spectrum anti-germ agent with antifungal, antimicrobial, and antibacterial activities [25]. The increasing use of silver in textile products poses a risk to human health and a threat to the ecosystem. Fabrics finished with silver antimicrobials tend to release silver when they come in contact with fluids like sweat and saliva. The possibility of silver ingestion through touch is higher in case of compromised skin like burns, scalds, and wounds [26]. There have also been previous reports of argyria, a blue-grey skin coloring brought on by extended skin exposure to silver ions [27]. Geochemical studies have detected silver nanoparticles in soil. Ag-np is proven to impair the nitrification process of the nitrogen cycle. Silver nanoparticles affect the microbes involved in N, P, and C metabolisms, in turn hampering soil nutrition [28].

Triclosan was used for textile antigerm finishing. The recently known toxicity concerns of Triclosan and awareness about its environmental persistence have limited its uses. Studies have shown it is toxic to aquatic creatures and has endocrine-disruptive characteristics. Triclosan is marked as an emerging pollutant [29]. The antimicrobial agent- Triclosan by itself is assumed to be non-persistent, but its transformation compound methyl-triclosan is highly persistent and bio-accumulative [30].

Quaternary ammonium compounds {$R_4N^+X^-$} are another class of antimicrobial chemicals especially used for finishing polyamide fibers. These finishes leach out easily post-laundering. The positively charged nitrogen bonds to an anionic fiber substrate via weak ionic interactions. These weak interactions are responsible for the easy leaching of such finishes [31]. This positively charged antimicrobial gets adsorbed easily on the negative cell wall of algae and soil [32] Polyhexamethylene biguanide (PHMB) is another frequently used antimicrobial agent for protective textile finishing. Higher dosage of active compounds is required for getting optimum results. This makes the application of PHMB an undesirable option in antigerm finishing despite its eco-friendly and comparatively safer environmental profile.

N-halamines are also used in the antimicrobial finishing of textiles. They are the only class of antimicrobial finishing agents that can be regenerated by chlorination. Exposure to halogens for their regeneration results in unpleasant odour and shade change. They are often coupled with wrinkle recovery finishes whose free formaldehyde release is another environmental concern to address [33]. The increasing use of antimicrobials has led to a threatening phenomenon known as Antimicrobial Resistance (AMR). The overuse of antimicrobials makes microorganisms immune to antimicrobial action. Antimicrobial treatments become ineffective on such micro-organisms, leading to their uncontrolled growth. Excessive antibiotics used in pharmaceuticals, personal care, and hygiene products like disinfectants, soaps, body washes, sanitizers, and disinfectants are often discussed as the main cause of antimicrobial resistance. However, the AMR resulting from using antimicrobial finishing agents on clothes has not been investigated.

2.3 Challenges in Biodegradability and Sustainability of Textile Finishing Treatments

Apart from the specialty finishes discussed above, some other finishing chemicals remain unaddressed with biodegradability and sustainability concerns. In recent years, customer demands and stringent regulations have driven the textile industry toward sustainable and cleaner production. However, certain finishing processes remain a bottleneck toward sustainability. For instance, free formaldehyde is released during durable press finishing treatments. Even the cross-linkers used for curing specific finishes are based on formaldehyde, a carcinogen [34]. Softeners are amongst the highly consumed finishes. Cyclic siloxanes offer versatile properties like lubrication, softness, smoothening, and water repellency. Octamethyl cyclotetrasiloxane (D4), Decamethyl cyclopentasiloxane (D5), and Dodecamethyl cyclohexasiloxane (D6) are the three cyclosiloxanes which are under vigilant scrutiny of GOTS, ZDHC and other regulatory bodies. The concentration of these cyclosiloxanes in finishing chemicals, especially silicone softeners, is regulated because of their high persistence and bio-accumulative characteristics. The silicone-based finishing agents also pose a risk of choking reverse osmosis (RO) membranes at the effluent treatment stage. The unexhausted finishing chemicals that are retained in the bath ultimately enter the waste streams. These silicate-containing streams can clog the RO membranes. To avoid these problems, some finishing processors prefer non-silicone-based softeners.

Another class of widely consumed softeners is cationic softeners. Cationic softeners used for textile applications are condensates of fatty acids derived from plant-based fats and oils. Most cationic softeners are synthesized from castor oil or other plant-based oils. The felling of trees and deforestation, as well as procuring raw materials for synthesizing these chemicals and then labeling them as 'biobased finishing' chemicals, is an irony. Coated textile materials find applications in healthcare, sports, space, construction, transportation, defense, and other industries. Polyvinyl chloride (PVC) and polyurethane (PU) are the major polymers used in coated textiles. Despite the hazardous production process of PVC, it is widely used. PVC becomes unstable above 80 °C and evolves hydrogen chloride. Stabilizers are added to avoid the degradation of PVC. The commonly used stabilizers are made from lead, which is highly toxic to human health and the environment. Prolonged exposure to PVC can cause dermatitis, skin irritation, and conjunctivitis. After its production, PVC can retain some vinyl chloride monomer, known as a class 1 category carcinogen.

Waste fabrics are either landfilled or incinerated. If fabrics coated with PVC are burnt, toxic dioxins may be released. Phthalates are used as plasticizers where soft and flexible coatings are desired on textile materials. Plasticizers can detach from the PVC coatings as they are not directly bonded to PVC molecules. Owing to these plasticizing compounds' high-fat solubility, they can easily enter the human food cycle. Eco-efficiency studies have shown that PVC plastisol, which has less than

5 ppm free vinyl chloride monomer with non-toxic plasticizers, can be a sustainable coating material. The same study revealed that commercial coating formulations containing polyacrylic resins dispersed in water are more eco-friendly than polyacrylic and polyurethane resins dispersed in organic solvents [35].

Among other polymers, polyurethane is extensively utilized for coating in the textile sector due to its superior flexibility, wear, and weather resistance. Polyurethane coatings can greatly improve a number of textile properties, such as wear comfort, tensile & tear strength, wrinkle resistance, water contact angle, antibacterial qualities, etc. [36]. Polyurethane (PU) is also extensively used to render a surface finish on fabrics intended to make artificial leather. Traditional solvents like Dimethyl Formamide (DMF), used for PU-based textile coatings, are not eco-friendly and are banned by many regulations. Alternative solvents like (N-Methyl-2-pyrrolidone, NMP, DMAc (dimethyl acetamide and dimethylsulfoxide DMSO have been suggested. Some of these ultimately ended up under the scrutiny of regulations. NMP was proposed for scrutiny under Annex XVII of REACH, and DMAc was proposed for Authorisation in Annex XIV of REACH. Other solvents like DMSO are not compatible because of their physical characteristics like Odor, color, solid state, and higher boiling point, which require higher energy for removal after the application is completed. Investigating different solvent types, including bio-based solvents, ionic liquids, and deep eutectic solvents, can offer viable solutions.

Coating manufacturers like Stahl and Coats have developed innovative formulations that are DMF-free and compliant with OEKO-TEX and bluesign systems. Deep eutectic solvents are promising solutions because of their low toxicity and biodegradability. The industrial performance and cost-effectiveness of ionic liquids are under investigation. Solvents derived from bio-based raw materials may have less impact on the environment. Using renewable sources for production is the key to enhancing sustainability. Extensive research studies are necessary to standardize and determine newly developed products' efficacy, chemical compatibility, and ecological hazards.

2.4 Sustainable Textile Finishing and Its Limitations

The finishing stage of textiles uses a huge amount of chemicals and can release toxic materials into the environmental systems. The textile industry has prioritized sustainability from fiber to the finishing stage. Low volatile organic compound (VOC) content, high efficiency, low toxicity, low energy needs, little waste, durability, recyclability, and renewable feedstocks are some sustainability indicators the textile industry is currently focussing. For example, processors have started using powder-based coatings recognized for their low toxicity, recyclability, and zero volatile organic compounds. Scientists have synthesized binder systems using sustainable feedstocks [37].

Emerging trends have shown the use of sustainable chemicals, such as natural materials and plant extracts, for textile finishing. These can be attributed to growing concerns about pollution control and environmental preservation. Natural oils possessing fragrance have gained wide interest in textiles' aroma and wellness finishing segment. Along with rejuvenating effects, such aroma finishes can also add antimicrobial properties to textiles to a certain extent. Other plant-derived natural extracts of citrus fruits, such as jasmine, lavender, mint, pine coalescence, rosemary, sandalwood, tea tree oil, and vanilla, have been proven to produce refreshing effects. Achieving durability in washing and resilience of applied finishes to repeated laundering is a major challenge in the aroma finishing of textiles with natural products [38].

Microencapsulation is believed to be the most effective application technique for retaining long-lasting fragrances on textile materials. Microencapsulation involves isolating active substances to create nano-sized capsules where a membrane safeguards the core material from external conditions. A continuous film of polymeric material surrounds tiny droplets of liquid or solid material. This outer polymeric layer is a barrier, protecting the active ingredients from environmental factors such as humidity, pH, and oxidation. The biodegradability of microencapsulated finishes prepared using core–shell technology is also debatable. Durability and biodegradability don't go hand in hand. The protective outer film of polymeric materials, which encases the active ingredient, is chemically synthesized using melamine–formaldehyde, phenol–formaldehyde, or urea–formaldehyde resins. These thermosetting polymers cannot be recycled, and additionally, they release formaldehyde.

As an alternative to these synthetic polymers, researchers have developed biopolymers as an alternative for sustainable microencapsulation. Gelatine alginate, chitosan, and cellulose are preferred biopolymers for textile applications because of their availability, efficiency, and ease of application. Biodegradation is an environmental challenge associated with microcapsules prepared using biopolymers. Unlike conventional polymers, biopolymers are believed to discharge large numbers of microcapsules into water bodies [39]. The sustainable and innovative techniques are represented in Fig. 2.3.

With the advent of white biotechnology, the use of enzymes has become quite popular in the textile industry. Enzymes have benefitted the textile industry in terms of energy saving. Incorporating enzymes in textile wet processes enables operations at lower temperatures with fewer chemicals. The substrate-specific action of enzymes produces no by-products and minimizes effluent load. Overall, enzymatic processes are considered more sustainable than conventional textile processes because of the biodegradable nature of enzymes. Bio-polishing fabrics, especially denim, is done using cellulase enzymes. Enzymatic bio polishing has substituted the use of pumice stones, previously used for getting an abraded salt and pepper effect on indigo-dyed denim. Pumice stones caused higher levels of suspended solids in finishing effluent. Besides the washed-down appearance, other properties like anti-pilling, softness, smoothness, and luster are enhanced after enzyme bio-polishing. Denim bleaching is conventionally done using sodium hypochlorite or potassium permanganate, followed by neutralization processes that produce salts that increase effluent treatment costs. The use of laccases has been reported in several studies for the bio-bleaching of denim [40]. An alternative for sustainable surface modification of polyester fabrics using cutinase enzymes was studied. Esterases have been suggested

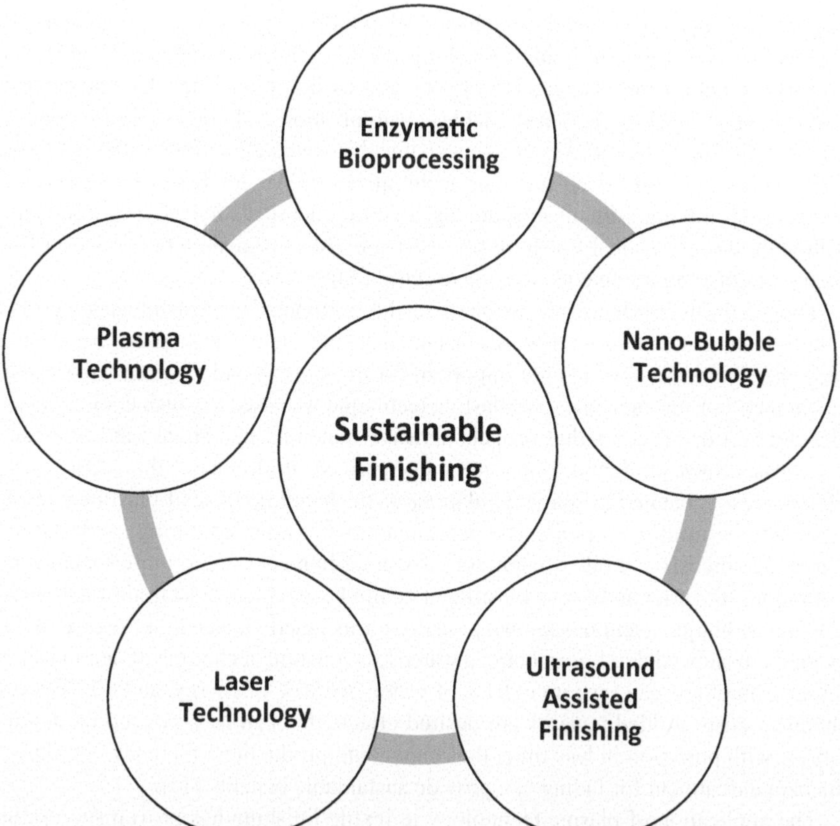

Fig. 2.3 Sustainable and innovative techniques for textile finishing

for eco-friendly weight reduction of polyester, which can avoid harsh chemicals like caustic [41].

Woolen fabrics tend to shrink when machine washing. To avoid this, they are given anti-shrink treatment. The most common process used for the anti-shrink finishing of wool is the chlorine-Hercosett process. Due to the environmental issue of absorbable organic halogens (AOXs), chlorine-based finishing of wool is banned in many developed countries. Other treatments using polymeric coating have also been explored to make wool shrink-resistant. The polymeric coating (resins) used for anti-felting wool is non-biodegradable. Sustainable finishing techniques for wool using proteolytic enzymes have been developed. However, without the polymeric coating, enzymatic treatment alone cannot achieve the necessary shrink resistance [42].

Waterless and dry finishing processes are thought to be resource-saving. Mechanical finishing processes impart desired impacts on fabrics without harmful chemicals. Mechanical processes like calendaring, raising, brushing, shearing, and embossing achieve desired effects such as smoothness, texture, luster, and dimensional stability.

The finished appearance and desired hand feel of fabric can be obtained with reduced reliance on chemicals. Mechanical finishing also minimizes the potential health risks of workers and operators engaged in textile processing mills. The added advantage of mechanical finishing is that it enables customization and allows precise control over the finishing effects. Waterless mechanical systems allow for further resource and cost savings. Advanced finishing techniques like Air Jet Texturing can create crimping or bulking without introducing harmful chemicals into the environment. Although the mechanical finishing processes provide sustainable benefits over the chemical finishing treatments, they have certain limitations.

Though the use of chemicals can be somewhat eliminated in mechanical processes, a significant amount of energy is consumed in producing heat, pressure, steam, and other parameters, which are important for the smooth and efficient working of machinery. For instance, the sandblasting technique was used to finish denim. Sandblasting is a dry process that propels abrasive materials like silica sand to create a distressed look on denim. Since it's a dry process, it eliminates the drying step. This process is banned by many regulations as the handling of sand-blasting guns is extremely harmful to workers. The particulate matter released during sandblasting can cause silicosis and other respiratory diseases. In recent years, more sustainable techniques like laser and ozone blasting of denim have replaced the traditional sandblasting technique. Denim is known globally for its faded vintage look. A case study in Vietnam showed that Laser, being a water-less finishing technique of denim using Laser technology, can save up to 97% of water, 70–90% of energy, and 60–70% on chemical costs. Added benefits are desired effects that can be produced on denim fabrics with precision in less time, thus enhancing production efficiency. Similarly, the ozone treatment for fading can provide sustainable benefits [43].

The application of plasma technology in textile finishing has proven successful in surface modification of textile substrates. Functional effects like water repellence, flame retardancy, and antimicrobial effect can be achieved on plasma-treated fabrics, which was otherwise possible only through chemical finishing. Enhancement in other surface properties like softness, hydrophilicity, and reduced pilling of fabrics by using plasma is also reported [44]. Plasma processes can be energy-intensive, requiring specialized operation and maintenance, which adds to production costs. Reproducibility can be compromised as achieving uniform effects across large substrates can be challenging. The main limitation of these techniques is the complex setup, initial investment cost of machinery which limits their commercialization on a wide scale. There are chances of fabric damage as well, depending on the sensitivity of fabrics. Moreover, plasma treatment causes only the surface modification of fabric,it might not be efficient in applications where deeper penetration of finishes is desired. The durability of these surface modifications can vary and the finished layer can diminish over time. Harmful gaseous emissions like ozone can be released during plasma treatment. It involves the use of high voltage, which may pose a safety risk to operators. For this reason, specialized operation knowledge is needed, which is rare in the current industrial setup. Similar limitations are witnessed while implementing ultrasound-assisted finishing in bulk industrial setups. High-frequency sound waves can create noise and hearing issues, which require soundproofing measures.

Another innovative finishing technique is nano-bubble technology, which was launched by a Spanish company, Jeanologia. Nano bubble technology uses only a small amount of water and the procedure produces no waste. In this process, atmospheric air is subjected to an electro-mechanical shock in an electro-flow reactor that generates a flow of wet air and nanobubbles. These nano-bubbles can produce a smooth surface of fabric. The exhaust application process of finishing is carried out using higher material-to-liquor ratios (MLR) ranging from 1:7 to 1:10. Higher MLR implies higher consumption of water, chemicals, and costs. Recent studies have suggested nano bubble technology can facilitate the use of MLR as low as 1:1 for the application of finishes like softener, resin, oil and water repellent, soil release and antimicrobial finishes onto the cotton fabric [45]. Challenges can be encountered in the production of uniform bubbles, which may lead to uneven finishing. The energy involved in the production of gases is also a concern. Also, these newer technologies are not standardized much. In a nutshell, sustainable alternatives for conventional finishing systems do exist but lack performance or cost-effectiveness. Despite these challenges in commercialization of technological advancements, research efforts continue to make sustainable finishing processes a viable option for textile manufacturers.

2.5 Role of Regulations

Some chemicals involved in textile finishing can be carcinogenic, mutagenic, and reprotoxic agents. The non-biodegradable nature of some chemicals makes them persistent in the environment. These non-degradable chemicals lead to bioaccumulation and biomagnification in the ecosystems. Many countries worldwide have set up regulations on textile products to ensure the wearer's safety and environmental protection. Finishing chemical manufacturers must abide by stringent safety norms. Processors should follow the standards for minimum water use and reduced emissions. Textile companies are being urged to reconsider their product and process management by organizations like the Sustainable Apparel Coalition, Roadmap to Zero Discharge of Hazardous Chemicals (ZDHC) group, and Affirm group, which aim for decreased environmental impacts from the textile, apparel, and footwear industries. The textile processing products are supposed to comply with the standards of regulatory certification bodies like OEKO TEX, bluesign, REACH, etc. (Table 2.2).

Generally, the main challenge of finishing chemical manufacturers lies in abiding by the compliances without compromising on the functional performance of the finish. Sustainability comes with a price. Finishing formulators will focus on achieving a balance between sustainability and cost-effectiveness. Further sustainability in textile finishing can be achieved with digitalization, innovation, and collective efforts from all stakeholders in the textile value chain, including customers. The fast-changing regulations and the obligation to comply with several regulations are also major challenges in the textile finishing sector.

Table 2.2 Biodegradability and sustainability standards for textile finishes

Standards/labels/certification and regulatory systems	Objectives
OECD Guidelines for Testing Finishing Chemical formulations	• OECD 301 for biodegradability of chemical finishes • OECD 201/202/203 for aquatic toxicity, applicable to finishes like natural polymer-based softeners
ISO 14024 Type I Environmental Labelling	• Biodegradable coatings or treatments that facilitate decomposition in natural environments • Energy-efficient finishing processes
ISO 14040/14044 Life Cycle Assessment (LCA)	• International standards for conducting life cycle assessments • Applicable for biodegradable finishes such as enzyme-based finishes
Cradle-to-Cradle Certified	• Assesses products for environmental and social performance across five categories: material health, material reutilization, renewable energy and carbon management, water stewardship, and social fairness • Ensures textile finishes like water repellents or flame retardants are safe, biodegradable and designed for a circular economy
GreenScreen for Safer Chemicals	• Assesses finishing chemicals based on their hazards and promotes safer alternatives
ISO 14851/14852	• Determines the aerobic biodegradability of textile materials
EN 13,432	• European standard for compostable and biodegradable packaging • Applicable to biodegradable textile coatings
ASTM D5988	• Used to evaluate the biodegradability of textile materials when disposed in soil environments
Ecolabel by the European Union	• Sets rigorous environmental performance standards, ensuring biodegradability and low toxicity of textile chemicals
ISO 9001	• Ensures consistency and traceability in textile finishing processes
Global Organic Textile Standard (GOTS)	• Covers the processing, manufacturing, packaging, labelling, trading, and distribution of organic textiles • Prohibits the use of toxic and persistent chemicals during the finishing process
Bluesign System	• Eliminates harmful substances from the beginning of the manufacturing process • Covers chemical management, safety, emissions, worker health and safety in textile finishing

(continued)

Table 2.2 (continued)

Standards/labels/certification and regulatory systems	Objectives
REACH (Registration, Evaluation, Authorization, and Restriction of Chemicals)	• Applicable to manufacturers and importers of the European Union to identify risks associated with the finishing chemicals
OEKO-TEX Standard 100	• Tests for harmful substances in textiles at all processing stages, including finishing

The permissible limits of various compliance management organizations and regulatory bodies can be similar. However, the criteria of each certifying body can be different as well. Chemical regulatory criterion varies according to geographical regions. Some countries have their own regulations that impose restrictions on the importation of some chemicals. The need for a single finishing product to comply with several regulations and different certification criteria is challenging for finishing chemical manufacturers. Setting up a single textile certification body for global compliance can help solve the confusion of abiding by so many restricted substance lists and save the costs of the already fragmented textile sector (Table 2.3).

2.6 Conclusion

Finishing plays a crucial role in textile processing by enhancing the functional properties and commercial value of fabrics. However, the environmental impact of finishing chemicals, particularly their biodegradability, persistence, and potential for bioaccumulation, must be carefully considered. As finishing agents often remain on textiles throughout their lifecycle and even after disposal, their long-term effects on ecosystems cannot be overlooked. This chapter highlights the need to assess the sustainability of commercially available finishing chemicals and explore viable alternatives with improved environmental profiles. The discussion on sustainable finishing agents demonstrates that effective substitutes exist, which can provide the desired functional properties while minimizing harm to the environment. The adoption of biodegradable and non-toxic finishing formulations is a necessary step toward reducing the ecological footprint of textile production. Moving forward, the textile industry must prioritize research and development in green finishing technologies, ensuring that sustainability becomes an integral part of finishing processes. Regulatory frameworks, industry standards, and consumer awareness will also play key roles in accelerating this transition. By addressing the challenges associated with finishing chemicals, the textile sector can contribute to a more sustainable and responsible manufacturing ecosystem, balancing performance with environmental responsibility.

Table 2.3 Regulatory threshold values of finishing chemicals

Some regulated chemicals of textile finishes	Standards	Permissible limits/ Toxicity Indicator
Formaldehyde	OEKO TEX STD 100 Edition 2.2024	Class I- 16 mg/kg
		Class II-75 mg/kg
		Class III-150 mg/kg
		Class IV-300 mg/kg
	Bluesign Restricted substance List Version 14	Type A-15 mg/kg
		Type B-75 mg/kg
		Type C-300 mg/kg
	GOTS 7.0	Baby wear < 16 mg/g
		Other products < 75 mg/kg (Skin Contact)
		Other products < 150 mg/kg (NonSkin Contact)
Phthalates in PVC Coatings		
Phthalates such as BBP, DBP, DCHP, DEHP, DEP, DHNUP, DHP, DHxP, DIBP, DIDP, DIHP, DIHxP, DINP, DMEP, DMP, DNOP, DNP, DPP, DPrP	GOTS 7.0	<100 mg/kg
Phthalates in Plasticisers DnPP, DIPP, DnHP, DCHP, DIHxP, DnOP, DIOP, DEHP, DNP, DIDP	Bluesign Restricted substance List Version 14	Limit Type A, B, C-50 mg/kg
DEHP, DBP, BBP, DIBP	Annex XIV of REACH	Listed as SVHC
Cyclosiloxanes in silicone finishes		
D4 (Octamethylcyclotetrasiloxane)	Bluesign Restricted substance List Version 14	Limit Type A, B, C-30 mg/kg
D5 (Decamethylcyclopentasiloxane)		Limit Type A, B, C-200 mg/kg
D6 (Dodecamethylcyclohexasiloxane)		Limit Type A, B, C-200 mg/kg
Flame retardants		
Antimony	Bluesign Restricted substance List Version 14	Type A- 5 mg/kg Type B- 10 mg/kg Type C–10 mg/kg
Tetra-, penta-, hexa-, hepta-, decabromodiphenyl ether / Tetra-, Penta-, Hexa-, Hepta-Decabromodiphenylether	OEKO-TEX ECO Passport threshold values	10 mg/kg

(continued)

Table 2.3 (continued)

Some regulated chemicals of textile finishes	Standards	Permissible limits/ Toxicity Indicator
	Limit value for certification with restriction	10 mg/kg
	Limit values for undiluted products according to OEKO-TEX® ECO PASSPORT	10 mg/kg
Brominated alkyl alcohols like 2,2-Bis(bromomethyl)-1,3-propanediol—(BBMP);2,3-Dibromopropan-1-ol -(2,3-DBPA);1-Propanol, 2,2-dimethyl-, tribromo deriv	Bluesign system substances List Version 14	5 mg/kg
Chlorinated Paraffin waxes		5 mg/kg
Hexabromocyclododecan, all isomers—group for all major diastereoisomers, Polybrominated diphenyl ethanes, Polybrominated diphenyl ethers		5 mg/kg
Fluorocarbons for durable water & oil repellence, soil release finishes		
PFAS (Poly- and perfluoroalkyl substances)	bluesign system substances list (BSSL) I v14.0 I 01 July 2023	Limit Type A, B, C 50 mg/kg
Perfluorobutane sulfon amides, Perfluorobutane sulfon amido ethanols, Perfluorobutane sulfon amidoethyl (meth)acrylate, Perfluorobutane sulfon halides, Perfluorobutane sulfon polymers		15 mg/kg
Perfluorohexane sulfonic acid and its derivatives		20 μg/kg
Perfluoroalkyl sulfonic acid and its derivatives F(CF2) n [n > 8]		20 μg/kg
Branched Perfluoroalkyl compounds		0.05 mg/kg
Perfluorocarboxylic acid (C9-C14) related Substances		260 μg/kg
Perfluorobutanoic acid, Perfluorohexanoic acid Perfluorooctanoic acid related substances		1000 μg/kg
Antimicrobial finishes		
Permethrin	OEKO-TEX ECO Passport threshold values	250 mg/kg
Triclosan		
Permethrin	Limit value for certification with restriction	1250 mg/kg

(continued)

Table 2.3 (continued)

Some regulated chemicals of textile finishes	Standards	Permissible limits/ Toxicity Indicator
Triclosan		
Permethrin	Limit values for undiluted products according to OEKO-TEX® ECO PASSPORT	250 mg/kg
Triclosan		
Alkylphenol (AP) and Alkylphenol Ethoxylates (APEOs) in wetting agents, softeners		
Nonylphenol (NP), Octylphenol (OP), Octylphenol ethoxylates (OPEO), Nonylphenol ethoxylates (NPEO)	ZDHC MRSL for Textiles and Coated fabric processing	Group B chemical supplier commercial formulation limit-250 ppm
Residual solvent content post-finishing		
NMP, DMAc, DMF	GOTS 7.0	Baby wear and other products -0.05%
Formamide		0.02% by weight
Nonylphenol ethoxylates		100 mg/kg
Chlorinated benzenes & Toluenes		1 mg/kg
Quinoline		100 mg/kg

Declaration

The authors declare that they have no known competing financial interests or personal relationships that could have appeared to influence the work reported in this chapter. Generative AI and AI-assisted technologies have been used in the writing process to improve the readability and language of the work only. After using this tool, the authors carefully reviewed and revised the content to ensure accuracy and quality. The authors take full responsibility for the final version of this publication.

References

1. Roy Choudhury AK (2013) Green chemistry and the textile industry. Text Prog 45(1):3–143
2. Le Marechal AM, Krianec B, Vajnhandl S et al. (2012) Textile finishing industry as an important source of organic pollutants. In: Puzyn T (ed.) Organic pollutants ten years after the stockholm convention—environmental and analytical update. InTech. Available at: http://www.intechopen.com/books/organic-pollutants-ten-years-after-the-stockholm-convention-environmental-and-analytical-update/textile-finishing-industry-as-an-important-source-of-organic-pollutants. Accessed 15 May 2024
3. Leal Filho W, Perry P, Heim H et al (2022) An overview of the contribution of the textiles sector to climate change. Front Environ Sci 10:973102

4. Islam MdS, Islam JMM (2021) Sources and fates of textile solid wastes and their sustainable management. In: Baskar C, Ramakrishna S, Baskar S, et al. (eds) Handbook of solid waste management. Singapore: Springer Singapore, pp. 1–21. Available at: https://doi.org/10.1007/978-981-15-7525-9_109-1. Accessed 15 May 2024
5. Mehta S (2023) Biodegradable textile polymers: a review of current scenario and future opportunities. Environ Technol Rev 12(1):441–457
6. Zambrano MC, Pawlak JJ, Daystar J et al (2021) Impact of dyes and finishes on the aquatic biodegradability of cotton textile fibers and microfibers released on laundering clothes: correlations between enzyme adsorption and activity and biodegradation rates. Mar Pollut Bull 165:112030
7. Lykaki M, Zhang Y-Q, Markiewicz M et al (2021) The influence of textile finishing agents on the biodegradability of shed fibres. Green Chem 23(14):5212–5221
8. Kulkarni S (2014) Alkyl halides. In: Encyclopedia of Toxicology. Elsevier, pp 144–145. Available at: https://linkinghub.elsevier.com/retrieve/pii/B9780123864543002293. Accessed 15 May 2024
9. Rickard BP, Rizvi I, Fenton SE (2022) Per- and poly-fluoroalkyl substances (PFAS) and female reproductive outcomes: PFAS elimination, endocrine-mediated effects, and disease. Toxicology 465:153031
10. Holmquist H, Schellenberger S, Van Der Veen I et al (2016) Properties, performance and associated hazards of state-of-the-art durable water repellent (DWR) chemistry for textile finishing. Environ Int 91:251–264
11. Chambers WS, Hopkins JG, Richards SM (2021) A review of Per- and Polyfluorinated Alkyl substance impairment of reproduction. Frontiers Toxicol 3:732436
12. Li F, Duan J, Tian S et al (2020) Short-chain per- and polyfluoroalkyl substances in aquatic systems: occurrence, impacts and treatment. Chem Eng J 380:122506
13. Van Der Veen I, Schellenberger S, Hanning A-C et al (2022) Fate of Per- and Polyfluoroalkyl substances from durable water-repellent clothing during use. Environ Sci Technol 56(9):5886–5897
14. Schellenberger S, Liagkouridis I, Awad R et al (2022) An outdoor aging study to investigate the release of per- and Polyfluoroalkyl substances (PFAS) from functional textiles. Environ Sci Technol 56(6):3471–3479
15. Mohsin M, Sarwar N, Ahmad S et al (2016) Maleic acid crosslinking of C-6 fluorocarbon as oil and water repellent finish on cellulosic fabrics. J Clean Prod 112:3525–3530
16. Cheng Y, Wang S, Xu Z et al (2023) Non-fluorine oil repellency: To what extent can it substitute perfluoroalkyl substances? Prog Org Coat 183:107726
17. Sharif R, Mohsin M, Ramzan N et al (2022) Synthesis of bio-based non-fluorinated oil and water repellent finishes for cotton fabric by using palmitic acid, succinic acid, and maleic acid. J Nat Fibers 19(16):14077–14088
18. Bashari A, Salehi KAH, Salamatipour N (2020) Bioinspired and green water repellent finishing of textiles using carnauba wax and layer-by-layer technique. J Textile Inst 111(8):1148–1158
19. Ziya Özek H (2018) Silicone-based water repellents. In: Waterproof and water repellent textiles and clothing. Elsevier, pp 153–189
20. Schellenberger S, Gillgard P, Stare A et al (2018) Facing the rain after the phase out: Performance evaluation of alternative fluorinated and non-fluorinated durable water repellents for outdoor fabrics. Chemosphere 193:675–684
21. Chirila L, Danila A (2021) Hydrophobic and oleophobic finishes for textiles. In: Rather LJ, Haji A, Shabbir M (eds) Innovative and emerging technologies for textile dyeing and finishing. 1st ed. Wiley, pp. 325–372. Available at: https://doi.org/10.1002/9781119710288.ch12. Accessed 15 May 2024
22. Tang B, Poma G, Bastiaensen M et al (2019) Bioconcentration and biotransformation of organophosphorus flame retardants (PFRs) in common carp (Cyprinus carpio). Environ Int 126:512–522
23. Horrocks AR (2011) Flame retardant challenges for textiles and fibres: New chemistry versus innovatory solutions. Polym Degrad Stab 96(3):377–392

24. Functional Finishes for Textiles (2015) Elsevier. Available at: https://linkinghub.elsevier.com/retrieve/pii/C20130163738. Accessed 16 May 2024
25. Bruna T, Maldonado-Bravo F, Jara P et al (2021) Silver nanoparticles and their antibacterial applications. Int J Mol Sci 22(13):7202
26. Trop M, Novak M, Rodl S et al (2006) Silver-coated dressing Acticoat caused raised liver enzymes and Argyria-like symptoms in burn patient: the journal of trauma: injury. Infection, and Critical Care 60(3):648–652
27. Vlachou E, Chipp E, Shale E et al (2007) The safety of nanocrystalline silver dressings on burns: a study of systemic silver absorption. Burns 33(8):979–985
28. Yonathan K, Mann R, Mahbub KR et al (2022) The impact of silver nanoparticles on microbial communities and antibiotic resistance determinants in the environment. Environ Pollut 293:118506
29. Yin Y, Wu H, Jiang Z et al (2022) Degradation of Triclosan in the water environment by microorganisms: a review. Microorganisms 10(9):1713
30. Contardo-Jara V, Meinecke S, Feibicke M et al (2021) Fate, bioaccumulation and toxic effects of triclosan on a freshwater community—a mesocosm study. Environ Adv 5:100100
31. Mohapatra S, Yutao L, Goh SG et al (2023) Quaternary ammonium compounds of emerging concern: Classification, occurrence, fate, toxicity and antimicrobial resistance. J Hazard Mater 445:130393
32. Hora PI, Pati SG, McNamara PJ et al (2020) Increased use of quaternary ammonium compounds during the SARS-CoV-2 pandemic and beyond: consideration of environmental implications. Environ Sci Technol Lett 7(9):622–631
33. Li R, Sun M, Jiang Z et al (2014) N-Halamine-bonded cotton fabric with antimicrobial and easy-care properties. Fibers and Polymers 15:234–240
34. Reitz TA, Reeder M, Dang A et al (2022) Formaldehyde released from clothing articles and pillowcases. AATCC J Res 9(4):205–212
35. Bidoki SM, Wittlinger R (2010) Environmental and economical acceptance of polyvinyl chloride (PVC) coating agents. J Clean Prod 18(3):219–225. https://doi.org/10.1016/j.jclepro.2009.10.006
36. Ghonia JR, Savani NG, Prajapati V, Dholakiya BZ (2024) A review on polyurethane based multifunctional materials synthesis for advancement in textile coating applications. J Polym Res 31(3):95. https://doi.org/10.1007/s10965-024-03941-5
37. Czachor-Jadacka D, Biller K, Pilch-Pitera B (2024) Recent development advances in bio-based powder coatings: a review. J Coat Technol Res 21(2):435–444. https://doi.org/10.1007/s11998-023-00849-5
38. Natarajan G, Rajan TP, Das S (2022) Application of sustainable textile finishing using natural biomolecules. J Nat Fibers 19(11):4350–4367. https://doi.org/10.1080/15440478.2020.1857895
39. Peng X, Umer M, Pervez MdN, Hasan KMF, Habib MA, Islam MdS, Lin L, Xiong X, Naddeo V, Cai Y (2023) Biopolymers-based microencapsulation technology for sustainable textiles development: a short review. Case Stud Chem Environ Eng 7:100349. https://doi.org/10.1016/j.cscee.2023.100349
40. Panwar V, Sheikh JN, Dutta T (2020) Sustainable denim bleaching by a novel thermostable bacterial laccase. Appl Biochem Biotechnol 192(4):1238–1254. https://doi.org/10.1007/s12010-020-03390-y
41. Wu J, Cai G, Liu J, Ge H, Wang J (2014) Eco-friendly surface modification on polyester fabrics by esterase treatment. Appl Surf Sci 295:150–157. https://doi.org/10.1016/j.apsusc.2014.01.019
42. Hassan MM, Carr CM (2019) A review of the sustainable methods in imparting shrink resistance to wool fabrics. J Adv Res 18:39–60. https://doi.org/10.1016/j.jare.2019.01.014
43. Nayak R, George M, Jajpura L, Khandual A, Panwar T (2022) Laser and ozone applications for circularity journey in denim manufacturing—a developing country perspective. Curr Opin Green Sustain Chem 38:100680. https://doi.org/10.1016/j.cogsc.2022.100680

44. Haji A, Kan C-W (2021) Plasma treatment for sustainable functionalization of textiles. In: Green chemistry for sustainable textiles. Elsevier, pp 265–277
45. Mohsin M, Sardar S, Hasan M, Akhtar KS, Anam W, Ijaz S, Hassan A (2022) Water efficient, eco-friendly and effluent free nano bubble finishing of cotton fabric. J Nat Fibers 19(16):12586–12595. https://doi.org/10.1080/15440478.2022.2073496

Chapter 3
Bioactive Finishes for Sustainable and Functional Textiles

Sayed Hasan Mahmud and Md. Reazuddin Repon

Abstract This chapter explores the dynamic area of bioactive finishes applied for the enhancement of sustainable and functional textiles. With the rise in consumer awareness about environmental concerns, there is a growing need for textiles that possess both exceptional performance qualities and adhere to eco-friendly principles. Bioactive finishes, derived from natural sources such as plants, animals, fungi, microorganisms, or renewable raw materials, are emerging as a potential alternative to satisfy these requirements. The chapter presents a thorough summary of several bioactive substances, such as antibacterial, UV-protective, flame-retardant, and moisture-wicking chemicals, that may be integrated into textile treatments. These bioactive finishes enhance the durability and use of textiles while also addressing concerns regarding their environmental effect and influence on human health. Considerations like biodegradability, less chemical use, and total life cycle analysis are looked at to see how sustainable these finishes are. Moreover, the chapter explores the methods and techniques such as padding, spraying, coating, or microencapsulation used to incorporate bioactive finishes into textiles, with a focus on the significance of effective and eco-friendly procedures. Furthermore, this chapter focuses on the notable progress achieved in creating bioactive coatings for textiles, with a particular emphasis on their capacity to both enhance sustainability and functionality. As the textile industry pursues new and environmentally aware solutions, bioactive finishes stand out as a viable path for satisfying the increasing expectations of modern customers while minimizing the environmental imprint of textile manufacturing.

Keywords Bioactive finishes · Functional textiles · Microencapsulation · Biodegradability · Eco-friendly

S. H. Mahmud
Department of Textile Engineering, National Institute of Textile Engineering & Research, Dhaka, Bangladesh

Md. R. Repon (✉)
Department of Bioproducts and Biosystems, School of Chemical Engineering, Aalto University, Vuorimiehentie 1, 02150 Espoo, Finland
e-mail: reazmbstu.te@gmail.com

S. Biranje et al. (eds.), *Sustainable Finishing Techniques in Textiles*,
https://doi.org/10.1007/978-981-96-4860-3_3

3.1 Introduction

The textile industry has experienced a notable transition in recent times toward sustainability and functionality, mostly due to increased consumer consciousness and regulatory demands on the consequences on the ecosystem and public health. Conventional textile treatments, which frequently involve the use of artificial chemicals, have drawn critique for their negative effects on the environment and possible health risks [1–3]. As a result, interest in bioactive finishes—creative remedies made from renewable raw materials or natural sources including fungus, microbes, plants, and animals—is growing. These bioactive treatments provide a viable way to create textiles that adhere to eco-friendly principles and high-performance criteria at the same time [4–6, 2, 7]. Bioactive finishes are a wide range of materials intended to give textiles particular qualities that improve their use while upholding sustainability criteria. For example, antibacterial treatments use natural substances like silver nanoparticles or chitosan from crustaceans to prevent microbial growth. This prolongs the life of textiles and decreases the need for frequent washing, saving water and energy [1, 7–11] UV-protective finishes often employ natural UV absorbers like flavonoids or lignin, providing textiles with enhanced protection against harmful ultraviolet radiation while avoiding the environmental and health risks associated with synthetic UV blockers [5, 12–14].

The integration of these bioactive finishes into textiles can be achieved through various methods, including padding [15], spraying, coating [16], and microencapsulation [8]. Every technique offers distinct benefits and obstacles concerning the effectiveness of the application, the longevity of the coating, and the ecological implications. For instance, microencapsulation extends the textile's useful life by protecting the bioactive chemicals throughout processing and enabling regulated release [17, 18].

In addition to their practical advantages, bioactive finishes are essential for mitigating the environmental effect of textile fabrication. Biodegradability is enhanced by the use of natural and renewable resources. Furthermore, the use of bioactive finishes might lessen the requirement for several chemical treatments, lowering the total chemical load and related emissions during the textile manufacturing process [2, 4, 5, 19].

Bioactive finishes emerge as a critical tactic in striking a balance between performance and sustainability as the industry searches for novel ways to satisfy the changing needs of customers and regulatory agencies. This chapter provides a critical discussion into the dynamic field of bioactive finishes used in sustainable and functional textiles. It looks at the different kinds of bioactive materials that are used, how they are applied, and how they affect both human health and the environment. This chapter seeks to provide a full grasp of the potential of bioactive finishes in altering the textile industry through a thorough examination of recent advancements and prospective future developments [7, 14, 17, 19].

Recent research has made significant strides in the development of bioactive finishes for sustainable and functional textiles. Anupama Rajput et al. [6] review antimicrobial coatings on infant diapers, highlighting issues with synthetic materials like non-biodegradability and rashes. Natural coatings (Curcumin, Aloe Vera, Tulsi, Neem) show similar antimicrobial effects compared to synthetic ones (silver and zinc nanoparticles, Titanium Dioxide). T Inprasit et al. [20] used Neem (*Azadiracta indica*) to extract a tannin-rich dye from the bark and an antibacterial agent from the leaves. The dye, applied to hemp fabrics, yielded optimal results at 5% concentration, 100 °C, for 60 min, producing reddish-brown fabrics with excellent color fastness. The antibacterial agent, extracted using methanol, achieved a 99.99% reduction in *Staphylococcus aureus* on treated hemp fabrics. Joshi et al. [21] described natural antimicrobial agents like chitosan, natural dyes, neem extract, and herbal products for textiles, highlighting their environmental benefits. It discusses active ingredients, antimicrobial mechanisms, challenges, and future potential for eco-friendly textile applications. A biofunctional polycotton has been developed using an antimicrobial agent from neem seeds and it's showed over 90% inhibition [22]. Antimicrobial effects lasted up to five washes, and treated fabrics had improved crease recovery with slight reduction in tensile strength. A natural antimicrobial agent has been explored from actinobacteria for eco-friendly finishing on cotton textiles [23]. The treated cotton showed strong antibacterial activity, anti-adhesive properties, and effectiveness against biofilm formation, lasting up to five washes without a binder.

3.2 Bioactive Finishes for Functional Textiles

Many advancements in textile science and engineering can be attributed to the search for functional fabrics. Beyond the conventional decorative and cozy use of textiles, functional textiles are intended to offer additional value by providing qualities such as antibacterial properties [10], UV protection [13], flame retardancy [24], and moisture management [25]. Bioactive finishes, derived from natural sources, present a sustainable and effective means to achieve these functionalities. This section describes various types of bioactive finishes used to enhance textile functionality, exploring their sources, mechanisms of action, and applications in contemporary textile products (Fig. 3.1).

3.2.1 Bioactive Finishes for Antibacterial Textiles

The proliferation of bacteria on textiles [26] can lead to unpleasant odors, stains, and health risks, particularly in environments requiring high hygiene standards, such as hospitals, sportswear, and everyday apparel. Traditional antibacterial treatments [27] often rely on synthetic chemicals, which may pose environmental and health concerns. Bioactive coatings made of natural materials have become a viable way to give textiles antibacterial qualities in response to these difficulties. Chitosan is a biopolymer that is derived from the exoskeletons of crustaceans and is well-known

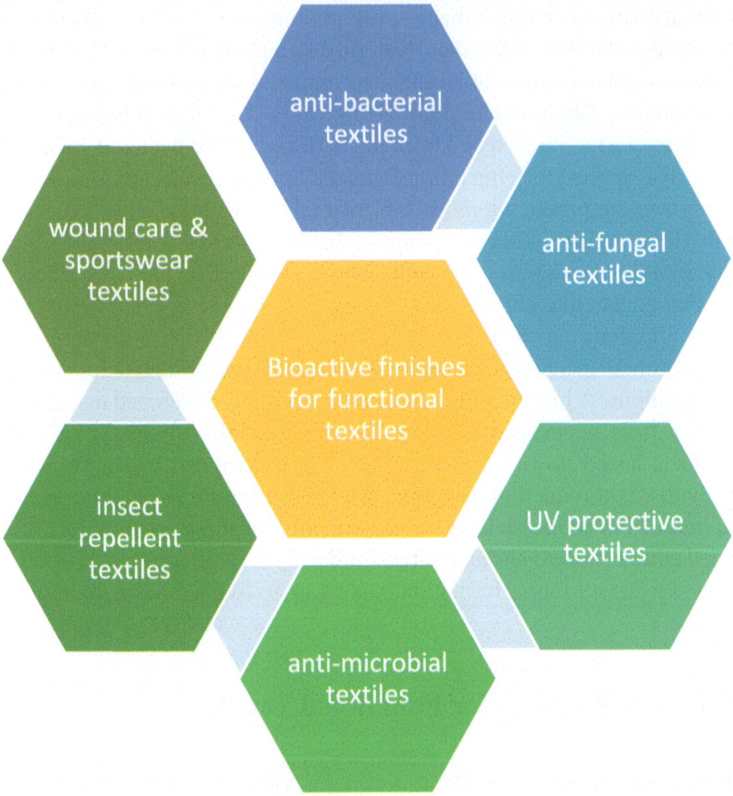

Fig. 3.1 Bioactive finishes for functional textiles

for having strong antibacterial qualities. It works by attaching itself to the negatively charged bacterial cell walls, causing structural disruption and eventual cell death [12]. These nanoparticles cause structural damage and cellular function inhibition by releasing silver ions that pierce bacterial cell walls. This technique works especially well for long-lasting antibacterial protection [28]. Essential oils and extracts from plants like tea tree, eucalyptus, and thyme contain active ingredients such as terpenes and phenolics, which exhibit broad-spectrum antibacterial activity (Table 3.1).

Bacterial cell walls are contacted by substances such as chitosan and silver nanoparticles, which increases permeability and ultimately causes cell lysis [19, 41]. Silver nanoparticles initiate reactive oxygen, which causes oxidative harm to bacterial cells, disrupting their metabolic functions and leading to cell death [42]. Some plant extracts contain compounds that interfere with bacterial enzyme systems, inhibiting their growth and reproduction [43].

Using padding process, the fabric is submerged in a solution that contains the bioactive ingredient, then it is squeezed and allowed to dry. Although it guarantees even distribution, large amounts of the finishing solution can be needed [15].

Table 3.1 Plant extract for anti-bacterial textiles

Plant	Refs.
Neem	[29]
Tulsi	[30]
Harmal	[31]
Guajava	[32]
Chamomile	[33]
Prickly chaff flower	[34]
Ash	[35]
Madder	[36]
Henna	[37]
Turmeric	[38]
Arjun	[39]
Aloe Vera	[40]

Spraying and coating methods can be more effective in using the available materials and enable tailored applications. They work especially well when using plant extracts and silver nanoparticles [16]. The process of encapsulating the bioactive chemicals in microscopic capsules and bonding them to textile fibers is known as microencapsulation. By providing a regulated release and protecting the bioactive compounds during processing, this method increases the textile's useful life [8].

3.2.2 *Bioactive Finishes for Anti-Fungal Textiles*

The growth of fungi on textiles can lead to several undesirable effects, including discoloration, unpleasant odors, deterioration of fabric strength, and potential health issues such as skin infections and allergies [44]. Natural antifungal chemicals are produced by many plants as defense systems. Strong antifungal action is exhibited by terpenoids, flavonoids, and phenolics, which are active components found in essential oils and extracts from plants including oregano, neem, and tea tree [3]. Chitosan is a versatile biopolymer that is derived from crab shells and is well-known for its antifungal qualities. By breaking down the fungal cell walls and preventing spore germination, it works as an antifungal agent. Because of its biocompatibility and biodegradability, chitosan is a desirable choice for long-term antifungal therapies [45]. Certain bacteria and fungi produce metabolites with antifungal properties. For example, Bacillus species produce lipopeptides that can inhibit fungal growth. These microbial metabolites can be harnessed and applied as bioactive finishes to textiles [46].

Some bioactive agents prevent the germination of fungal spores, thereby inhibiting the initial stages of fungal colonization on textiles [47]. Certain plant extracts contain compounds that interfere with the enzymatic activities necessary for fungal growth and metabolism [3].

3.2.3 Bioactive Finishes for UV Protective Textiles

Exposure to ultraviolet (UV) radiation poses significant risks to both textile materials and human health. UV radiation can degrade textile fibers, leading to reduced fabric strength and discoloration, while prolonged exposure to UV rays can cause skin cancer and other health issues in humans. Bioactive finishes derived from natural sources are gaining attention as a sustainable alternative for providing UV protection [48–50].

Numerous plants generate organic substances with the ability to either reflect or absorb UV rays. Commonly present in fruits, vegetables, and even flowers, flavonoids and polyphenols have potent UV-absorbing abilities. For example, substantial concentrations of these chemicals are found in plant extracts from green tea, pomegranates, and turmeric, which have been effectively used to UV-protective textile treatments [51, 52]. Lignin is an organic polymer that is exhibited in plant cell walls and is well-known for its superior UV absorption properties. It is a waste product from the paper and pulp industries that can be used to protect fabrics from UV rays in an environmentally beneficial way [52]. Carotenoids such as beta-carotene and lycopene can absorb UV radiation, thereby protecting the underlying textile fibers from UV-induced damage [53].

UV light can be absorbed by substances like lignin and flavonoids, stopping it from entering the textile fibers. Through this process of absorption, harmful UV photons are transformed into less hazardous energy sources, including heat, which are then released [5]. The quantity of UV light that reaches the textile surface can be decreased by some bioactive compounds' ability to scatter or reflect UV radiation. Minerals like zinc oxide and titanium dioxide are frequently utilized for this purpose [54]. Many plant extracts contain antioxidants that neutralize free radicals generated by UV radiation. This not only protects the textile fibers but also offers additional skin protection benefits to the wearer [49].

3.2.4 Bioactive Finishes for Antimicrobial Textiles

Antimicrobial textiles are essential in various applications, from healthcare to sportswear, due to their ability to inhibit the growth of microorganisms such as bacteria, fungi, and viruses [19, 55]. The antibacterial qualities of silver have long been known. Plant extracts can be used to reduce silver ions, which is an environmentally beneficial way to create silver nanoparticles or AgNPs. These nanoparticles leak silver ions that breach microbial cell walls, damaging structural integrity and impeding biological processes [56]. Many plants produce antimicrobial substances as defensive strategies. Broad-spectrum antimicrobial action is exhibited by terpenes and phenolics, which are active chemicals found in essential oils and plant extracts such as thyme, eucalyptus, and tea tree [19]. Agents like chitosan and silver nanoparticles interact with microbial cell walls, leading to increased permeability and eventual cell lysis [19, 41]. Certain plant extracts contain substances that impede the growth and reproduction of microorganisms by interfering with their enzymatic systems [43].

3.2.5 Bioactive Finishes for Insect Repellent Textiles

For a variety of uses, such as camping gear, military uniforms, and outdoor apparel, insect-repellent textiles are necessary to shield users from bug bites and the spread of vector-borne diseases [57, 58]. Citronella oil, which is derived from the leaves and stems of various Cymbopogon (lemongrass) species, is well recognized for its ability to ward off insects. Mosquitoes and other insects are successfully repelled by their active constituents, which include citronellal, citronellol, and geraniol [59]. Neem oil, which is extracted from the seeds of the Azadirachta indica neem tree, contains substances such as azadirachtin, which has potent insecticidal and repellant qualities. Neem oil is effective against a wide range of insects, including mosquitoes, ticks, and fleas [60]. Eucalyptus oil, particularly from the lemon eucalyptus (Eucalyptus citriodora), contains p-menthane-3,8-diol (PMD), a compound known for its mosquito-repellent properties. Eucalyptus oil is effective and provides a natural alternative to synthetic repellents [9]. Lavender oil, derived from the flowers of Lavandula angustifolia, contains linalool and linalyl acetate, which are known to repel mosquitoes and other insects [61].

A lot of bioactive insect repellents function by covering up or changing the host's aroma, which makes it harder for insects to find their intended victim. By interfering with insects' ability to detect scent, essential oils such as eucalyptus and citronella discourage them from landing on the treated fabric [18]. Some bioactive substances, such as neem oil, include chemicals that are harmful to insects upon touch or consumption. These substances prevent insects from growing and reproducing, which lowers their number [57]. Certain bioactive agents can affect the behavior of insects, making them less likely to bite or settle on treated surfaces. For instance, the compounds in lavender oil can reduce the feeding activity of mosquitoes [62].

3.2.6 Bioactive Finishes for Wound Care Textiles

Wound care textiles are specialized materials designed to promote healing, prevent infection, and improve patient comfort [4, 63]. Chitosan promotes cell proliferation and tissue regeneration, and its antimicrobial properties help prevent infection. Chitosan also has hemostatic properties, aiding in blood clotting and reducing bleeding [12]. Aloe vera extract is widely used in wound care for its soothing, anti-inflammatory, and antimicrobial properties. It promotes wound healing by stimulating fibroblast activity and collagen synthesis, which are crucial for tissue repair [40]. Since ancient times, honey—especially Manuka honey—has been used to treat wounds because of its strong antibacterial and restorative qualities. In addition to reducing inflammation and offering a moist healing environment, it also possesses broad-spectrum antibacterial activity against viruses, bacteria, and fungus [64].

Bioactive agents like chitosan, honey, and silver nanoparticles exhibit broad-spectrum antimicrobial properties, preventing infections by inhibiting the growth of bacteria, fungi, and viruses [10]. Aloe vera and honey are two examples of compounds that lower inflammation and foster a healing environment. By regulating the body's inflammatory response, they aid in the reduction of pain and swelling [65]. Bioactive substances that promote cell proliferation and collagen production, such as chitosan and aloe vera, hasten tissue healing and wound closure. These substances promote the development of new tissue, speeding up the healing process [66]. Chitosan has hemostatic properties, promoting blood clotting and reducing bleeding, which is crucial for acute wound management [53].

Using **impregnation** technique, the textile is soaked in a solution that contains the bioactive chemical, and then it is dried. The process of impregnation enables the bioactive ingredient to be distributed evenly throughout the textile matrix [67]. Electrospinning is a technique used to create nanofibrous mats that can incorporate bioactive agents. These mats provide a large surface area for interaction with the wound, enhancing the delivery of bioactive compounds and promoting healing [68].

3.2.7 Bioactive Finishes for Sportswear

Sportswear demands high-performance materials that offer comfort, durability, and enhanced functionality to support athletic activities [69, 70]. To meet these requirements, various natural substances, plant extracts, and biopolymers are being explored for their ability to enhance the performance and user experience of sportswear. Essential oils and plant extracts, such as those derived from eucalyptus, tea tree, and lavender, possess antimicrobial and anti-inflammatory properties. When incorporated into sportswear, these substances can provide odor control and skin-soothing benefits, particularly useful for athletes who experience extended periods of sweat and friction during physical activity [7]. These natural compounds reduce odor-causing bacteria and help maintain healthy skin, improving the overall comfort of sportswear. Aloe vera extract has a reputation for being calming and hydrating. Because it has a cooling effect and lessens skin irritation, it can be utilized to improve the comfort of sportswear [40]. A natural substance with well-known moisture-wicking and deodorizing qualities is bamboo charcoal. It can assist athletes in staying hydrated and rejuvenated during strenuous exercise [71]. In order to keep athletes dry and comfortable, agents like chitosan and bamboo charcoal improve fabrics' ability to wick away moisture from the skin by removing perspiration and encouraging rapid evaporation [72].

Some plant extracts and biopolymers offer UV protection by absorbing or reflecting UV radiation, protecting the skin from harmful UV rays [48].

3.3 Impact of Bioactive Finishes on Sustainability and Environment

The textile industry has long been evaluated for its negative effects on the environment, including excessive water use, chemical pollution, waste production, and energy use. Because of this, the industry is seeing a rise in the need for eco-friendly products and sustainable methods [62]. Bioactive agents are often derived from renewable resources such as plants, algae, and microbial sources. This reduces dependency on non-renewable petroleum-based chemicals and supports sustainable agricultural practices [73]. Many bioactive finishes are biodegradable, breaking down into harmless components after their useful life. This reduces the environmental burden of textile waste and prevents the accumulation of persistent pollutants [7]. Natural bioactive agents typically exhibit lower toxicity compared to synthetic chemicals. This minimizes the risk of harmful effects on aquatic and terrestrial ecosystems when these substances are released into the environment [62].

3.4 Future Prospects

The future of bioactive finishes in textiles is promising, with ongoing research and innovation aimed at overcoming current challenges and maximizing their sustainability potential. Advances in biotechnology, such as genetic engineering and synthetic biology, are expected to play a significant role in developing novel bioactive agents with enhanced properties. [1, 74]. Collaboration between academia, industry, and regulatory bodies is essential to develop standardized methods for testing and certifying bioactive finishes. This will ensure their safety, efficacy, and environmental performance, facilitating wider adoption [75]. Certifications highlighting the environmental benefits and functional performance of bioactive finishes can be a key motivator for consumers to support sustainable textiles. Increasing consumer awareness and demand for sustainable products can drive the adoption of bioactive finishes in textiles. Transparent labeling and certification can help consumers make informed choices and support sustainable practices. Interdisciplinary research combining textile science, material engineering, and green chemistry is essential for developing scalable, cost-effective, and sustainable bioactive finishes. Efforts to reduce reliance on fossil-based materials in textile finishes and integrate renewable raw materials will contribute to lowering the environmental footprint of bioactive finishes.

3.5 Conclusion

In the textile sector, bioactive finishes provide a revolutionary solution that balances high-performance functionality and environmental sustainability. These finishes, which derive from organic sources like microbes and plants, offer advantages, including wound healing, UV protection, antibacterial and antifungal properties, and are environmentally benign substitutes for conventional synthetic treatments. The effectiveness of bioactive compounds such as essential oils, silver nanoparticles, chitosan, and plant extracts improves the comfort, safety, and longevity of textiles used in sporting, daily wear, and medicinal applications. Because of their renewable sources, they are more biodegradable and less toxic, supporting environmental sustainability by consuming less energy and fewer chemicals throughout the course of a textile's lifetime. But there are still issues with durability, scalability, pricing, and regulatory compliance. To get over these obstacles, biotechnology research and application strategies like microencapsulation are essential. The development of standardized testing procedures requires cooperation between academics, business leaders, and government agencies. Additionally, the textile industry will benefit from growing customer demand for sustainable products, which will drive the use of bioactive finishes.

Declaration The authors declare that they have no known competing financial interests or personal relationships that could have appeared to influence the work reported in this chapter. AI tools have been used in the writing process to improve the readability and language of the work only. After using this tool, the authors carefully reviewed and revised the content to ensure accuracy and quality. The authors take full responsibility for the final version of this publication.

References

1. Eid BM, Ibrahim NA (2021) Recent developments in sustainable finishing of cellulosic textiles employing biotechnology. J Clean Prod 284:124701
2. Thakker AM, Sun D (2021) Sustainable plant-based bioactive materials for functional printed textiles. J Text Inst 112(8):1324–1358
3. Yıldırım FF, Avinc O, Yavas A, Sevgisunar G (2020) Sustainable antifungal and antibacterial textiles using natural resources. Sustain TextE Appar Ind: Sourc Nat Raw Mater, 111–179
4. Nazir F, Javed S, Nazir A, Hussain T, Abid S (2022) Natural plant extract-treated bioactive textiles for wound healing. In Med Text Nat Resour (pp 137–166): Elsevier
5. Nosheen A, Hussain MT, Khalid M, Javid A, Aziz H, Iqbal S, Ali S (2022) Development of protective cotton textiles against biohazards and harmful UV radiation using eco-friendly novel fiber-reactive bioactive agent. Process Saf Environ Prot 165:431–444
6. Rajput A, Ramachandran M, Gotmare VD, Raichurkar PP (2017) Recent bioactive materials for development of eco-friendly dippers: an overview. J Pharm Sci Res 9(10):1844–1848
7. Zhou Q, Rather LJ, Ali A, Wang W, Zhang Y, Haque QMR, Li Q (2020) Environmental friendly bioactive finishing of wool textiles using the tannin-rich extracts of Chinese tallow (Sapium sebiferum L.) waste/fallen leaves. Dye Pigment, 176, 108230
8. Dridi D, Bouaziz A, Gargoubi S, Zouari A, B'chir F, Bartegi A, Boudokhane C (2021) Enhanced antibacterial efficiency of cellulosic fibers: microencapsulation and green grafting strategies. Coatings 11(8):980

9. Endris J, Govindan N (2021) Dyeing and finishing of cotton fabric with eucalyptus leaves extracts. Res J Text Appar 25(3):193–208
10. Jabar JM (2021) Antimicrobial functional textiles. Text Funct Appl, 209
11. Purwar R, Joshi M (2004) Recent developments in antimicrobial finishing of textiles—a review. AATCC review, 4(3)
12. Mondal MIH, Saha J (2019) Antimicrobial, UV resistant and thermal comfort properties of chitosan-and Aloe vera-modified cotton woven fabric. J Polym Environ 27:405–420
13. Shabbir M, Mohammad F (2018) Multifunctional AgNPs@ Wool: Colored, UV-protective and antioxidant functional textiles. Appl Nanosci 8:545–555
14. Xiang B, Liu Q, Sun Q, Gong J, Mu P, Li J (2022) Recent advances in eco-friendly fabrics with special wettability for oil/water separation. Chem Commun 58(97):13413–13438
15. Lee HJ, Yeo SY, Jeong SH (2003) Antibacterial effect of nanosized silver colloidal solution on textile fabrics. J Mater Sci 38:2199–2204
16. Ye Z, Li S, Zhao S, Deng L, Zhang J, Dong A (2021) Textile coatings configured by double-nanoparticles to optimally couple superhydrophobic and antibacterial properties. Chem Eng J 420:127680
17. Singh N, Sheikh J (2020) Microencapsulation and its application in production of functional textiles
18. Valdés A, Ramos M, Beltran A, Garrigos MC (2018) Recent trends in microencapsulation for smart and active innovative textile products. Curr Org Chem 22(12):1237–1248
19. Murugesh Babu K, Ravindra KB (2015) Bioactive antimicrobial agents for finishing of textiles for health care products. J Text Inst 106(7):706–717
20. Inprasit T, Motina K, Pisitsak P, Chitichotpanya P (2018) Dyeability and antibacterial finishing of hemp fabric using natural bioactive neem extract. Fibers Polym 19:2121–2126
21. Joshi M, Ali SW, Purwar R, Rajendran S (2009) Ecofriendly antimicrobial finishing of textiles using bioactive agents based on natural products
22. Joshi M, Ali SW, Rajendran S (2007) Antibacterial finishing of polyester/cotton blend fabrics using neem (Azadirachta indica): a natural bioactive agent. J Appl Polym Sci 106(2):793–800
23. Reshma A, Brindha Priyadarisini V (2021) Eco friendly antimicrobial finishing of textiles using bioactive metabolite from endophytic Streptomyces fradiae CQLW against bio deterioration. J Text Inst 112(6):902–913
24. Gaan S, Salimova V, Rupper P, Ritter A, Schmid H (2011) Flame retardant functional textiles. In Funct TextEs Improv Perform, Prot Health (pp 98–130): Elsevier
25. Bal K, Das B (2023) Moisture management and wicking behavior of textiles. In Funct Tech Textes (pp 453–498): Elsevier
26. Freney J, Renaud FNR (2011) Textiles and microbes. In intelligent textiles and clothing for ballistic and NBC protection: technology at the cutting edge (pp 53–81): Springer
27. Ilieş A, Hodor N, Pantea E, Ilieş DC, Indrie L, Zdrîncă M, Ghergheles C (2022) Antibacterial effect of eco-friendly silver nanoparticles and traditional techniques on aged heritage textile, investigated by dark-field microscopy. Coatings 12(11):1688
28. Simončič B, Klemenčič D (2016) Preparation and performance of silver as an antimicrobial agent for textiles: A review. Text Res J 86(2):210–223
29. Rajendran R, Radhai R, Balakumar C, Ahamed HAM, Vigneswaran C, Vaideki K (2012) Synthesis and characterization of neem chitosan nanocomposites for development of antimicrobial cotton textiles. J Eng Fibers Fabr 7(1):155892501200700130
30. Ravindra KB, Murugesh Babu K (2016) Study of antimicrobial properties of fabrics treated with Ocimum sanctum L (tulsi) extract as a natural active agent. J Nat Fibers 13(5):619–627
31. Adeel S, Anjum F, Zuber M, Hussaan M, Amin N, Ozomay M (2022) Sustainable extraction of colourant from harmal seeds (Peganum harmala) for dyeing of bio-mordanted wool fabric. Sustain 14(19):12226
32. Zayed M, Ghazal H, Othman H, Hassabo AG (2022) Psidium guajava leave extract for improving ultraviolet protection and antibacterial properties of cellulosic fabrics. Biointerf Res Appl Chem 12(3):3811–3835

33. El-Zawahry MM, El Khatib HS, Shokry GM, Rashad HG (2022) One-Pot robust dyeing of cotton fabrics with multifunctional chamomile flower dyes. Fibers Polym 23(8):2234–2249
34. Yasotha P, Pachiyappan KM, Aafrin Banu Mu M, Bhanindhraa B (2019) Natural antimicrobial agents for textile materials. World J Pharm Res, 8(8), 167–176
35. Bukit BF, Frida E, Humaidi S, Sinuhaji P (2022) Selfcleaning and antibacterial activities of textiles using nanocomposite oil palm boiler ash (OPBA), TiO2 and chitosan as coating. S Afr J Chem Eng 41:105–110
36. Güzel ET, Karadag R, Alkan R (2020) Durability, antimicrobial activity and HPLC analysis of dyed silk fabrics using madder and gall oak. J Nat Fibers
37. Yusuf M, Ahmad A, Shahid M, Khan MI, Khan SA, Manzoor N, Mohammad F (2012) Assessment of colorimetric, antibacterial and antifungal properties of woollen yarn dyed with the extract of the leaves of henna (Lawsonia inermis). J Clean Prod 27:42–50
38. Ghoreishian SM, Maleknia L, Mirzapour H, Norouzi M (2013) Antibacterial properties and color fastness of silk fabric dyed with turmeric extract. Fibers Polym 14:201–207
39. Rather LJ, Azam M, Shabbir M, Bukhari MN, Shahid M, Khan MA, Mohammad F (2016) Antimicrobial and fluorescence finishing of woolen yarn with Terminalia arjuna natural dye as an ecofriendly substitute to synthetic antibacterial agents. RSC Adv 6(45):39080–39094
40. Canche-Escamilla G, Colli-Acevedo P, Borges-Argaez R, Quintana-Owen P, May-Crespo JF, Cáceres-Farfan M, Vera-Ku BM (2019) Extraction of phenolic components from an Aloe vera (Aloe barbadensis Miller) crop and their potential as antimicrobials and textile dyes. Sustain Chem Pharm 14:100168
41. Kamonwannasit S, Nantapong N, Kumkrai P, Luecha P, Kupittayanant S, Chudapongse N (2013) Antibacterial activity of Aquilaria crassna leaf extract against Staphylococcus epidermidis by disruption of cell wall. Ann Clin Microbiol Antimicrob 12:1–7
42. Mourenza Á, Gil JA, Mateos LM, Letek M (2020) Oxidative stress-generating antimicrobials, a novel strategy to overcome antibacterial resistance. Antioxids 9(5):361
43. Haase H, Jordan L, Keitel L, Keil C, Mahltig B (2017) Comparison of methods for determining the effectiveness of antibacterial functionalized textiles. PLoS ONE 12(11):e0188304
44. Riaz S, Ashraf M (2020) Recent advances in development of antimicrobial textiles. Adv Funct Finish Text, 129–168
45. Butola BS (2019) Recent advances in chitosan polysaccharide and its derivatives in antimicrobial modification of textile materials. Int J Biol Macromol 121:905–912
46. Hamed AA, Soldatou S, Qader MM, Arjunan S, Miranda KJ, Casolari F, Eshelli M (2020) Screening fungal endophytes derived from under-explored Egyptian marine habitats for antimicrobial and antioxidant properties in factionalised textiles. Microorg 8(10):1617
47. Taha AS, Ibrahim IHM, Abo-Elgat WAA, Abdel-Megeed A, Salem MZM, El-Kareem MSMA (2023) GC–MS, quantum mechanics calculation and the antifungal activity of river red gum essential oil when applied to four natural textiles. Sci Rep 13(1):18214
48. Mavrić Z, Tomšič B, Simončič B (2018) Recent advances in the ultraviolet protection finishing of textiles. Tekstilec, 61(3)
49. Rather LJ, Shabbir M, Li Q, Mohammad F (2019) Coloration, UV protective, and antioxidant finishing of wool fabric via natural dye extracts: cleaner production of bioactive textiles. Environ Prog Sustainable Energy 38(5):13187
50. Samanta KK, Basak S, Chattopadhyay SK (2015) Specialty chemical finishes for sustainable luxurious textiles. Handb Sustain Lux Textes Fash: 1:145–184
51. Kamboj A, Medha K, Gupta V, Jose S (2024) Ultraviolet protection of textiles with herbal dyes: A contemporary review. Sustain Chem Pharm 41:101689
52. Lv S, Liang S, Zuo J, Zhang S, Wang J, Wei D (2023) Lignin-based anti-UV functional materials: recent advances in preparation and application. Iran Polym J 32(11):1477–1497
53. Janarthanan M, Senthil Kumar M (2018) The properties of bioactive substances obtained from seaweeds and their applications in textile industries. J Ind Text 48(1):361–401
54. Emam HE, Ahmed HB (2023) Ultraviolet protection finishing agents in textile functionalization. In Adv Healthc Prot TextEs (pp 423–446): Elsevier

55. Simončič B, Tomšič B (2017) Recent concepts of antimicrobial textile finishes. Text Finish: Recent Dev Futur Trends, 1–68
56. Zhou Y, Tang R-C (2018) Facile and eco-friendly fabrication of colored and bioactive silk materials using silver nanoparticles synthesized by two flavonoids. Polym 10(4):404
57. Martinez MA, Gende LB, Alvarez VA, Gonzalez JS (2022) Biofunctional textiles: functional polymer-carriers with antiviral, antibacterial, antifungal, and repellent activity. In Biobased Mater: Recent Dev Ind Appl (pp 227–258): Springer
58. Zannat A, Uddin MN, Mahmud ST, Mia R, Ahmed T (2024) Natural dyes and pigments in functional finishing. In Renewable Dyes and Pigments (pp. 271–287): Elsevier
59. Cerempei A (2017) Aromatherapeutic textiles. Act IngredS Aromat Med Plants 853:439–446
60. Ibrahim NA, Eid BM, El-Zairy ER (2011) Antibacterial functionalization of reactive-cellulosic prints via inclusion of bioactive Neem oil/βCD complex. Carbohyd Polym 86(3):1313–1319
61. Türkoğlu GC, Erkan G, Karavana SY, Sarıışık AM, Çetmeli Bakadur A, Ütebay B, Popescu A (2023) Spray-Dried oregano oil and lavender oil microcapsules for antibacterial sports and leisurewear. AATCC J Res 10(6):332–345
62. Raj A, Chowdhury A, Ali SW (2022) Green chemistry: its opportunities and challenges in colouration and chemical finishing of textiles. Sustain Chem Pharm 27:100689
63. Schoukens G (2019) Bioactive dressings to promote wound healing. In Adv Textes Wound Care (pp 135–167): Elsevier
64. Freitas AS, Oliveira R, Ribeiro A, Almeida-Aguiar C (2024) Biofunctional textiles: antioxidant and antibacterial finishings of cotton with propolis and honey. Int J Mol Sci, 25(15)
65. Maleki SJ, Crespo JF, Cabanillas B (2019) Anti-inflammatory effects of flavonoids. Food Chem 299:125124
66. Gorgiev S, Zemljić LF, Strnad S, Kokol V (2018) Textile-based biomaterials for surgical applications. In Fundam Biomater: Polym (pp 179–215): Elsevier
67. Abramiuc D, Ciobanu L, Muresan R, Chiosac M, Muresan A (2013) Antibacterial finishing of cotton fabrics using biologically active natural compounds. Fibers Polym 14(11):1826–1833
68. Akram MW, Hoque MMU, Miah MS, Shahid MA, Hossain MF, Mahmud SH (2023) Fabrication and characterization of antimicrobial wound dressing nanofibrous materials by PVA-betel leaf extract. Heliyon, 9(7)
69. Gupta D, Chaudhary H, Gupta C (2015) Sericin based bioactive coating for polyester fabric. Indian J Fibre & TextE Res (IJFTR) 40(1):70–80
70. Hassabo AG, Elmorsy H, Gamal N, Sediek A, Saad F, Hegazy BM, Othman H (2023) Applications of nanotechnology in the creation of smart sportswear for enhanced sports performance: Efficiency and comfort. J TextEs, Color Polym Sci 20(1):11–28
71. Narzary I, Mahato RK, Middha SK, Usha T, Goyal AK (2024) Valorization of bamboo charcoal as a low-cost adsorbent for waste water treatment: a mini review. Adv Bamboo Sci, 100067
72. Gocek İ, Duru SC (2019) Investigating the effects of wicking and antibacterial finishing treatments on some comfort characteristics of Meryl skinlife for seamless activewear/sportswear. J Eng Fibers Fabr 14:1558925019852790
73. Chatha SAS, Asgher M, Asgher R, Hussain AI, Iqbal Y, Hussain SM, Iqbal HMN (2019) Environmentally responsive and anti-bugs textile finishes–Recent trends, challenges, and future perspectives. Sci Total Environ 690:667–682
74. Radhakrishnan S (2014) Application of biotechnology in the processing of textile fabrics. Roadmap to sustainable textiles and clothing: eco-friendly raw materials, technologies, and processing methods, 277–325
75. Baptista-Silva S, Borges S, Brassesco ME, Coscueta ER, Oliveira AL, Pintado M (2022) Research, development and future trends for medical textile products. In Med TextEs Nat Resour (pp 795–828): Elsevier

Chapter 4
Wellness Finishes of Textiles for Aroma, Health, and Hygiene: Integrating Functional Textile Finishing Techniques

Janmay Singh Hada and Chet Ram Meena

Abstract Functional finishes, designed to enhance the performance properties of textile materials, have been widely applied in various technical textiles, including medical textiles. In the healthcare and hygiene sector, textile finishing is important, encompassing medicine and various techniques for functional purposes. Fabrics play a crucial role as a barrier between environmental microorganisms and human skin. Improper washing post-use can lead to health-related issues, as microorganisms can degrade fabric quality, causing discoloration, stains, fiber damage, unpleasant odors, and a slimy feel. The dyeing and finishing with herbs are another area free from toxic and artificial chemicals and provide eco-friendly wellness in apparel and home furnishing. Antimicrobial textiles enhance infection control, serve as barrier materials, and find applications in health and hygiene products. This chapter explores the wellness finishes in textiles for enhancing aroma, health, and hygiene. It emphasizes functional finishing techniques, including antimicrobial treatments, to protect wearers and fabrics from infections while addressing issues like discoloration, odors, and fiber damage caused by microorganisms. Eco-friendly approaches, such as herbal extracts and microencapsulation, are highlighted for their durability and antibacterial efficacy. The chapter also investigates the integration of essential oils, leveraging textiles' proximity to the skin as a medium for aromatherapy. By combining traditional knowledge with scientific advancements in textile finishing, the chapter focuses on enhancing textile functionality, improving quality of life, and optimizing therapeutic benefits through innovative wellness finishes.

Keywords Functional finishes · Wellness · Aroma therapy · Microencapsulation · Hygiene · Different approach

J. S. Hada (✉) · C. R. Meena
Textile Design Department, National Institute of Fashion Technology, Jodhpur, Rajasthan, India
e-mail: janmay.hada@nift.ac.in

C. R. Meena
e-mail: chetram.meena@nift.ac.in

S. Biranje et al. (eds.), *Sustainable Finishing Techniques in Textiles*,
https://doi.org/10.1007/978-981-96-4860-3_4

4.1 Introduction

Rising consumer concerns about health and environmental safety drive the textile industry toward more sustainable practices. Key strategies to meet this demand include adopting green technologies, using eco-friendly textile auxiliaries, and implementing cleaner chemical finishing processes to reduce environmental impact, enhance product safety, and align with global efforts for sustainability [1, 2]. As fabric comes off the loom, its appearance and performance are often not up to consumer standards. Before it reaches the market, the fabric undergoes several finishing processes to enhance these qualities [3]. The choice of finishes depends on the fiber type and the intended end-use of the fabric (Fig. 4.1).

The fibers are being engineered in great variety that they are finer than human hair. It reveals how a remarkable surface can house an intricate structure with a hollow center that performs great feats of climate control or other well-being benefits. Fabric structures are built on the performance of these fibers, extending capabilities through their construction. The range and scale are breathtaking, from the tiniest medical stent invisible to the naked eye to the large composites of the aerospace industry. Surface such as embedding, coating, and finishing further dimension to these materials and form the main function. Antimicrobial, antistatic, anti-odor, and flame-retardant treatments are being used individually and increasingly in combination to provide multiples of performance to the fibers. Some finishes are applied broadly to improve the fabric's visual appeal, while others are tailored to provide specific performance characteristics. Finishing can involve either physical treatment or chemical reactions. Certain finishes permanently alter the fabric's structure and last throughout its life, while others may wash away during laundering and need reapplication to maintain their effectiveness [4, 5]. Fabrics designed for medical applications are referred to as medical textiles and functional finishes are often

Fig. 4.1 Requirement of finishing products

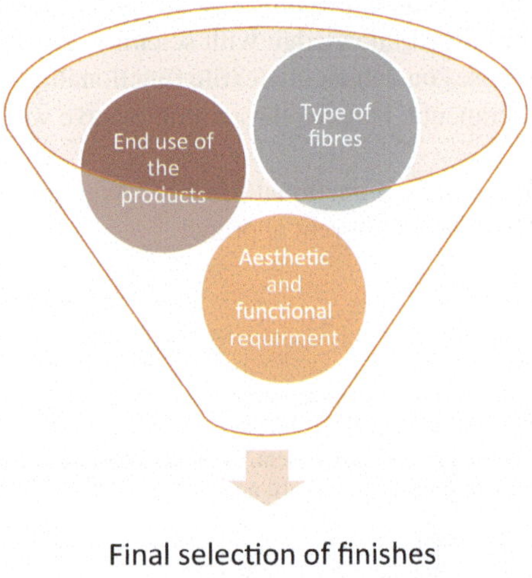

applied to enhance their performance and meet specific healthcare requirements. These finishing processes improve the properties of untreated fabrics and add value by imparting essential functional features tailored for medical use [6, 7]. More advanced finishing treatments can accommodate changing environmental conditions. Finishing treatments can be divided into two phases: pretreatment to clean and free the impurities of the fabrics so that further treatments like dyeing, printing, and finishing are done effectively. Performance enhancement finishes are applied to improve functional characteristics like flame resistance, stain, and water resistance, as well as a durable press, antistatic, soil release, etc.

Integrative medicine, emphasizing the convergence of complementary therapies and conventional medicine, is gaining prominence. Aromatherapy, a centuries-old tradition, is one such complementary therapy. Yet, its therapeutic benefits lack comprehensive scientific understanding, especially regarding essential oil dosage, delivery mechanisms, and overall efficacy on human physiology. The inhaled aroma from essential oils is widely used to stimulate the function of the brain by olfactory receptors through herbs like lemongrass, lavender fennel, etc. Textiles, which remain in direct contact with the skin for extended periods, can be effective carriers for essential oils, facilitating the gradual release and absorption of their therapeutic benefits [8]. This interaction leverages the fabric's proximity to the body, offering a unique method for integrating wellness practices into daily life through clothing, bedding, or wearable accessories. Essential oils, renowned for their aromatic and bioactive qualities, are vital in multiple sectors: food, nutrition, pharmaceuticals, healthcare, cosmetics, and textiles. Understanding their chemical composition and biological properties is crucial to harnessing their full potential and enabling the development of novel, high-value applications. The objective of finishing is to remove deficiencies like the first category: yarn and fabric defects due to mechanical stress) and the second category (impurities, short fibers, thick and thin places, neps and pills, etc.). The finishing process can add new performance characteristics not inherently present in the grey fabrics. It includes performance characteristics like flame resistance, wrinkle resistance, hydrophilicity, hydrophobicity, soft hand, fragrance finish, aroma finish, microencapsulation, etc. Also, it can add changing environmental conditions like self-cleaning and a thermally adapted finish.

4.2 Functional Finishes

Performance-enhancing finishes play a crucial role in improving the functional properties of textiles, addressing a wide range of consumer needs and application requirements. These finishes enhance flame resistance, stain and water repellence, fabric feel (hand), durable press, and soil release. For instance, water-repellent finishes are vital in creating fabrics that resist moisture absorption, making them suitable for various applications, from outdoor gear to medical textiles. Moisture management, in particular, is a key consideration in fabrics designed for water repellence, as it helps regulate moisture levels, ensuring comfort and functionality in everyday and specialized products. In medical textiles, products designed for baby care, adult incontinence, and feminine hygiene play an essential role in the daily lives of millions of people worldwide. These product categories may serve different demographics but share common consumer needs, clinical performance standards, skin care considerations, and technological requirements [9]. As these products must meet functional and comfort expectations, performance finishes that offer moisture control, breathability, and protection against skin irritation are critical. For example, water-repellent fabrics may be employed in incontinence products to manage moisture while maintaining skin dryness, reducing the risk of irritation or infection. Advanced textiles, with their innovative designs and materials, hold the potential to revolutionize various sectors, including healthcare, transportation, environmental sustainability, and personal well-being. These textiles are at the forefront of emerging technologies that enhance the quality of life. For example, medical textiles have diverse applications in healthcare, grouped into four major segments: healthcare and hygiene products, non-implantable medical textiles, implantable medical textiles, and extracorporeal devices. Each category serves a distinct purpose but is interconnected in addressing the broader needs of patient care, hygiene, and treatment. Healthcare and hygiene products encompass wound dressings, incontinence products, and antimicrobial fabrics designed to offer protection and comfort and promote healing. Non-implantable medical textiles are used in products such as surgical gowns, bandages, and drapes, critical in preventing infection and ensuring safe medical procedures. Implantable medical textiles, which include items like surgical mesh and heart valves, are designed to integrate with the human body to support or enhance biological functions. Extracorporeal devices, such as dialysis filters and blood oxygenators, rely on textiles that can handle biological fluids and support life-saving medical treatments. With the advancement of textile technologies, these specialized fabrics can offer significant improvements in clinical performance, comfort, and longevity. Advanced medical textiles improve the functionality of products, address skincare and infection prevention, enhance patient comfort, and are used in wound care, surgical implants, prosthetics, diagnostic devices, and bandages, all while contributing to the development of innovative healthcare solutions, highlighting their significant promise in the medical field [10, 11].

4.3 Wellness and Aroma Therapy

While most fabrics can perform protective or beneficial functions, some are designed for well-being. Fragrance, an increasingly essential part of daily life, can uplift mood, enhance the quality of life, have beneficial effects, and boost work efficiency. Aromatherapy, a centuries-old tradition and one of the complementary therapies has been used in various cultures and societies either as an alternative to or alongside conventional medicine. However, there is limited scientific understanding of its therapeutic benefits, particularly regarding the correct dosage of essential oils, their delivery mechanisms, and their overall efficacy on human physiology [8]. The term 'aromatherapy' was first used by Rene-Maurice Gattefosse in the 1920s to treat patients in military and civilian hospitals [12]. 'Armachology' is defined as the scientific analysis of olfactory effects on mood, physiology, and behavior. Essential oils have long been utilized for their therapeutic properties, particularly through aromatherapy [13]. The precise mechanisms through which these oils exert their effects involve complex interactions with multiple physiological systems. Essential oil molecules, when inhaled, activate systemic pathways and induce neurological responses to promote analgesia and overall well-being. Upon inhalation, volatile molecules of essential oils interact primarily with the olfactory system. These molecules can also influence the gastrointestinal and integumentary systems depending on their biochemical composition and pathway of entry. The olfactory activation triggers signal transduction pathways that influence the central nervous system. Key among these responses is the release of neurotransmitters, particularly endorphins, which act on opioid receptors to produce a calming and analgesic effect.

Textiles (second skin), being in close contact with the skin—have the potential to act as an effective delivery system for the therapeutic benefits of essential oils. These compounds are widely used in perfumes, food, and beverages due to their ability to invoke pleasurable sensations and shift emotional levels from earlier history, as shown in Table 4.1. Acting as chemical messengers, fragrances interact with olfactory receptors in the nose. While naturalness is often preferred in aromas, the creativity of perfumers plays a critical role in crafting unique blends. Fragrance and flavor substances are strong-smelling organic compounds characterized by typically pleasant odors or fragrances (pleasant smell or odor) (Table 4.2).

Aromatic plants are a unique category of flora known for their fragrant compounds, commonly referred to as essential oils (EOs). These oils can be extracted from various plant parts, including petals, flowers, grasses, seeds, stems, leaves, needles, rinds, fruits, roots, rhizomes, woods, and resins. Chemically, EOs are complex, volatile liquids composed of diverse compounds such as terpenes, sesquiterpenes, oxygenated derivatives, aldehydes, oxides, phenols, ethers, acids, and ketones [14, 15]. To be perceived, fragrances must be volatile, with molecular mass typically around 200 and rarely exceeding 300. Due to the sensitivity of olfactory receptors, certain compounds are detectable at extremely low concentrations, with the threshold concentration representing the minimum level distinguishable from a blank under

Table 4.1 The historical evolution of aromatherapy

Time	Particulars
1200 BC	Ancient Hindu scriptures in the Vedas highlight the use of essential oils like sandalwood and cinnamon for promoting health and wellness
1800 BC	A clay tablet from Babylon, dating back to 1800 BC, mentions the use of oils such as cedar, myrrh, and cypress
3500 years ago	The ancient Egyptians, followed by the Romans and Greeks, used plants for medicine, healing, surgery, food preservation, and mummification

Table 4.2 Descriptive terms for various odors

Term	Describe odours
Aldehydic	Fresh, fatty aroma resembling ironed laundry or seawater, commonly found in synthetic and natural scents
Animal	Intense, musky notes derived from natural animal sources like musk, civet, and ambergris
Balsamic	Sweet, warm scents like vanilla, cocoa, and cinnamon, often used for their calming properties
Camphoraceous	Sharp, cooling aroma reminiscent of camphor, frequently used for its refreshing effect
Citrus	Fresh, invigorating scents of citrus fruits like lemon and orange, promoting a revitalizing fee
Earthy	Humus-like aroma evoking the scent of soil, associated with grounding and relaxing effects
Fatty	Mild aroma resembling animal fat or tallow, often used in base compositions
Floral	Diverse scents of flowers, offering a wide range of moods from relaxing (e.g., lavender) to uplifting (e.g., jasmine)
Fruity	Sweet and tangy scents of fruits, imparting a playful and cheerful fragrance
Green	Freshly cut grass and leaves aroma, symbolizing freshness and vitality
Herbaceous	Complex aroma of green herbs such as mint, sage, and eucalyptus, known for their refreshing and therapeutic effects
Medicinal	Sharp, antiseptic-like aroma reminiscent of disinfectants, used for their perceived cleanliness
Metallic	Unique metallic aroma, resembling copper or brass, adding distinctiveness to fragrance profiles
Minty	Cooling, invigorating aroma of peppermint, often associated with alertness and energy
Mossy	Forest and seaweed-inspired scents, evoking nature and tranquility
Powdery	Soft, delicate aroma associated with talcum powders, creating a sense of comfort and cleanliness
Resinous	Rich, aromatic scents of tree exudates, often used for their deep and lasting fragrance
Spicy	Warm, pungent aromas of spices like cinnamon and clove, providing energy and stimulation
Waxy	Subtle scent resembling candle wax, often used for neutral and soothing profiles
Woody	Deep, earthy aromas of woods like cedarwood and sandalwood, offering warmth and sophistication

standard conditions. It is applied using padding mangle with chitosan (by microencapsulation) or silicone softener with Pad dry cure methods. The oils are solubilized by using ethylene glycol.

Aroma compounds, when applied to textiles through an aroma finish, significantly enhance the fabric's sensory appeal and therapeutic properties. As illustrated

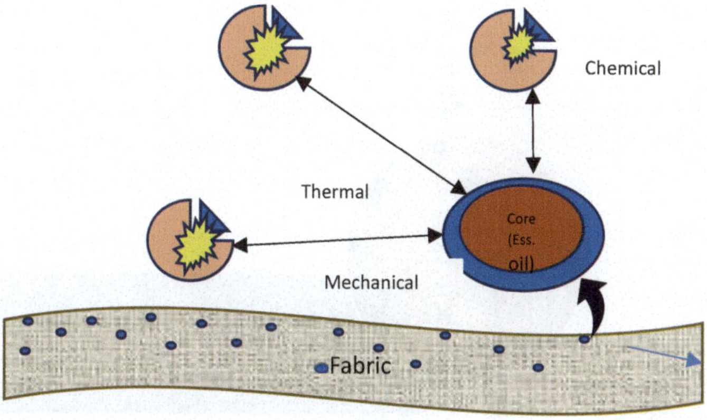

Fig. 4.2 Mechanism of release of oils (essential) in aroma textiles

in Fig. 4.2, these compounds are released gradually through mechanical, chemical, and thermal mechanisms. This controlled release creates a pleasant and memorable experience for the wearer, combining functionality with wellness benefits.

The other aspect is that textile materials are designed to absorb odors from daily human activities like smoking, cooking, and perspiration. These odors negatively affect both the environment and human health, disrupting a healthy lifestyle, as shown in Fig. 4.3. Traditional methods for eliminating absorbed malodours in textiles include absorption/adsorption of volatile organic compounds, hot or steam water extraction, and fragrance finishes to control odor-causing agents. However, these techniques have drawbacks, such as high costs, non-reusability of absorptive chemicals, fabric degradation, and excessive water consumption during hot and steam treatments.

While fragrance finishes using scented powders and sprays offer a temporary solution by providing a pleasant smell, the malodour often returns after use. Microencapsulation has emerged as a more effective technique for controlling odours caused by microbial activity. Essential oils, widely used in this method, can be coated onto textiles to create fabrics that control and reduce odors. Developing textile-based substrates to combat odour-causing organisms is essential for addressing indoor air pollution and maintaining a healthier environment [16]. In 1982, the concept of 'aromachology' emerged, which studies the relationship between fragrance and psychology, exploring how different scents evoke specific emotions and sensations, such as relaxation, happiness, and overall well-being [8, 17].

Common agents used for aroma finishes include essential oils such as lavender, basil, rose, clove, and jasmine, which contain aromatic compounds [18]. These oils are applied to fabrics using different techniques, one of which is the microencapsulation method with the help of chitosan. In this process, the essential oils are encapsulated in tiny capsules bonded to the fabric, allowing for a gradual fragrance

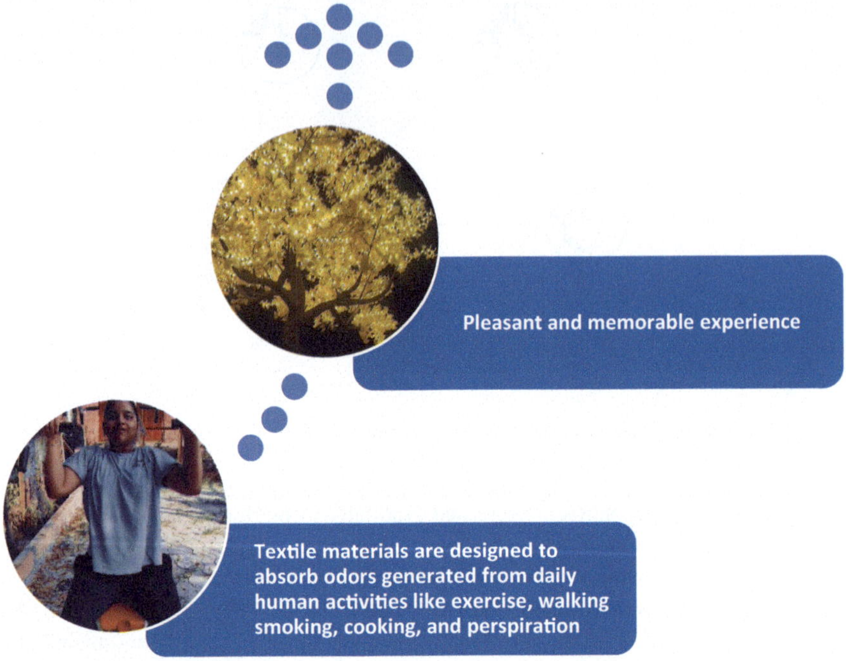

Fig. 4.3 Objective of armoa finish

release. Another method, the padding method, uses a silicone softener for application. Since essential oils are not water-soluble, propylene glycol is used as a solubilizer to dissolve the oils in water, enabling effective application [16]. Microencapsulation effectively manages fragrance longevity by controlling scent release and maintaining potency over time, leveraging essential oils' indispensable properties across various industries [19]. The encapsulation of essential oils addresses their volatile and delicate nature, providing a protective shell that ensures their stability and controlled release. This technique enhances their functionality in textiles, allowing for a sustained release under specific physiological conditions. Encapsulated essential oils expand the scope of textile applications, as shown in Fig. 4.4.

Antimicrobial textiles guard against harmful microorganisms, offering an invisible shield of protection. Cosmetic textiles go a step further, nourishing your skin with slow-releasing skincare benefits as you wear them. Meanwhile, medical textiles work quietly to provide therapeutic effects, supporting health and wellness. And for those who venture outdoors, mosquito-repellent textiles act as a reliable defense against pesky insect bites. Together, these innovative textiles transform everyday garments into tools for comfort, care, and protection (Fig. 4.5).

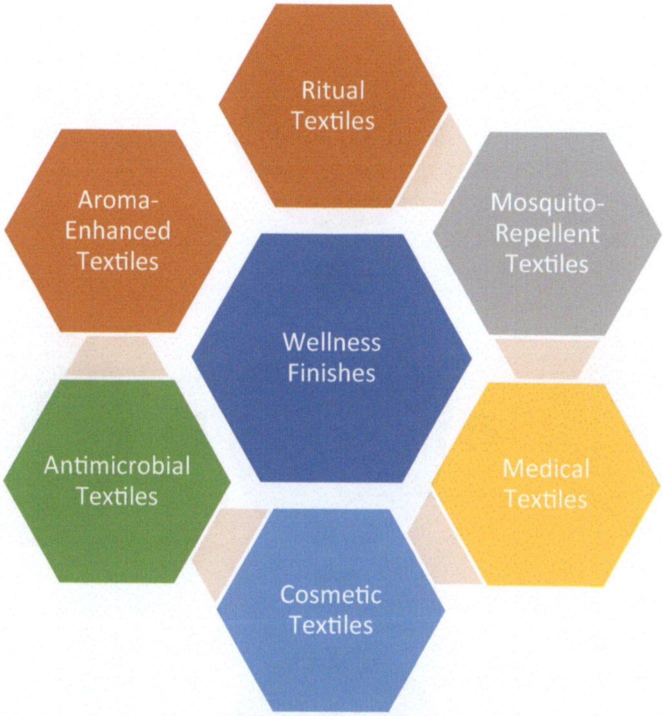

Fig. 4.4 Aroma-enhanced textiles: transforming fabrics, transforming lives

4.4 Microencapsulation

Microencapsulation is a technology that uses microcapsules to add functional properties to textiles and is an important element in developing functional textile finishes [20]. The problem with the traditional finish is the uncontrolled release of the active principle becomes ineffective whereas microencapsulation control increases the duration and effectiveness of the finishing process with functionality, stability, and chemical release. Microencapsulation is a process in which tiny particles or droplets in liquid/solid/gaseous in metrically size or form of an active substance, such as fragrances, drugs, or dyes, are enclosed within a coating or shell material to form small capsules of biopolymers instead of synthetic polymers. Microencapsulation provides a protective barrier and controlled release mechanism, preventing unwanted chemical interactions, preserving the core material, and tailoring the delivery of active ingredients to meet specific needs. This encapsulation helps protect the core material from environmental factors (like heat, moisture, or light), control its release over time, and improve its stability [21]. Depending on the intended use, the coating can be made from various materials, including polymers, lipids, or natural substances (Table 4.3). Microencapsulation offers a distinct combination

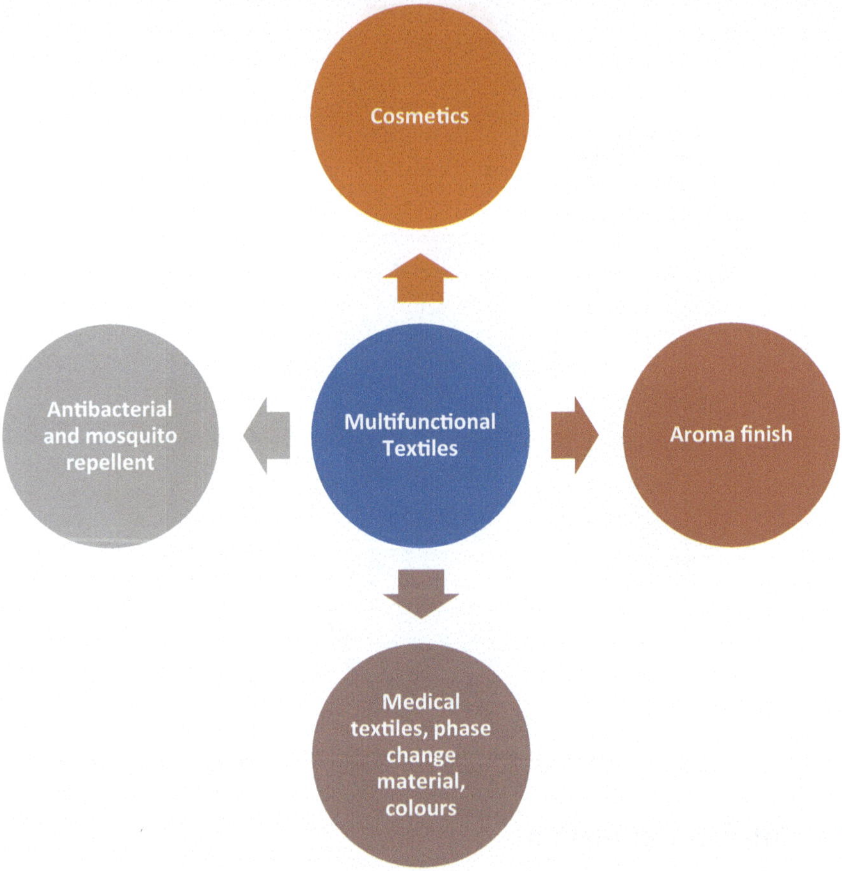

Fig. 4.5 Applications of multifunctional textiles: from cosmetics and aroma finishes to medical uses and protective features

of benefits: eco-friendly biodegradability, cost-effective production, and enhanced accessibility. In textiles, microencapsulation is used to embed fragrances, antibacterial agents, or other functional additives in fabrics [22, 23]. When the fabric is worn or exposed to certain conditions (like heat, moisture, or pressure), the microcapsules break or slowly release their contents, providing long-lasting effects [24]. Microencapsulation is a technique that enhances the thermal stability of fragrances by enveloping them within microscopic capsules. Encapsulating fragrances prevents their degradation due to heat exposure, oxidation, or other external factors, thereby extending their shelf life and maintaining their original potency. Using composite wall materials specially chosen to withstand varying temperatures and environmental conditions further improves the performance. The composite materials selected for the capsule walls ensure that the fragrance remains stable and retains its intended characteristics over time [25, 26]. Combining different materials creates stronger protection, making the microcapsules more resistant to heat, pH, and oxidation and reducing fragrance loss. Fragrances are made of components that evaporate quickly and don't hold up well to heat [27], so using microcapsules to protect them is a better approach. However, when exposed to high heat during processing, the fragrance in standard microcapsules often fades. Over the years, researchers have been trying to develop microcapsules that can handle heat better [28]. Compared to single-layer microcapsules, double or multi-layered ones offer better protection against heat. The double-layer microcapsules are especially helpful because they can fill in shell gaps, preventing fragrance from escaping [29]. Composite shells, made from a mix of organic and inorganic nanoparticles, form strong bonds that make them more stable, helping to lock in the fragrance [30]. The heat resistance of these composite materials depends on how the nanoparticles interact with the polymer [31]. The various materials and mechanisms used for microencapsulation in textiles, showing how they can improve fragrance retention and stability under different conditions, are shown in Table 4.3.

Inorganic–organic composites combine the best qualities of both types of materials, making the production of microcapsules easier, safer, and more versatile. These can be used to encapsulate a wide variety of core materials, including fragrances, drugs, dyes, and more. When used on cotton fabrics, coatings improve properties like conductivity, antibacterial function, and UV protection. For example, the longer

Table 4.3 Materials and mechanisms for microencapsulation in textiles for enhanced fragrance retention and stability

Material type	Mechanism of microencapsulation	References
Gelatin	Gelatin forms a film around the core material (fragrance), which solidifies upon cooling. This creates a protective layer that can release the fragrance slowly over time	Zhang et al. [28]
Chitosan	Chitosan forms a biodegradable and biocompatible shell via ionic gelation, which is responsive to pH changes. This allows for controlled fragrance release under specific conditions	Du et al. [30]
Alginate	Alginate forms biocompatible hydrogels ideal for encapsulating and controlled release of active agents, Alginate for core encapsulation and gel structure	Peng [33]
Cellulose	Cellulose and its derivatives act as protective coatings, enhancing capsule stability and release control. Cellulose derivatives for outer coating and release control	Peng [33]
Polyurethane	Polyurethane creates a strong, flexible coating by polymerizing around the fragrance core, providing a durable encapsulation that offers slow, sustained release under friction or heat	Santos et al. [29]
Silica	Silica nanoparticles form an inorganic shell, creating a barrier that improves thermal stability and reduces fragrance loss by reducing permeability	Jang and Wilkie [31]
Cyclodextrin	Cyclodextrin molecules trap fragrance molecules in their hydrophobic cavity, protecting them from environmental factors and releasing them gradually upon exposure to moisture or heat	Hsieh, Chang, and Gao [27]

the fabric is soaked in a special coating to block electromagnetic interference, the better its shielding properties become [32]. Biopolymer-derived microencapsulated materials are identified as promising for textiles [33].

4.5 Hygiene, Different Approach

Antimicrobial activity encompasses all mechanisms and agents that inhibit the growth of microorganisms, prevent microbial colony formation, or destroy microorganisms entirely. These agents target bacteria, fungi, and other pathogens, ensuring enhanced hygiene and safety in various applications, including textiles, as shown in Table 4.4. It is a situation where an active agent only affects the bacteria or fungi and harms the vitality of the organisms [34]. Mucha defined antimicrobial activity as the interaction where an active agent adversely affects the vitality of microorganisms, creating an environment that limits their growth or survival. These antimicrobial agents find significant applications in textiles, where they enhance fabric performance by providing protection against microbial contamination and maintaining hygiene.

Specific antimicrobial actions:

- **Antibacterial Activity**: Targets and inhibits bacteria specifically.
- **Antimycotic Activity**: Aims to prevent the growth or vitality of fungi.

The application of an antimicrobial finish is microbial degradation and avoiding bacterial infection. It protects fibrous products and the effects of microbes like fungi, like school uniforms, tents, geotextiles, curtains, and baths. Liquid additives like sanitizers are used in the final laundry skins, where cloth is used as apparel and worn against the skin. The category of antimicrobial products is shown in Fig. 4.6. It is used for outdoor and sporting pursuits to prevent the odor on clothing. Antimicrobial fibrous products can be divided into two categories: passive and inactive. The key criteria for these microbial agents are low toxicity risk and washing and dry cleaning resistance.

Table 4.4 Terminology in antimicrobial activity finishing

Antibacterial/antimycotic activity	Active agents that specifically inhibit the growth or vitality of bacteria
Bactericidal	Agents capable of killing living microorganisms, including bacteria and fungi
Bacteriostatic	Agents that prevent bacterial growth without necessarily killing the bacteria
Biostatic	Agents that inhibit the growth of microorganisms without causing immediate death

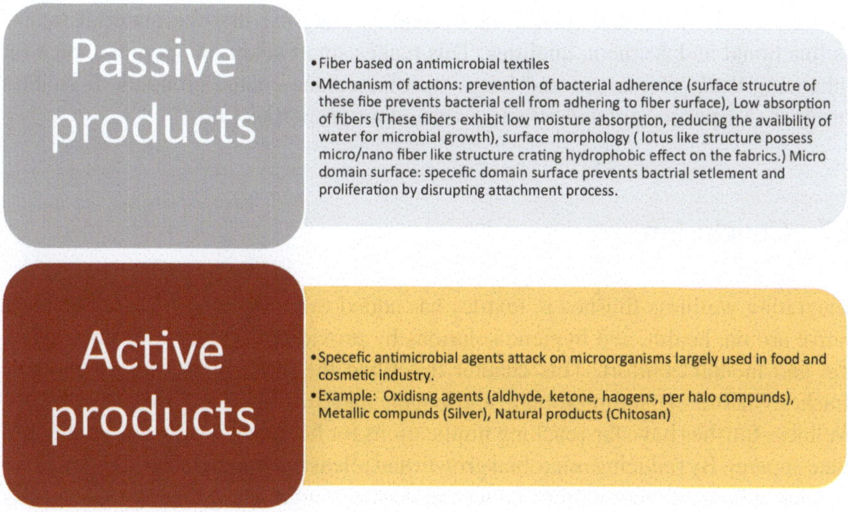

Fig. 4.6 Classification of antimicrobial products

4.6 Silver as Nano Finish/Nano Technology

Nanotechnology presents a wide range of exciting applications, one of which involves harnessing the unique properties of silver at the nanoscale. When reduced to nano-sized particles, silver exhibits remarkable anti-odor and antimicrobial capabilities. These properties make nanosilver highly effective in combating bacteria, fungi, and other microorganisms, which is why it is increasingly used in various industries. From medical equipment and textiles to personal care products and food packaging, nanosilver enhances hygiene and prolongs freshness by inhibiting the growth of harmful microbes. This innovative use of nanotechnology addresses health and cleanliness concerns and contributes to developing more sustainable and efficient solutions in everyday life [35, 36]. Silver is regarded as a natural and safer alternative to other toxic chemicals, offering non-toxic solutions. However, silver does have some limitations, including its high cost, the shedding of silver particles, diminished effectiveness over time, and potential environmental concerns related to effluent discharge [37]. Spinneret clogs after applying silver nanopowder; it does not bond with polymers like cotton and airborne silver dust [38]. Nano Horizons introduced this smart silver as a permanent anti-Odor/anti-microbial additive that can be used with cotton, nylon, polyurethane, polyester, etc. According to Nano Horizons, smart silver can be used with T-shirts, shoe lining, socks, gloves, carpets, and other coatings. Urethane-based materials are widely utilized across diverse applications, including upholstery and rainwear, paints, caulks, and specialized coatings. Integrating advanced silver-based technologies, such as smart silver, significantly enhances these materials by imparting odor resistance and antimicrobial properties that inhibit the growth of mildew and molds. Notably, adding smart silver does not adversely affect key fabric attributes such as elasticity, texture (hand), or dyeability, ensuring that the material retains its functional and aesthetic qualities. This makes smart silver an optimal choice for enhancing the performance and longevity of urethane-coated products. It enables designers for permanent and effective anti-odour properties [39].

4.7 Conclusion

Integrating wellness finishes in textiles has added extra advantages, offering innovative aroma, health, and hygiene solutions by providing wearers with more physical and mental comfort. This chapter described the efficacy of various wellness finishes, including antimicrobial, antifungal, and fragrance-releasing treatments. Wellness finishes have far-reaching implications for healthcare, activewear, and intimate apparel. By reducing microbial growth and releasing beneficial fragrances, these finishes enhance functionality like wearer comfort, confidence, and overall quality of life. In the healthcare and hygiene sectors, functional finishes play an important role in enhancing the performance of products. It enhances performance characteristics as per the end use of the products. Advanced textiles are key to innovative designs, with potential applications in human habitats, transportation, and personal health. Wellness textiles are designed for well-being, incorporating fragrance for improved quality of life. Aromatherapy in textiles uses fragrance compounds to evoke pleasurable sensations, often applied via microencapsulation, which increases the affinity and durability of the process. This technology embeds fragrances in fabrics, releasing them slowly upon wear or exposure to heat and moisture. Microencapsulation also enhances fragrance stability, protecting against heat and oxidation, though research continues to improve thermal resistance. This chapter highlights the importance of antimicrobial agents, terms used in antimicrobial finishing, and types of aroma material used in safeguarding both fabrics and wearers from infections, focusing on the category of active and passive products. The study demonstrates that microencapsulation of essential oils and herbal extracts improves antimicrobial efficacy and offers potential wellness benefits through aromatherapy. By integrating traditional knowledge with modern science, this research provides valuable insights into optimizing therapeutic applications and improving textiles' overall functionality, quality, and value of fabrics. Future research should focus on developing sustainable, eco-friendly wellness finishes and exploring their applications in diverse textile categories.

Declaration The authors declare that they have no known competing financial interests or personal relationships that could have appeared to influence the work reported in this chapter. Generative AI and AI-assisted technologies have been used in the writing process to improve the readability and language of the work only. After using this tool, the authors carefully reviewed and revised the content to ensure accuracy and quality. The authors take full responsibility for the final version of this publication.

References

1. Huma Tariq RA (2023) Recent progress in the microencapsulation of essential oils for sustainable functional textiles. Polym Bull 81(9):7585–7629
2. Rahaman MT, Pranta AD, Repon MR, Ahmed MS, Islam T (2024) Green production and consumption of textiles and apparel: importance, fabrication, challenges and future prospects. J Open Innov Technol Mark Complex 10(2):100280
3. Sekhri S (2011) Text book of fabric science. PHI Learning Pvt. Ltd, New Delhi
4. Joseph M (1988) Essentials of textiles, 6th edn. Rinehart and Winston Inc., Florida
5. Paul R (ed) (2014) Functional finishes for textiles: improving comfort, performance and protection. Elsevier
6. El Mogahzy Y (2009) Engineering textiles. Woohead publishing limited, Cambridge, England
7. Karim N, Afroj S, Lloyd K, Oaten LC, Andreeva DV, Carr C, Farmery AD, Kim ID, Novoselov KS (2020) Sustainable personal protective clothing for healthcare applications: a review. ACS Nano 14(10):12313–12340
8. Mehta S, MacGillivray M (2022) Aromatherapy in textiles: a systematic review of studies examining textiles as a potential carrier for the therapeutic effects of essential oils. Textiles 2(1):29–49
9. Wiesemann F, Adam R (2011) Absorbent products for personal health care and hygiene. In: Bartels, VT (eds) Handbook of medical textiles. Woodhead Publishing, pp 316–33
10. O'Mahony M (2011) Advanced textile for health and well being. Thames and Hudson, London, United Kingdom
11. Guru R, Kumar A, Kumar R (2022) Healthcare and hygiene products application in medical textile. In: Ibrahim, H (eds) Next-generation textiles. IntechOpen. https://doi.org/10.5772/intechopen.103662
12. Brahms JC (2004) Aroma, touch and well-being: following the mind to wellness. Int J Cosmet Sci 26(3):168–168
13. Thangaleela S, Sivamaruthi BS, Kesika P, Bharathi M, Kunaviktikul W, Klunklin A, Chanthapoon C, Chaiyasut C (2022) Essential oils, phytoncides, aromachology, and aromatherapy—a review. Appl Sci 12(9):4495
14. Ghayempour S, Montazer M (2013) Micro/nanoencapsulation of essential oils and fragrances: focus on perfumed, antimicrobial, mosquito-repellent and medical textiles. J Microencapsul 33(6): 497–510
15. Hanif MA, Nisar S, Khan GS, Mushtaq Z, Zubair M (2019) Essential oils. In: Malik S (eds) Essential oil research. Springer, Cham. https://doi.org/10.1007/978-3-030-16546-8_1
16. Hemamalini T, Dev VRG (2022) Fragrance finishing on textile substrate for odour control. In: Thilagavathi, G, Rathinamoorthy R (eds) Odour in textiles. CRC Press, pp 143–166
17. Neha (2020) Aromatic finishes. Textile value chain. https://textilevaluechain.in/news-insights/fabrics-processings-news/aromatic-finishes
18. Bauer K, Garbe D, Surburg H (2008) Common fragrance and flavor materials: preparation, properties and uses. Wiley
19. Kłosowska A, Wawrzyńczak A, Feliczak-Guzik A (2023) Microencapsulation as a route for obtaining encapsulated flavors and fragrances. Cosmetics 10(1):26
20. Nelson G (2002) Application of microencapsulation in textiles. Int J Pharm 242(1–2):55–62
21. Xu Y, Yan X, Zheng H, Li J, Wu X, Xu J, Zhen Z, Du C (2024) The application of encapsulation technology in the food Industry: classifications, recent advances, and perspectives. Food Chem X 21:101240
22. Boh Podgornik B, Šandrić S, Kert M (2021) Microencapsulation for functional textile coatings with emphasis on biodegradability—a systematic review. Coatings 11(11):1371
23. Bojana BP, Marica S (2016) Microencapsulation technology and applications in added-value functional textiles. Phys Sci Rev 1(1):20150003
24. Ghosh SK (2006) Functional coatings and microencapsulation: a general perspective. In: Ghosh SK (eds) Functional coatings: by polymer microencapsulation, pp 1–28

25. Nelson G (2001) Microencapsulation in textile finishing. Rev Prog Color Relat Top 31:57–64
26. Tariq H, Rehman A, Raza ZA, Kishwar F, Abid S (2024) Recent progress in the microencapsulation of essential oils for sustainable functional textiles. Polym Bull 81(9):7585–7629
27. Hsieh W-C, Chang C-P, Gao Y-L (2006) Controlled release properties of Chitosan encapsulated volatile citronella oil microcapsules by thermal treatments. Colloids Surf B Biointerfaces 53(2):209–214
28. Zhang T, Luo Y, Wang M, Chen F, Liu J, Meng K, Zhao H (2020) Double-layered microcapsules significantly improve the long-term effectiveness of essential oil. Polymers 12(8):1651
29. Santos ANB, dos Santos DJ, Carastan DJ (2021) Microencapsulation of reactive isocyanates for application in self-healing materials: a review. J Microencapsul 38(5):338–356
30. Du WY, Li JH, Liu Q (2020) Effect of temperatures on self-healing capabilities of concrete with different shell composition microcapsules containing toluene-di-isocyanate. Constr Build Mater 247:118575
31. Jang BN, Wilkie CA (2005) The thermal degradation of polystyrene nanocomposite. Polymer 46(9):2933–2942
32. Ghosh S, Ganguly S, Das P, Das TK, Bose M, Singha NK, Das AK, Das NC (2019) Fabrication of reduced graphene oxide/silver nanoparticles decorated conductive cotton fabric for high performing electromagnetic interference shielding and antibacterial application. Fibers Polym 20:1161–1171
33. Peng X, Umer M, Pervez MN, Hasan KF, Habib MA, Islam MS, Lin L, Xiong X, Naddeo V, Cai Y (2023) Biopolymers-based microencapsulation technology for sustainable textiles development: a short review. Case Stud Chem Environ Eng 7:100349
34. Mucha H, Hofer D (2002) Antimicobial finishes and modifications. Melliand Int 148–151
35. Butola BS, Mohammad F (2016) Silver nanomaterials as future colorants and potential antimicrobial agents for natural and synthetic textile materials. RSC Adv 6(50):44232–44247
36. Rai M, Birla S, Ingle AP, Gupta I, Gade A, Abd-Elsalam K, Marcato PD, Duran N (2014) Nanosilver: an inorganic nanoparticle with myriad potential applications. Nanotechnol Rev 3(3):281–309
37. Drexler KE (1990) Engines of creation: The coming era of nano technology. Anchor Books, New York
38. Silver S (kein Datum) Nanohorizons. Von Abgerufen. www.nanohorizons.com.
39. El Mogahzy YE (2009) Finishing processes for fibrous assemblies in textile product designing. In: Engineered textiles. Woodhead Publishing, pp 316–335

Chapter 5
Biotechnological Innovations for Sustainable and Environment-Friendly Textile Production

Mahwish Salman, Shahid Adeel, Muhammad Usama, Afifa Zaka, Zafer Özomay, and Rony Mia

Abstract Textile production and finishing is a vital area in the textile sector due to its advancements in various functional applications. The textile production sector uses multiple chemicals to process natural and synthetic fabrics and discharges toxic materials into the natural environment. Consequently, it is necessary to reevaluate and improve the methods used as well as the production procedures. Biotechnology has made manufacturing processes more effective and eco-friendlier, which has a significant impact on the textile sector. Reducing the consumption of natural resources, hazardous chemicals, waste material formation, resulting costs, and energy consumption are all made possible by utilizing sustainable production techniques. This development in the textile sector has been accelerated because of the efficient and promising results of biotechnology. Many chemical processes used in textile production have been replaced with more promising and environmentally friendly biological treatments. To achieve a sustainable bio-based economy, the use of enzymatic methods has been widely considered in the textile sector. Different biological procedures have been employed to remove toxic and hazardous materials from the effluent emitted into the environment. In this chapter, current advancements and the application of biotechnological methods in textile production have been discussed. It

M. Salman · M. Usama
Department of Biochemistry, Government College University Faisalabad, Faisalabad, Pakistan

S. Adeel (✉) · A. Zaka
Department of Applied Chemistry, Government College University Faisalabad, Faisalabad, Pakistan
e-mail: shahidadeel@gcuf.edu.pk

Z. Özomay
Faculty of Applied Technologies, Department of Printing Technologies, Marmara University, Istanbul, Turkey

R. Mia (✉)
Wilson College of Textiles, North Carolina State University, Raleigh, NC, USA
e-mail: mroni_mia@yahoo.com

S. Biranje et al. (eds.), *Sustainable Finishing Techniques in Textiles*,
https://doi.org/10.1007/978-981-96-4860-3_5

has also aimed to provide new guidelines for traders, academic researchers, and the textile community to take steps toward the use of biotechnology in textile production.

Keywords Biotechnology · Bio-functionalization · Sustainability · Bioprocessing · Bioeconomy

5.1 Introduction

Ensuring an adequate supply of raw materials is crucial for the shift toward a bio-based economy. Therefore, it is necessary to acknowledge the availability of resources from both supply and demand aspects [1]. The biomass that will be used for energy production comes from a variety of sources, such as food and feed production, waste, and leftovers from crops, ocean, and forest areas [2]. Animal byproducts and waste, and domestic garbage are further possible sources for bio-based products after recycling [3]. Biotechnological tools offer more sustainable and eco-friendly alternatives to conventional procedures to make textile production environment-friendly [4]. Nowadays, enzymes are widely used in textile processing, such as bio-desizing, bio-polishing, and bio-scouring, to replace the expensive and dangerous chemical methods [5, 6]. Biotechnological advancement also improves the textile processing from colouring to finishing aspects [7, 8]. The cost and time effect approaches of enzyme technology has spread the ecological awareness among the global community for in their many-fold benefits [9], which have been expressed in Fig. 5.1.

Modern bio-aided technology is the process of employing microbes or cell constituents to produce beneficial industrial products [10]. This technology is used to produce bio-based products through a sustainable raw material supply with minimal waste production. This technology produces eco-friendly products using sustainable biological processes, systems, and organisms [11, 12]. Through these technologies, specialized genetic engineering approaches have been employed to modify the enzyme functionalities. Most of the enzymes are isolated from genetically modified organisms [13]. The use of enzymes in the technical textile industry is rapidly gaining global interest. This utilization has given green products at the global level with pollution-free effects [14, 15]. Using this technology, the reduction in water consumption has been reduced from 30 to 50%, which shows that this tool has the potential to move toward low-emission discharge [16, 17]. Thus, sustainable approaches have gained fame around the globe for their widespread use to generate products of people's interest [18].

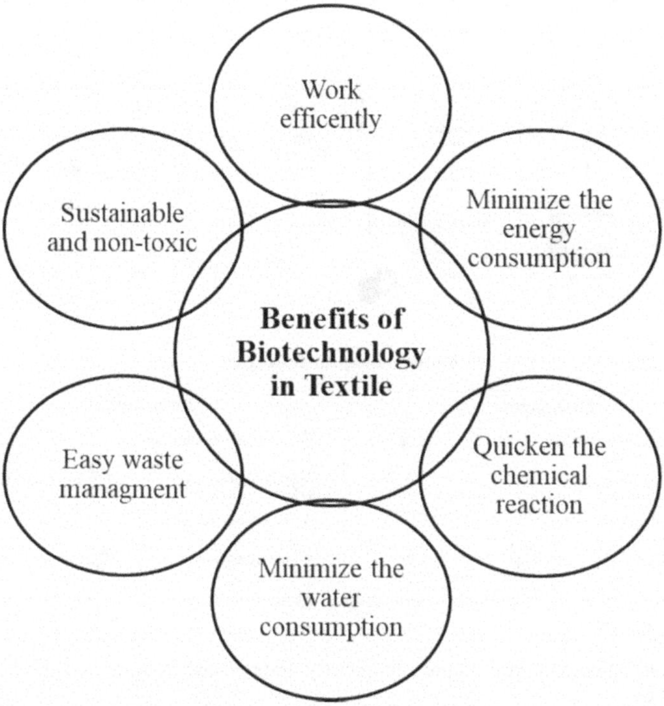

Fig. 5.1 Innovative approaches of enzymes in the textile industry

5.2 Textile Processing

The primary goal of the textile sector is to manufacture yarn, fiber, and fabric, which after the dyeing and finishing process leads to garments [19]. Many toxic and hazardous chemicals, including lubricants, solvent zing agents, cleaning agents, defoamers, surfactants, and dyes, are employed excessively at the various stages of textile manufacturing [20]. This process causes an enormous number of harmful effluents which pollute the environment. According to a recent report, China uses 49% of chemicals in the textile industry whereas other Asian countries use 12%, European countries use 10% and North America contributes 11% in total chemical consumption. In the Indo-Pak region, this contribution lasts up to 10%, which is very alarming for global health [21]. To overcome the damaging effect of such hazardous chemicals, the textile industries now approaching the use of a variety of enzymes for different processes such as bleaching, finishing, dyeing, and effluent management to make the process more sustainable [22]. The application of enzyme biotechnology in textile chemical treatments is quickly gaining attention globally because of its non-toxic and environment-friendly qualities. This is due to the increasing demand for manufacturers of textiles to minimize pollution throughout the textile manufacturing process [23]. Enzymes are biological catalysts, where they can accelerate the process

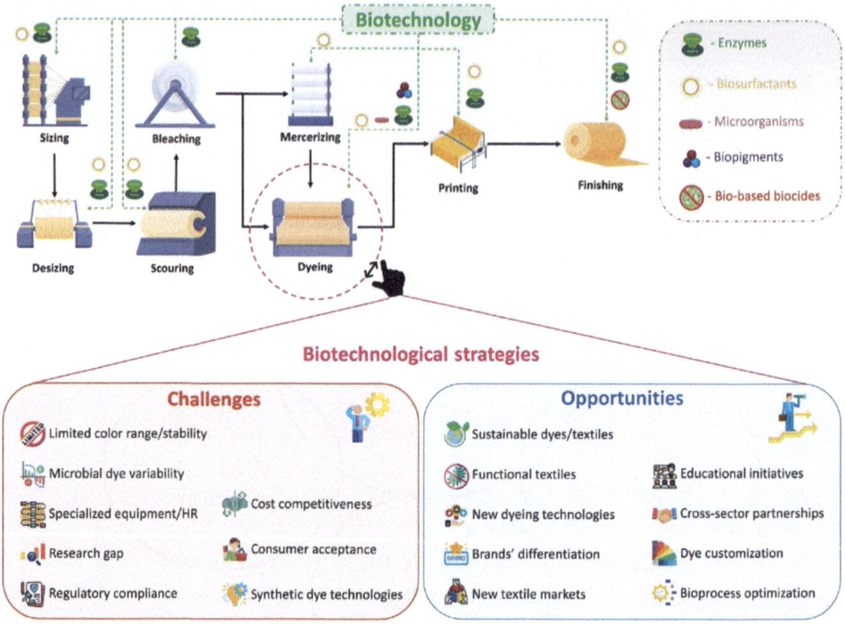

Fig. 5.2 Biotechnology used in the textile industry, Reproduced from the Ref. [26], Copyright 2024, Cell Press

with minimal pollution load [24]. Due to these advanced sustainable technologies, the global market of enzymes for fabric processing will reach up to 801.3 million (USD) by 2032 [25].

The application of biotechnology has the potential to revolutionize the way of dye garments and fabrics. Figure 5.2 illustrates how the partial or full integration of new biotechnological products can improve the sustainability of the entire textile industry [26]. This can be accomplished by addressing a wide range of environmental concerns and manufacturing challenges through the utilization of biopigments, microorganisms, enzymes, and biosurfactants, among other bio-based products [27]. A significant reduction in environmental impact, water and energy consumption, and waste generation can be achieved through the use of these compounds, which have the potential to successfully replace unsustainable compounds utilized in conventional textile manufacturing methods [28, 29].

5.2.1 *Biotechnological Advancements in Cotton Processing*

For the wet-processing of natural and manmade fabrics, it is essential to remove the starch-based protective layer. Different conventional methods such as inorganic acids and oxidizing agents were used to seize the fabric [30]. However, these methods cause damage to the fabric and reduce its tensile strength [31]. The enzymes isolated from different microbial sources such as *Bacillus licheniformis, Bacillus amyloliquefaciens*, and *Bacillus stearothermophilus* have been employed for this purpose [32]. These enzymes work by degrading the starch molecules into dextrin and maltose subunits which can easily wash out in hot water after enzymatic application [13]. This enzymatic desizing is better than the traditional method due to the specificity of amylase toward starch molecules [33]. A temperature of 30–60 °C and pH of 5.5–6.5 is ideal for amylase desizing because at these conditions minimum amount of water is used [34]. The goal of the scouring procedure for cellulosic fabric is to eliminate any added or natural contaminants for hydrophobic properties [35]. This process also uses a high proportion of chemicals, water, and energy after processing and emits carcinogenic water into our ecosystem [36]. To replace the traditional alkaline-based method, a broader range of eco-friendly enzymes, such as pectinases, lipases, proteases, and cellulases alone or in the form of a binary mixture has been studied [37]. Bio-polishing is a very important technique for cotton processing which gives whiteness to the treated fabric with or without removing the colorant [38]. In conventional bleaching approaches, H_2O_2 has been employed which requires a lot of energy and time [39]. These enzyme-based approaches are more sustainable and eco-friendly as compared to the conventional approaches [40]. Cellulase enzyme has been employed for the removal of micro and fuzzy fibrils from the surface of textile fabric. The enzymatic treatment provides a better hand feel and reduces the formation of pills [41]. The previous studies reported that the biopolishing before dyeing process gives better results because pre-treatment of bio-polishing enhances the smoothness of the surface by removing the microfibrils [42]. For biopolishing, an acid and neutral cellulase bath is maintained at the acidic (4.5–5 pH) or neutral (7 pH) level starting from 55 to 85 °C [43]. Hence, enzymatic approaches are eco-friendlier and sustainable and are considered as one of the alternatives to conventional approaches [44]. New enzymes are now used to minimize the energy consumption and adverse effects of industrial effluents in the ecosystem [45].

5.3 Biological Action of Textiles

Textile fabrics composed of both manmade and natural fibers can be ready to color after the finishing step [46]. In textile production, this procedure has received a lot of attention and become a bio-tool to add value to fabric sustainability.

5.3.1 Antimicrobial Activity of Fabrics

The use of antimicrobial fabrics has been grown to reduce the bioburden in the clinical sector and consequently minimize the risk of hospital-acquired infections [47]. Wearable textiles are a suitable place for microbes to grow due to their large surface area and ability to retain moisture [48]. Antimicrobial textile products help to prevent the transmission of deadly pathogens. The selection of antimicrobial agents to get bactericidal and bacteriostatic characteristics depends on the required efficacy toward specific pathogen microbe [49]. The coating of inorganic nanoparticles has gained more interest due to their high bactericidal activity as compared to many antibiotics, but their high concentration can be toxic and can cause adverse effects on human cells. These inorganic nanoparticles work by producing reactive oxygen species (ROS) which disrupt the microbial cell membrane and lead to microbial death. However, the addition of plasma or chemical actions makes the process costly [50, 51].

Cellulase derivatives have been broadly examined through practical procedures due to their chemical stability, biocompatible and biodegradable potential. These derivatives are eco-friendly, non-toxic, and suitable for enzymatic activity maintenance [52]. Moreover, the presence of hydroxyl group on the surface of cellulase, makes them favorite for immobilization. Different functionalized molecules have been added to obtain strong covalent immobilization [53, 54]. To get antimicrobial products, different enzymes are used both individually and in combination with other enzymes, along with metals, organic and inorganic nanoparticles, antibiotics, and antimicrobial peptides. Different enzymes such as laccase and oxidoreductase have been used for the manufacturing of antimicrobial products [55, 56]. They are used to enhance the attachment of biomolecules by modifying the polymeric surface and increasing the coating efficiency of textile material by polymerization of polyphenols. Similarly, lysozyme is used for getting antibacterial products [57, 58].

5.3.2 Bio-functionalization of Textile

Bio-functionalization of textiles is now a day warm topic under experimentation and observation. The effect of humidity, temperature, pH, microbes, and wind can be engineered by overcome through functionalized materials [59, 60]. These functional fabrics made of biomaterial have the potential to deliver more than one functionality apart from the protection of the community. Due to the advancements in biotechnology, various approaches have been reported to develop a thin uniform film of bioactive material for the functionalization of fabrics. Mostly enzyme-based nanomaterial is employed for fabric functionalization [61]. The production of functionalized textile fabric is divided into two groups. One is to functionalize the fiber by modifying the fiber with additives. The other way is to modify the textile surface with functional resins, biomolecules, and bio-finishing techniques [62]. The application of laccase on woolen fabric can improve the water-repellent properties [63]. Moreover, the coating of lysozyme can also enhance the antimicrobial properties as well as the antioxidant potential of fabrics. For example, enzymatic phosphorylation with kinases gives flame-retardant properties while the application of laccase and protease can improve the functional characterization of fabrics [64, 65].

5.3.3 Color Stripping

Since the last decade, the work for stripping colors using biomaterials is on the way. Particularly degraded textile effluents having carcinogens are of great importance [66, 67]. The complex enzymes derived from *Ganoderma lucidum* show such capacity to strip the color [68, 69]. The concentrated extract with three key enzymes (laccase, lignin peroxidases, and manganese peroxidase) has worked well for the stripping of textile effluents. To eliminate color from woven, dyed cotton fabric, this enhanced enzymatic stripping method has the potential to replace costly chemical methods [70]. Hence, bio-stripping can probably be used to treat colored wastewater to make the process green and clean [71, 72].

5.4 Denim Bio-wash

Indigo dyes are used to dye the warp fiber of cotton in the denim industry [73]. The denim stuff was washed to give them a worn appearance and they generally stay at the fiber surface after dyeing due to low substantively of indigo dye for cellulose. These ring-dyed textiles are treated with stone or enzymes to remove the color from the fabric-abraded areas, exposing white surfaces [74]. Denim that has been warp-dyed is used to produce the trendy faded jeans look (Fig. 5.3). A microscopic study revealed that the pH of the bath affects how much colorant penetrates the cotton yarn cross-section during indigo dyeing. Ring-dyeing of the yarn occurs gradually as the pH of the dye bath is lowered from 13 to 11 [75]. More color yield is attained on the yarn surface in combination with the increasing ring dyeing process. The pH range of 10.8–11.2 provides the best color output in denim dyeing. Stonewashing is the traditional method of fading blue denim using pumice stones with abrasive impact [26, 76]. Since their invention in the 1980s, cellulases have been used in up to 80% of denim finishers to give denim a worn-in appearance [77]. Stones and cellulases can be combined to achieve this effect with improved dyeing.

Cellulases function through a process called biostoning, which involves the release of denim indigo dye [78]. Knit and woven cellulosic materials, including blends of cotton, viscose, and linen, can be treated using a stone washing tool [79]. A particular setup is needed for this process. An average denim finisher produces 18 tonnes of sludge every week when processing 100,000 garments with stones [80]. This should be removed from wastewater since it can block the drains. When jeans were subjected to an environmental evaluation, it was found that the bio-stoning method has the potential to produce very little sludge which shows its sustainable operational nature. Thus, stone washing in the denim industry using biomaterials is the prime demand of consumers [81, 82].

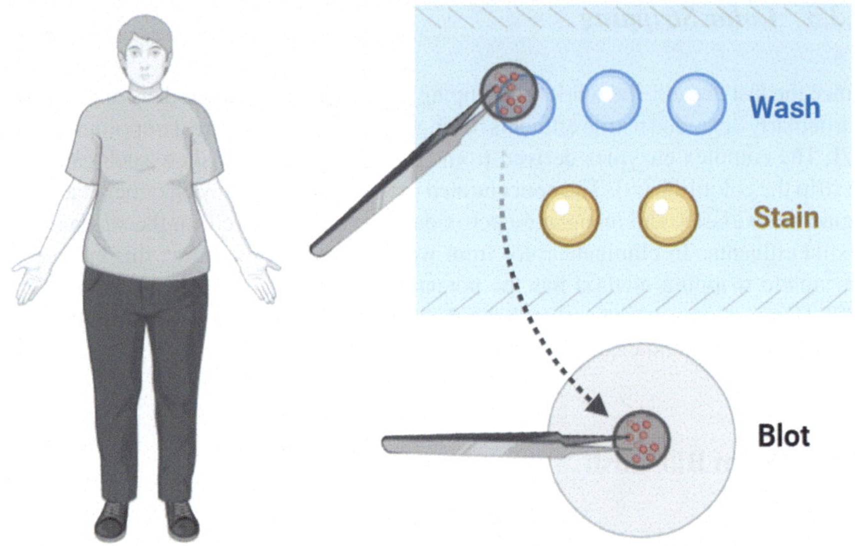

Fig. 5.3 Elements of denim bio-washing

5.5 Degumming

5.5.1 Degumming of Silk

China is the major consumer of enzymes used for silk degumming [83]. Trypsin, papain, and pepsin are the recommended enzymes for proteolytic hydrolysis derived from animal sources. These enzymes break down peptide linkages made of arginine and lysine carboxyl groups [84]. Moreover, natural wax, grime, and lubricating oils must be removed with a light alkali treatment. The pancreas secretes the proteolytic enzyme called trypsin, which is the most efficient in the pH range of 7–9 and concentration of 2% [85].

Papain employed for degumming the silk fabric, extracted from *Papyrus latex*, which showed excellent activity at pH 5–7.5 and temperature 70–90 °C. Various degumming agents (acidic, alkaline, and neutral proteases) have been investigated that dissolve sericin without altering the protein in silk fiber. Although these agents are rarely used commercially, yet the alkaline proteases appear to be the best at eliminating sericin and enhancing the surface properties of silk [86]. According to previous studies, a bacterial enzyme alkalase completely hydrolyzes sericin at 60 °C and pH 7 [87]. Degumming at 50 °C and pH 7–9 offers several advantages such as saving energy. The water consumption can be low down using this procedure with less effluent production [88]. Hydrolysis of sericin by protease revealed the presence of small molecular peptides of 1139.7 and 616.2 Dalton. The enzymatic degumming process has minimum risk as compared to alkaline soap degumming while weight loss of fabric can easily be reduced by adjusting the temperature, pH, and concentration of enzyme. Low-temperature requirements in the enzymatic process lower energy consumption as well as prevent fiber weakness [89].

5.5.2 *Degumming of Ramie*

Ramie fiber is extensively utilized in the biomass sectors (biocomposite and biofuel) and the textile industries. However, ramie fiber can only be put into practice if degumming removes excessive non-cellulosic components [90]. The conventional degumming process notably harms fiber quality. Moreover, the use of excess chemicals in the form of polluted loads harms the environment also [91]. According to a study, *Pectobacterium sp.* CXJZU-120 was the only strain of bacteria that could eliminate more than 90% of the gum in untreated ramie in just 6 h [92]. In comparison to the conventional chemical degumming method, this enzymatic process is more appropriate for extracting ramie fibers from various raw materials while maintaining fiber quality [93]. However, this process can lower the cost by up to 20.5%, increase resource utilization by more than 50%, and lower pollution loads by more than 80%. In another studies it has been found that the removal of the gum, hemicellulase, and lignin left the treated fibers with a typical cellulose structure suitable for textile purposes [94]. There has been observed a noticeable improvement in the fabric fineness, breaking durability, and brightness which has proved that the enzyme degumming process is environmentally friendly and applicable in textiles [95].

5.6 Wool Functionalization

Different alkaline and emulsion solutions are used in carbonization to remove the impurities from woolen fabric [96]. Enzymes are eco-friendly and alternative solutions to replace these toxic and hazardous chemicals in wool functionalization. The enzymatic application makes the finishing process more sustainable and eco-friendly by producing minimal pollution load as well as reducing water and energy consumption [97]. Different enzymes such as pectinases and lipases are employed for wool functionalization. The proteases hydrolyzed the scaly surface of the woolen fabric and released the embedded leftovers from the fiber core. These protease-scoured woolen fabrics have shown excellent fastness properties as well as enhanced dyeability in the dyeing process. The enzymatic treatment on woolen fabric provides better cleaning and shrink-proofing qualities [98].

The enzymatic surface modification of woolen fabric enhances the anti-shrinkage properties as well as wearing comfort. In traditional methods, different oxidative reagents such as potassium permanganate and hypochlorite have been employed to modify the woolen fabric surface [99]. The protease treatment has replaced these conventional approaches and provided exceptional quality of softness by removing fibrous contaminants. Previous studies reported that the combination of lipase, cutinase, and protease has improved the shrink resistance and water-repelling properties of woolen fabric. The application of laccase with its suitable mediator also provides excellent shrink-proofing qualities onto woolen fabric whereas improper enzymatic approaches can cause unwanted fabric damage. Hence, this enzymatic approach should be employed carefully [100]. Usually, chemical treatment can affect the fabric strength and its sorption behavior. Thus, proteinase treatment has taken place due to its environmental nature. However, proteinase treatment breaks down the proteins, which weakens the fiber and reduces fabric quality. Numerous studies have also demonstrated that increasing the size of the enzyme through chemical cross-linking with glutaraldehyde results in the reduction of fabric strength and weight loss. Hence, proteinase treatment has some restrictions for its use in proteinic fabric [101, 102].

In Fig. 5.4, the scale layer that was present on the surface of the wool was partially hydrolyzed by incubating the wool fabric with keratinase (K) and a protease of Savinase (S). The diffusion resistance of dye molecules in wool fibers was reduced as a result of the partial elimination of the scale barrier, which allowed for the realization of low-temperature dyeing. Additionally, glutamine (Gln) was combined with ε-polylysine (ε-PLL) in order to acquire a modified-PLL (mPLL). This modified-PLL was then grafted onto wool fibers by means of TG, resulting in the formation of a network structure that encompasses the fiber surface through cross-linking and spot-welding process [103].

Li et al. [104] proposed the use of milder oxidants such as H_2O_2 followed by protease treatment to improve the woolen shrink resistance. Pill-resistant wool can be synthesized significantly without losing fiber strength by bleaching with H_2O_2 in the presence of imino disuccinic acid sodium salt (IDAS), a novel environmentally friendly chelating agent, and then treated with lipoprotein lipase at 50 °C and pH 7 for 1 h. The enzymatic treatment has improved the whiteness and wettability characteristics of pre-oxidized wool fabrics [105]. After treating wool for 30 min with 1.4% sodium sulfite and 1% enzyme subtilisin serine alkaline protease, the results showed that although fabric strength was retained yet its weight decreased by 3.71%. A new class of cuticle-specific protease-producing bacteria was used to replace the current protease application for wool. Previous studies reported that the usage of mediator and laccase from *Tinea versicolor* enhanced the wool resistance to shrinkage [106]. Additionally, the work done by Yuan et al. [107] reported the use of laccase for the activation of wool fibers in the presence of an appropriate mediator.

The impact of xylanase, pectinase, and resinase on wool structural, chemical, and physical characteristics was assessed by Choudhury et al. [108]. The treatments with xylanase and pectinase were just as effective at cleaning as traditional soap scouring of wool. This enzymatic treatment does not harm the physical properties of the fabric. However, resinase performance as a scouring agent was not very satisfactory. In all dyeing techniques that have been studied, an appropriate amount of these three proteolytic enzymes are present in the dye bath, and the amount of dye absorbance increases [109]. It was also noticed that by lowering the dye temperature, dye absorbance increases significantly until the temperature approaches the enzyme's maximum activity [110].

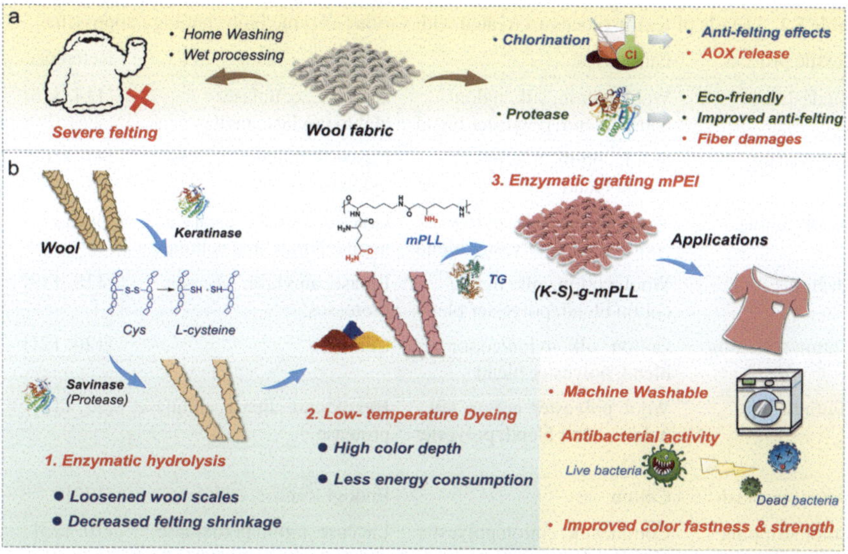

Fig. 5.4 **a** Contrasting the effects of traditional chlorine-based treatments and enzyme-based approaches for reducing wool shrinkage, **b** Schematic representation of a novel process utilizing enzymes to create machine-washable wool with antibacterial properties, employing a method that involves both removal and addition of elements. Reproduced from the Ref. [103], Copyright 2024, American Chemical Society

5.7 Bio-finishing of Cotton

Cotton, ramie, linen, lyocell, and viscose are natural cellulosic stuff. These fibers have a propensity to generate fuzz formation at the surface of the fabric [111]. A method known as biopolishing allows for the complete elimination or reduction of fibrillation, which improves softness, increases the fiber volume, and creates the peach-skin handle. The loss of strength during stretching by the breakdown of crystalline cellulose is one of the serious issues with cellulases [112]. The idea of treating fabrics with cellulase binding domains (CBD) has been investigated as a potential solution to overcome this issue [113]. The applications of different enzymes in textile processing are shown in Table 5.1.

Table 5.1 Details of textile processes treated with various enzyme biotechnology approaches

Textile process	Fabric type	Enzymes	References
Biodesizing	Wool, cotton, silk, nylon/cotton blend, polyester blend	Amylase, cellulase, alpha-amylase, maltase	[114, 115]
Bioscouring	Wool, cotton, silk, nylon/cotton blend, polyester blend	Lipase, cellulase, pectinase, xylanase, protease	[41, 116]
Biobleaching	Wool, cotton, silk, nylon/cotton blend, polyester blend	Glucose oxidase, catalase, hemicellulase and cellulase	[39, 117]
Dyeing	Wool, cotton, silk, nylon/cotton blend, polyester blend	Lipase, amylase, laccases, proteases	[118, 119]
Cotton softening	Cotton, silk, nylon/cotton blend, polyester blend	Cellulase	[120, 121]
Biofinishing	Wool, polyester, cotton, silk, nylon/cotton blend, polyester blend	Hexokinase, lipase, cellulase, protease	[12, 111]
Denim biowash	Cotton	Endoglucanase, cellulase	[122]
Color stripping	Cotton, silk, cotton polyester blend, polyester	Laccase, lignin peroxidase, manganese peroxidase	[70, 123]
Biodegradation of textile effluents	Wastewater from various processing units	Ligninolytic enzymes, lignin peroxidase, manganese peroxidase	[124, 125]

5.8 Conclusion

In the upcoming decades, it is expected that the use of industrial biotechnologies by the textile sector will significantly increase the eco-efficiency of the textile processing sectors. To switch textile production in the future to a more environmentally friendly and sustainable process. To achieve this goal, new approaches and sophisticated instruments must be developed. The advancement in enzymatic technology for textiles has attracted global interest. The textile industry already uses several well-established enzyme-based methods that show how industrial biotechnology can successfully provide greener choices. The most notable applications of white biotechnology in the present textile industry are the economic viability of enzymatic desizing, denim washing, and biopolishing procedures. For practically all the main textile wet processing, several additional enzyme-based methods are emerging, offering prospective advantages including specificity, moderate operating circumstances, environmental viability, and financial support. Enzymatic biotechnologies have now become a part of textile manufacturing, from fiber processing to final consumer goods. By offering substitutes for limited conventional textile production techniques, the invention of textile advances in biotechnology as an industrial tool has enormous potential to revolutionize the current state of the textile industry.

5.9 Future Trends

A major benefit of industrial biotechnology is the combination of various scientific fields, including biochemistry, biology, molecular biology, chemistry, microbiology, chemical engineering and other related fields. Such combinations of knowledge can provide remarkable outcomes. The textile industry, one of the biggest causes of pollution, is a crucial area for the sustainable application of biotechnology. The textile sector can use innovative techniques to survive in the competitive marketplace where clean technologies and environmental friendliness are necessary. Nucleic acid (DNA/RNA), protein/peptide, metabolite, and cellular/subcellular level-related technologies are the sophisticated biotechnology approaches appropriate for the textile industry to make the process cleaner and safer for the community.

The production of novel enzymes with improved functionalities can be made possible by genetic engineering. In the future, it is anticipated that the usage of such genetically engineered microbial enzymes will have commercial importance that will attract researchers for their use in daily life fields. Further research is needed to put the enzyme-based methods for bio-modification of manmade and natural fibers into an industrial context. The goal of the research should be to identify novel enzyme-producing bacteria that can work effectively in extreme conditions. Many textile manufacturers will eventually switch from conventional textile techniques to eco-friendly enzyme technologies. New enzymes have been extensively explored for the textile production process. Hence, it is necessary to increase productivity and efficiency so that modern biotechnologies can compete with traditional methods economically. In the future, it is predicted that genetically modified microbial enzymes of commercial significance will be spread into numerous textile industry sectors, eventually replacing current chemical or mechanical approaches.

Declaration The authors declare that they have no known competing financial interests or personal relationships that could have appeared to influence the work reported in this chapter. Generative AI and AI-assisted technologies have been used in the writing process to improve the readability and language of the work only. After using this tool, the authors carefully reviewed and revised the content to ensure accuracy and quality. The authors take full responsibility for the final version of this publication.

References

1. Wu S, Snajdrova R, Moore JC, Baldenius K, Bornscheuer UT (2021) Biocatalysis: enzymatic synthesis for industrial applications. Angew Chem Int Ed 60(1):88–119. https://doi.org/10.1002/anie.202006648
2. Mia R, Sk MS, Oli ZBS, Ahmed T, Kabir S, Waqar MA (2021) Functionalizing cotton fabrics through herbally synthesized nanosilver. Clean Eng Technol 4:100227. https://doi.org/10.1016/j.clet.2021.100227

3. Zhang Y, Rather LJ, Li Q (2022) Recent advances in the surface modification strategies to improve functional finishing of cotton with natural colourants-a review. J Clean Prod 335:130313. https://doi.org/10.1016/j.jclepro.2021.130313
4. Zhara H, Adeel S, Özomay Z, Mia R (2024) Properties and performance relationship of biopolymers in textile industry. In: Biopolymers in the textile industry: opportunities and limitations. Springer, pp 87–121
5. Rajulapati V, Dhillon A, Kumar Gali K, Katiyar V, Goyal A (2020) Green bioprocess of degumming of jute fibers and bioscouring of cotton fabric by recombinant pectin methylesterase and pectate lyases from clostridium thermocellum. Process Biochem 92:93–104
6. Mia R, Mofasser A, Bhat MA, Howlader MI, Sayed NI, Bakar MA (2024) Utilizing neem wood waste as a natural dye in the development of bio-based functionalized colored jute fabric. Sustain Chem Pharm 38:101489. https://doi.org/10.1016/j.scp.2024.101489
7. Al-Dhabi NA, Esmail GA, Ghilan A-KM, Arasu MV, Duraipandiyan V, Ponmurugan K (2020) Characterization and fermentation optimization of novel thermo stable alkaline protease from Streptomyces sp. Al-Dhabi-82 from the Saudi Arabian environment for eco-friendly and industrial applications. J King Saud Univ-Sci 32(1):1258–1264
8. Mia R, Das S, Banna BU, Ahmed T, Bakar MA (2023) Enhancing antibacterial properties of organic cotton fabric using mahogany wood waste. Case Stud Chem Environ Eng 100387. https://doi.org/10.1016/j.cscee.2023.100387
9. Mia R, Bakar MA, Islam MR, Ahmed T (2023) Eco-friendly coloration from mahogany wood waste for sustainable dyeing of organic nonwoven cotton fabric. Results Eng 17:101032. https://doi.org/10.1016/j.rineng.2023.101032
10. Kida D, Konopka T, Jurczyszyn K, Karolewicz B (2023) Technological aspects and evaluation methods for polymer matrices as dental drug carriers. Biomedicines 11(5):1274. https://doi.org/10.3390/biomedicines11051274
11. Yuan H, Chen L, Cao Z, Hong FF (2020) Enhanced decolourization efficiency of textile dye reactive blue 19 in a horizontal rotating reactor using strips of BNC-immobilized laccase: optimization of conditions and comparison of decolourization efficiency. Biochem Eng J 156:107501
12. Eid BM, Ibrahim NA (2021) Recent developments in sustainable finishing of cellulosic textiles employing biotechnology. J Clean Prod 284:124701. https://doi.org/10.1016/j.jclepro.2020.124701
13. Gautam RL, Bharadwaj AK, Kumar S, Naraian R (2023) Microbial enzymes for the variable applications of textile industry processing. Valorization Biomass Bioprod Elsevier pp 297–321
14. Madhu A, Chakraborty JN (2017) Developments in application of enzymes for textile processing. J Clean Prod 145:114–133
15. Darwesh OM, Al-Balakocy NG, Ghanem A, Matter IA (2023) Application of microalgal-ZnO-NPs for reusing polyester/cotton blended fabric wastes after modification by cellulases enzymes. Waste Dispos Sustain Energy 5(4):471–482. https://doi.org/10.1007/s42768-023-00170-2
16. Unuofin JO, Okoh AI, Nwodo UU (2019) Aptitude of oxidative enzymes for treatment of wastewater pollutants: a laccase perspective. Molecules 24(11):2064
17. Eisazadeh B, Mirzajani F, Sefidbakht Y (2020) How is the effect of silver nanoparticles and lipase/cellulase enzymes on each other? Iran J Sci Technol Trans Sci 44(1):27–35
18. Ramasamy R, Subramanian RB (2023) Microfiber mitigation from synthetic textiles—impact of combined surface modification and finishing process. Environ Sci Pollut Res 30(17):49136–49149
19. Islam MR, Lotif MA, Hossain MS, Ullah M, Mia R (2023) Sustainable production and evaluation of the properties of polyester-okra blended knitted fabric. Results Eng 17:100923. https://doi.org/10.1016/j.rineng.2023.100923
20. Tapia-Tussell R, Pereira-Patrón A, Alzate-Gaviria L, Lizama-Uc G, Pérez-Brito D, Solis-Pereira S (2020) Decolorization of textile effluent by trametes hirsuta Bm-2 and lac-T as possible main laccase-contributing gene. Curr Microbiol 77(12):3953–3961

21. Roy R, Ashmika R (2024) Textile products in healthcare: innovations, applications, and emerging trends. In: Emerging technologies for health literacy and medical practice, pp 288–314
22. Chandra P, Singh ER, Arora PK (2020) Microbial lipases and their industrial applications: a comprehensive review. Microb Cell Factories 19:1–42
23. Sarkar S, Banerjee A, Chakraborty N, Soren K, Chakraborty P, Bandopadhyay R (2020) Structural-functional analyses of textile dye degrading azoreductase, laccase and peroxidase: a comparative in silico study. Electron J Biotechnol 43:48–54
24. Kahoush M, Behary N, Cayla A, Mutel B, Guan J, Nierstrasz V (2020) Influence of remote plasma on PEDOT: PSS-coated carbon felt for improved activity of glucose oxidase. J Appl Polym Sci 137(14):48521
25. Niyonzima FN et al (2023) Microbial enzymes used in textile industry. Biotechnol Microb Enzym Elsevier pp 649–684
26. Alves MB, Jorge AM, Pereira JF (2024) The biotechnology revolution in textile dyeing. Trends Biotechnol. https://doi.org/10.1016/j.tibtech.2024.02.010
27. Iglesias MS, Sequeiros C, García S, Olivera NL (2019) Eco-friendly anti-felting treatment of wool top based on biosurfactant and enzymes. J Clean Prod 220:846–852
28. Rahman M, Billah MM, Hack-Polay D, Alam A (2020) The use of biotechnologies in textile processing and environmental sustainability: an emerging market context. Technol Forecast Soc Chang 159:120204
29. Srivastava A, Rani RM, Patle DS, Kumar S (2022) Emerging bioremediation technologies for the treatment of textile wastewater containing synthetic dyes: a comprehensive review. J Chem Technol Biotechnol 97(1):26–41. https://doi.org/10.1002/jctb.6891
30. Schimper CB, Ibanescu C, Keckeis R, Bechtold T (2008) Advantages of a two-step enzymatic process for cotton–polyester blends. Biotech Lett 30:455–459
31. Maryan AS, Montazer M, Damerchely R (2015) Discoloration of denim garment with color free effluent using montmorillonite based nano clay and enzymes: nano bio-treatment on denim garment. J Clean Prod 91:208–215
32. Rahman M, Hack-Polay D, Billah MM, Un Nabi MN (2020) Bio-based textile processing through the application of enzymes for environmental sustainability. Int J Technol Manag Sustain Dev 19(1):87–106
33. Cheng Z, Xia Y, Zhou Z (2020) Recent advances and promises in nitrile hydratase: from mechanism to industrial applications. Front Bioeng Biotechnol 8:352
34. Shi W, Liu S-Q, Sun J-P, Long J-J (2018) A strategy for environmentally-friendly removal of impurities from cotton based on biocatalytic reaction in supercritical carbon dioxide. Cellulose 25:6771–6792
35. Hoque MT, Mazumder NUS, Islam MT (2021) Enzymatic wet processing. Sustain Pract Text Ind 87–110
36. Abou Taleb M et al (2022) Bioscouring of wool fibres using immobilized thermophilic lipase. Int J Biol Macromol 194:800–810
37. Gupta GK, Dixit M, Pandey D, Kapoor RK, Kango N, Shukla P (2023) Microbial enzyme bioprocesses in biobleaching of pulp and paper: technological updates. Microb Bioprocesses Elsevier pp 319–337
38. Ragab MM, Othman H, Hassabo AG (2022) Various extraction methods of different enzymes and their potential applications in various industrial sector (a review). Egypt J Chem 65(10):495–508
39. Kumar A (2021) Biobleaching: an eco-friendly approach to reduce chemical consumption and pollutants generation. Phys Sci Rev 6(4):20190044
40. Wei S, Liu K, Ji X, Wang T, Wang R (2021) Application of enzyme technology in biopulping and biobleaching. Cellulose 28:10099–10116
41. Shoily SS et al (2023) The pectinolytic activity of Burkholderia cepacia and its application in the bioscouring of cotton knit fabric. J Genet Eng Biotechnol 21(1):136
42. Bahtiyari Mİ, Körlü AE, Bilisik K (2021) Bioprocessing of natural textile fibres and clothes. Fundam Nat Fibres Texts Elsevier pp 221–262

43. Hamdan NT, Jasim HM (2021) First report of bio-polishing of jute fibers using cellulase purified from a novel isolate of trichoderma longibrachiatum in Iraq. World Bull Public Health 4:44–51
44. Marzoug IB, Cheriaa R (2023) Effects optimization of bio-polishing industrial process parameters. J Text Sci Technol 9(1):30–51
45. Jegannathan KR, Nielsen PH (2013) Environmental assessment of enzyme use in industrial production–a literature review. J Clean Prod 42:228–240
46. Kumar D, Bhardwaj R, Jassal S, Goyal T, Khullar A, Gupta N (2021) Application of enzymes for an eco-friendly approach to textile processing. Environ Sci Pollut Res 1–11
47. Kamboj A, Jose S, Singh A (2022) Antimicrobial activity of natural dyes–a comprehensive review. J Nat Fibers 19(13):5380–5394
48. Metwally RA, El Sikaily A, El-Sersy NA, Ghozlan HA, Sabry SA (2021) Antimicrobial activity of textile fabrics dyed with prodigiosin pigment extracted from marine Serratia rubidaea RAM_Alex bacteria. Egypt J Aquat Res 47(3):301–305
49. Hu J, Zhao Y, Meng Y, Su J, Han J (2022) Long-lasting antimicrobial activity achieved through the synergy of graphene oxide and cuprous oxide coating on PET fabrics. Synth Met 286:117033
50. Staneva D, Atanasova D, Nenova A, Vasileva-Tonkova E, Grabchev I (2021) Cotton fabric modified with a PAMAM dendrimer with encapsulated copper nanoparticles: antimicrobial activity. Materials 14(24):7832
51. Xu AY, McGillivray DJ, Dingley AJ (2021) Active antibacterial coating of cotton fabrics with antimicrobial proteins. Cellulose 28(12):8077–8094
52. Al-Zahrani SS, Bora RS, Al-Garni SM (2021) Antimicrobial activity of chitosan nanoparticles. Biotechnol Biotechnol Equip 35(1):1874–1880
53. Atta AM (2021) Immobilization of silver and strontium oxide aluminate nanoparticles integrated into plasma-activated cotton fabric: luminescence, superhydrophobicity, and antimicrobial activity. Luminescence 36(4):1078–1088
54. Olczyk J et al (2020) Antimicrobial activity and barrier properties against UV radiation of alkaline and enzymatically treated linen woven fabrics coated with inorganic hybrid material. Molecules 25(23):5701
55. Janusz G et al (2020) Laccase properties, physiological functions, and evolution. Int J Mol Sci 21(3):966
56. de Boer SR, Schäffer A, Moreira MT (2023) Towards oxidoreductase-based processes for the removal of antibiotics from wastewater. Rev Environ Sci Bio Technol 22(4):899–932
57. Ferraboschi P, Ciceri S, Grisenti P (2021) Applications of lysozyme, an innate immune defense factor, as an alternative antibiotic. Antibiotics 10(12):1534
58. Wang S-W, Wang T-Y (2022) Study on antibacterial activity and structure of chemically modified lysozyme. Molecules 28(1):95
59. Verma C et al (2023) Functional designing of textile surfaces for biomedical devices. In: Fiber and textile engineering in drug delivery systems. Elsevier, pp 443–460
60. Velmurugan P et al (2020) Bio-functionalization of cotton, silk, and leather using different in-situ silver nanoparticle synthesis modules, and their antibacterial properties. Res Chem Intermed 46:999–1015
61. Humayoun UB et al (2023) Harnessing bio-immobilized ZnO/CNT/chitosan ternary composite fabric for enhanced photodegradation of a commercial reactive dye. Molecules 28(18):6461
62. Cherian T et al (2023) Bio-functionalization, stabilization and potential functionalities of hyaluronate macromolecules capped copper oxide nanoparticles. J Environ Chem Eng 11(6):111051
63. Stanzione I, Pitocchi R, Pennacchio A, Cicatiello P, Piscitelli A, Giardina P (2022) Innovative surface bio-functionalization by fungal hydrophobins and their engineered variants. Front Mol Biosci 9:959166
64. Sui J, Liu S, Chen M, Zhang H (2022) Surface bio-functionalization of anti-bacterial titanium implants: a review. Coatings 12(8):1125

65. Szadkowski B, Rybiński P, Sarlin E, Marzec A (2024) Bio-functionalization of fillers with natural phytic acid for sustainable flame-retardant agents for natural rubber composites. Ind Crops Prod 222:119504
66. Kishor R et al (2021) Ecotoxicological and health concerns of persistent coloring pollutants of textile industry wastewater and treatment approaches for environmental safety. J Environ Chem Eng 9(2):105012
67. Mohan D, Shukla SP (2022) Hazardous consequences of textile mill effluents on soil and their remediation approaches. Clean Eng Technol 7:100434
68. Latif Z et al (2021) Development of efficient strain of Ganoderma lucidum for biological stripping of cotton fabric dyed reactive blue 21. Saudi J Biol Sci 28(12):7550–7560
69. Asgher M, Muzammil M, Qamar SA, Khalid N, Bilal M (2020) Environmentally friendly color stripping of solar golden yellow R dyed cotton fabric by ligninolytic consortia from ganoderma lucidum IBL-05. Case Stud Chem Environ Eng 2:100031
70. Chatha SAS, Asgher M, Ali S, Hussain AI (2012) Biological color stripping: a novel technology for removal of dye from cellulose fibers. Carbohyd Polym 87(2):1476–1481
71. Adeel S et al (2020) Biotechnology: an eco-friendly tool of nature for textile industries. In: Advances in functional finishing of textiles, pp 85–114
72. Pociecha M, Sircelj H, Lestan D (2009) Remediation of cu-contaminated soil using chelant and EAOP. J Environ Sci Health Part A 44(11):1136–1143
73. Maulik SR, Pandit P (2022) Advancement in indigo dyeing. In: Textile dyes and pigments: a green chemistry approach, pp 107–121
74. Roy S, Maiti S (2022) Advancement in denim dyeing. In: Textile dyes and pigments: a green chemistry approach, pp 123–136
75. Gürses A et al (2016) Dyeing and dyeing technology. Dye Pigment 47–67
76. Xin JH, Chong C, Tu T (2000) Colour variation in the dyeing of denim yarn with indigo. Color Technol 116(9):260–265
77. Periyasamy AP, Periyasami S (2023) Critical review on sustainability in denim: a step toward sustainable production and consumption of denim. ACS Omega 8(5):4472–4490
78. Arja M-O (2007) Cellulases in the textile industry. In Industrial enzymes: structure, function and applications. Springer, pp 51–63
79. Bajaj P, Mahajan R (2019) Cellulase and xylanase synergism in industrial biotechnology. Appl Microbiol Biotechnol 103(21):8711–8724
80. Ahmed J et al (2023) Characterization of an endo-beta-1, 4 glucanase gene from paper-degrading and denim bio-stoning cellulase producing aspergillus isolates. Biotechnol Appl Biochem 70(3):1057–1071
81. Yusuf M (2021) Current and future perspectives of enzyme treatments for cellulosic fibers: a review. In: Green chemistry for sustainable textiles, pp 69–76
82. Singhania RR, Adsul M, Pandey A, Patel A (2017) Cellulases. Curr Dev Biotechnol Bioeng Elsevier 73–101
83. Zhu L, Lin J, Pei L, Luo Y, Li D, Huang Z (2022) Recent advances in environmentally friendly and green degumming processes of silk for textile and non-textile applications. Polymers 14(4):659
84. Wang R et al (2023) A novel method for silkworm cocoons self-degumming and its effect on silk fibers. J Adv Res 53:87–98
85. Wang F, Cao T-T, Zhang Y-Q (2015) Effect of silk protein surfactant on silk degumming and its properties. Mater Sci Eng, C 55:131–136
86. Liu X et al (2023) Comparative study of the preparation of high-molecular-weight fibroin by degumming silk with several neutral proteases. Polymers 15(16):3383
87. Kim J, Kwon M, Kim S (2016) Biological degumming of silk fabrics with proteolytic enzymes. J Nat Fibers 13(6):629–639
88. Rastogi S, Kandasubramanian B (2020) Processing trends of silk fibers: silk degumming, regeneration and physical functionalization. J Text Inst 111(12):1794–1810
89. More SV, Chavan S, Prabhune AA (2018) Silk degumming and utilization of silk sericin by hydrolysis using alkaline protease from Beauveria Sp. (MTCC 5184): a green approach. J Nat Fibers 15(3):373–383

90. Cheng L et al (2020) Ramie-degumming methodologies: a short review. J Eng Fibers Fabr 15:1558925020940105
91. Jiang W et al (2018) A green degumming process of ramie. Ind Crops Prod 120:131–134
92. Liu Z et al (2012) A rapid process of ramie bio-degumming by pectobacterium sp. CXJZU-120. Text Res J 82(15):1553–1559
93. Ding J, Zhou C, Dong Z (2023) Trend of ramie industry development: a review of green degumming and the utilization of processing residues. J Clean Prod 384:135487. https://doi.org/10.1016/j.jclepro.2022.135487
94. Huang H et al (2022) Anthraquinone-assisted deep eutectic solvent degumming of ramie fibers: evaluation of fiber properties and degumming performance. Ind Crops Prod 185:115115
95. Chaudhuri SR et al (2020) Optimization of bio-chemical degumming of ramie fiber for improved strength and luster. Biotechnol Rep 28:e00532
96. Allafi FA, Hossain MS, Shaah M, Lalung J, Ab Kadir MO, Ahmad MI (2022) A review on characterization of sheep wool impurities and existing techniques of cleaning: industrial and environmental challenges. J Nat Fibers 19(14):8669–8687
97. de Oliveira CRS, da Silva Júnior AH, Ferreira AJS, Immich APS, Lopes TJ (2024) Treatments on wool and structural evaluation. In: Handbook of natural polymers Elsevier pp 177–204
98. Al-Sayed W, Abdelrahman SH (2021) Sustainable chemistry in textile processes (pretreatment, coloration and chemical finishing). In: Green chemistry for sustainable textiles. Elsevier, pp 93–111
99. Shamey R, Sawatwarakul W (2014) Innovative critical solutions in the dyeing of protein textile materials. Text Prog 46(4):323–450
100. Kaur A, Chakraborty J (2015) Controlled eco-friendly shrink-resist finishing of wool using bromelain. J Clean Prod 108:503–513
101. El-Sayed H (2022) The current status and future insight into the production of machine-washable wool. J Nat Fibers 19(15):10293–10305
102. Popescu C, Stanescu MD (2024) Eco-friendly processing of wool and sustainable valorization of this natural bioresource. Sustainability 16(11):4661
103. Li Z et al (2024) Enzymatic preparation of machine-washable antibacterial wool through a subtraction-addition technique using hydrolase and transglutaminase. ACS Sustain Chem Eng 12(44):16480–16493
104. Li B et al (2022) Development of environmentally friendly wool shrink-proof finishing technology based on L-cysteine/protease treatment solution system. Int J Mol Sci 23(21):13553
105. Mojsov K (2017) Enzymatic treatment of wool fabrics-opportunity of the improvement on some physical and chemical properties of the fabrics. J Text Inst 108(7):1136–1143
106. El-Gabry L, El-Nouby G, Allam O, El-Sayed H (2008) Effect of mechanical and enzymatic treatments on some properties of coarse wool. J Nat Fibers 5(4):461–475
107. Yuan M, Wang Q, Shen J, Smith E, Bai R, Fan X (2018) Enzymatic coloration and finishing of wool with laccase and polyethylenimine. Text Res J 88(16):1834–1846
108. Choudhury AKR (2020) Enzyme applications in textile chemical processing. In: Sustainable technologies for fashion and textiles. Elsevier, pp 91–115
109. Smith E, Shen J (2012) Enzymatic treatment of wool pre-treated with cetyltrimethylammonium bromide to achieve machine washability. Biocatal Biotransform 30(1):38–47
110. Park M, Kim H-Y, Jin F-L, Lee S-Y, Choi H-S, Park S-J (2014) Combined effect of corona discharge and enzymatic treatment on the mechanical and surface properties of wool. J Ind Eng Chem 20(1):179–183
111. Ali H, Hashem M, Shaker N, Ramadan M, El-Sadek B, Hady MA (2012) Cellulase enzyme in bio-finishing of cotton-based fabrics: effects of process parameters. Res J Text Appar 16(3):57–65
112. Yu Y, Yuan J, Wang Q, Fan X, Wang P, Cui L (2014) A promising approach for bio-finishing of cotton using immobilized acid-cellulase. Fibers and Polym 15:932–937
113. Refaee AA, El-Naggar ME, Mostafa TB, Elshaarawy RF, Nasr AM (2022) Nano-bio finishing of cotton fabric with quaternized chitosan Schiff base-TiO_2-ZnO nanocomposites for antimicrobial and UV protection applications. Eur Polymer J 166:111040

114. Korsa G, Konwarh R, Masi C, Ayele A, Haile S (2023) Microbial cellulase production and its potential application for textile industries. Ann Microbiol 73(1):13
115. ZA Raza, A Rehman, F Anwar, N Ahmad (2023) Multiresponse optimization in amylolytic desizing of viscose fabric in the copresence of ferrous ions and an anionic surfactant. Pigment Resin Technol
116. Bashir T, Iftikhar T, Majeed H (2024) Bulk industrial textile production of bio scouring for cellulosic fabric utilizing indigenous hot springs triacylglycerol acylhydrolases from bacillus toyonensis and Bacillus thuringiensis. Cellulose 31(2):1353–1381
117. Shatalov AA, Pereira H (2009) Impact of hexenuronic acids on xylanase-aided bio-bleaching of chemical pulps. Biores Technol 100(12):3069–3075
118. Mahboob M et al (2023) Microwave-assisted santalin extraction from Pterocarpus santalinus for mordanted woolen yarn dyeing. Sustain Chem Pharm 35:101224. https://doi.org/10.1016/j.scp.2023.101224
119. Prabu HG, Sundrarajan M (2002) Effect of the bio-salt trisodium citrate in the dyeing of cotton. Color Technol 118(3):131–134
120. Tang K, Fan J, Zhang J, Sarkar M, Kan CW (2013) Effect of softeners and crosslinking conditions on the performance of easy-care cotton fabrics with different weave constructions. Fibers and Polymers 14:822–831
121. Parvinzadeh M (2007) The effects of softeners on the properties of sulfur-dyed cotton fibers. J Surfactants Deterg 10(4):219–223
122. Lee IY, Jeong GE, Kim SR, Bengelsdorff C, Kim SD (2015) Effects of biowashing and liquid ammonia treatment on the physical characteristics and hand of denim fabric. Color Technol 131(3):192–199
123. Arooj F, Ahmed N, Shaikh IA (2020) Application of ozone in stripping of cotton fabric dyed with reactive dyes. Ozone Sci Eng 42(4):319–330
124. Vijayalakshmidevi S, Muthukumar K (2015) Improved biodegradation of textile dye effluent by coculture. Ecotoxicol Environ Saf 114:23–30
125. Khan S, Malik A (2018) Toxicity evaluation of textile effluents and role of native soil bacterium in biodegradation of a textile dye. Environ Sci Pollut Res 25:4446–4458

Chapter 6
Advances in Microbial Enzymes in Processing and Applications in Textiles

Aneesa Fasim, K. L. Vaishnavi, R. Maheshwari, G. Sriprada, S. M. Veena, and Sunil S. More

Abstract Microbial enzymes play a crucial role in textile processing. They have substituted numerous dangerous and harmful chemicals owing to their environmentally favorable characteristics, widespread availability, simple production process, lack of toxicity, and selectivity toward substrates. These enzymes enhance the overall quality of the textile by catalyzing reactions at a faster rate, resulting in the conservation of water and energy. The processing of textiles primarily relies on using key enzymes such as amylases, laccases, cellulases, esterases, catalases, proteases, pectinases, etc., that are mostly bacterial or fungal in origin. Recent advancements in enzyme-based textile processing include depilling, desizing, and scouring, which help eliminate impurities; biopolishing aids in enhancing the fabric quality; denim washing to improve color and removal of hydrogen peroxide; depilling and felting of wool to prevent it from shrinkage. However, traditional textile processing methods have been observed to adversely affect the stability of enzymes. Genetic engineering and protein processing offer opportunities to modify enzymes that enhance their suitability and adaptability for harsh processing treatments. Further, the implementation of enzyme immobilization technology has improved enzyme usage efficiency and simplified the separation process. As a result, these biocatalysts have enhanced the cost-effectiveness and environmental sustainability of textile processing and production. The present chapter discusses the diverse enzymes employed in textile processing, along with the latest developments in enzyme-based technology within the textile processing sector.

A. Fasim · K. L. Vaishnavi · R. Maheshwari · G. Sriprada · S. S. More (✉)
School of Basic and Applied Sciences, Dayananda Sagar University, Bangalore, India
e-mail: sunilsmorey@gmail.com

S. M. Veena
Department of Biotechnology, Sapthagiri College of Engineering, Bengaluru, India

S. Biranje et al. (eds.), *Sustainable Finishing Techniques in Textiles*,
https://doi.org/10.1007/978-981-96-4860-3_6

6.1 Introduction

The textile sector, one of the oldest and most pervasive worldwide, has relied on conventional techniques in textile processing, which heavily depend on abrasive chemicals and high energy consumption. Nevertheless, in the recent past, enzyme technology has revolutionized this sector due to its immense benefits. Enzymes signify a critical turn toward more environmentally friendly techniques, which are pertinent in light of the increasing global issues about sustainability and environmental impact. In addition to enhancing the quality of products, enzymes are biodegradable and highly specific in their actions, which reduces the need for harmful chemicals and minimizes energy and water consumption. This has made the textile industry more sustainable, eco-friendly, and efficient. Enzymes are now essential in several textile processes, such as desizing, scouring, bleaching, and finishing. Traditionally, these procedures were carried out by using chemicals such as urea, caustic soda, acids, bases, salts, and bleach, which are hazardous to humans and might result in environmental contamination if discarded after use. As a result, enzymes provide a considerably cleaner and more environmentally friendly alternative [21]. Fungi account for over 60% of industrial enzyme production, bacteria for 24%, yeast for 4%, and plants and animals for the remaining 10% [42]. Nevertheless, microbial enzymes are favored because of their high yields, activity, repeatability, cost-effectiveness, exponential growth, use of inexpensive media, simple optimization, and strain improvement [49].

The most critical phase of bioprocess development is the enhancement of strains through enzyme engineering. It is a fast-emerging technology for producing customized enzymes on a big scale, at a low cost, and with ease. Most natural enzymes cannot withstand the severe industrial conditions that impede effective bioprocessing. Currently, numerous laboratories employ enhanced enzyme engineering techniques to optimize biocatalyst performance in various textile industry processes. Site-directed mutagenesis and directed evolution have become increasingly popular methods for producing and selecting mutant enzymes with enhanced activity, selectivity, stability, and availability [21]. Additionally, the application of efficient downstream processing and immobilization techniques has greatly improved the stability, purity, and reusability of the enzymes. When enzymes are immobilized, they are attached to a carrier or support substance that facilitates their separation, operation, and possible reuse. Immobilized enzymes have a lot of promise for industrial use due to their increased tolerance to temperature, pH, solvents, and auxiliaries, as well as their possibility of recovery and repurposing [31]. Therefore, this chapter addresses the numerous textile processing methods utilized in the industry, emphasizing the enzymes and enzyme-based technologies involved. It also discusses current advances in increasing enzyme efficiency and productivity.

6.2 Textiles

The textile industry uses various raw materials to produce textiles, garments, and other textile-based products. Each type of raw material has its own unique properties and is used in different types of textile products, depending on the desired performance and aesthetic characteristics [52]. Different raw materials used in the textile industry include natural fibers and synthetic or man-made fibers (Fig. 6.1). Natural fiber refers to the fibrous plant material that is derived from natural resources such as plants, animals, and minerals through the process of photosynthesis, which is already found in the environment. Based on their source (origin), natural fibers can be categorized into three main classes: vegetable, animal, and mineral. Natural fibers such as cotton fiber, jute fiber, wool fiber, hemp, kenaf, jute, sisal fiber, banana, flax, oil palm, etc., have become increasingly popular due to their environmentally friendly and renewable nature. Additionally, natural fibers are biodegradable, offer superior mechanical and thermal qualities, and have a low density. Man-made fibers are created by humans through chemical synthesis. They are divided into synthetic and regenerated fibers. Synthetic fibers are spun using chemically based polymers, whereas regenerated fibers are made of a natural polymer, most often cellulose. Man-made fibers include aramid, carbon, nylon, viscose, acetate, acrylic, and polyester [27].

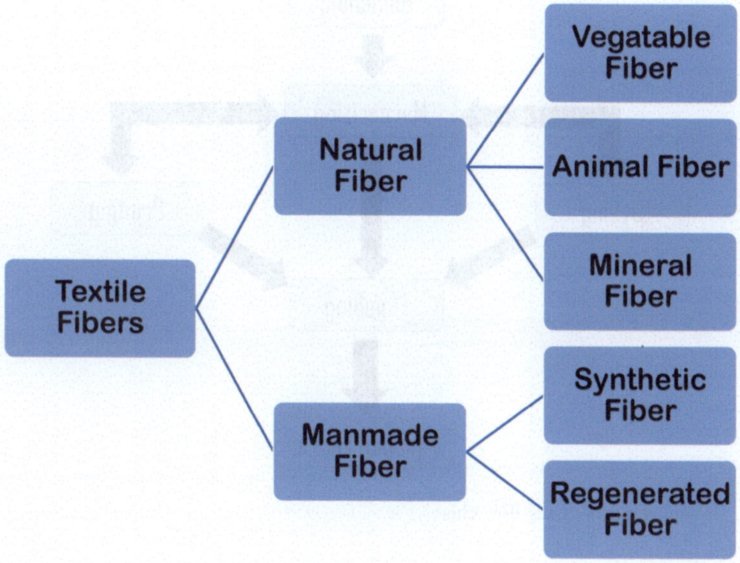

Fig. 6.1 Classification of textile fibers

6.3 Textile Processing

Textile processing involves a complex sequence of operations that transform raw fibers into finished goods like fabrics, clothing, and household textiles. The process of transforming unprocessed materials into finished textiles entails a variety of mechanical and chemical treatments intended to improve the fabric's functionality, durability, texture, and appearance. The quality of a textile product depends on fiber preparation, yarn manufacture, fabric creation, and finishing treatments [25]. A typical textile processing sequence is shown in Fig. 6.2.

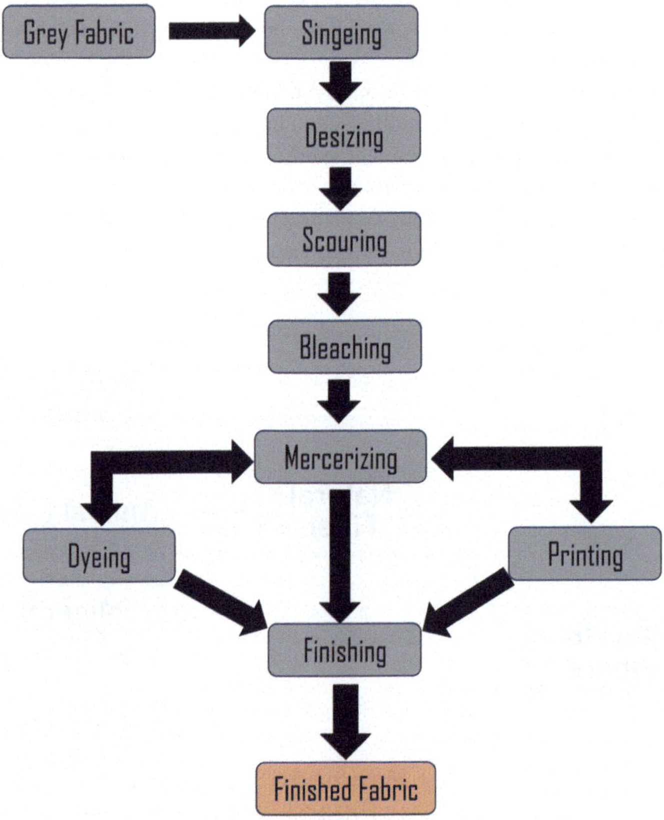

Fig. 6.2 Textile wet processing flow chart

6.3.1 *Singeing*

The singeing operation is conducted before the commencement of moist processing to eliminate hairy and protruding fibers from the fabric. The singeing process involves passing fabric over a gas flame at a distance and speed sufficient to burn only the protruding fibers while avoiding harm to the main fabric. The primary goal of the singeing process is to eliminate protruding filaments from the fabric surface, thereby reducing the likelihood of fabric pilling and achieving a clean fabric surface.

6.3.2 *Desizing*

De-sizing removes sticky chemicals, or sizes, from fabric threads placed during weaving to prevent yarn breakage. Chemicals like alkalis, acids, and oxidizing agents remove these particles from cloth as they hinder dyeing, printing, and finishing. Unfortunately, these chemicals remove non-cellulosic contaminants and weaken the cloth. These compounds raise wastewater COD, BOD, and TDS. Size-adding chemicals in textile finishing yield 50–80% COD wastewater [43].

6.3.3 *Scouring*

Wet textile processing begins with scouring raw cotton or other natural fibers to remove oils, waxes, and contaminants. Chemicals such as soda ash, sodium hydroxide, and hydrogen peroxide are used in the process of scouring.

6.3.4 *Bleaching*

Bleaching removes natural color from textiles or fabrics to whiten them. Chemicals like chlorine or hydrogen peroxide are used in the bleaching process. Commonly, hydrogen peroxide is used with catalysts like sodium silicate, sodium hydrosulfite, or sodium formaldehyde sulfoxylate. One of the first industrial bleaching chemicals was hypochlorite, which produces toxic chlorine gas and chlorinated organics. The use of absorbable organically bound halogen (AOX) has minimized the hazardous chlorinated organic byproducts during bleaching to a certain extent. In recent years, hypochlorite has been replaced by safe bleaching chemicals like peracetic acid, which completely decomposes into oxygen and acetic acid. Additionally, peracetic acid improves brightness without damaging fibers [5].

6.3.5 Macerisation

Mercerization improves cotton fabric shine, strength, and dyeability. The fabric is treated with a potent (18–24%) sodium hydroxide solution, which induces fiber swelling and enhances dye absorption. This solution has been shown to induce longitudinal shrinkage in cotton, which can be avoided by stretching or straining [30].

6.3.6 Dyeing

Dyeing is the process of incorporating color into the fabric. There are numerous methods for dyeing, including immersion dyeing, textile dyeing, and piece dyeing. Acid dyes are employed for protein fibers, while dispersed dyes are employed for synthetic fibers. The chromophore and auxo chrome groups of the dyes are responsible for the color, which contributes to environmental pollution [11].

6.3.7 Printing

Printing is a branch of dyeing that uses pigments or dyes to create images or patterns on fabrics. Screen printing, digital printing, roller printing, block printing, and rotary printing are among the numerous printing methods. Dyeing involves the application of color as a solution, whereas printing involves the application of color as a dense dye paste. Color fixation in printing is accomplished using appropriate post-treatment of the printed material. Recently developed screen-free printing techniques use electrical control to evenly disperse colour on fabric. Within this set of techniques are ink-jet and electrostatic printing. Implementing screen-free printing techniques appears to be an effective approach to reducing pollution [19].

6.4 Chemicals Used in Textiles

The textile industry is increasingly utilizing an extensive range of chemicals for various processes. Chemical treatments have the potential to considerably enhance the functionality, appearance, and wearability of fabrics. The textile industry relies on these chemicals in many stages of production and processing like bleaching, dying, printing, etc. Chemicals may also be incorporated into the final product to fulfill a variety of purposes, such as the application of antibiotics to sportswear, the application of flame-retardant treatments to fabrics, the impregnation of garments for doors, and the prevention of mold growth during storage and transportation, etc.

Therefore, the production of contemporary fabric would be exceedingly challenging in the absence of chemicals [36]. Apart from the chemicals mentioned in textile processing, a few of the other corrosive chemicals employed in the textile industry are.

6.4.1 Formaldehyde

Formaldehyde has many applications within the textile industry, including wrinkle prevention, dye penetration, preservation, wool protection, anti-mold, and cross-linking agents. Furthermore, it enhances both color fastness and waterproofness. Adverse effects of formaldehyde include allergic contact dermatitis, respiratory complications, skin irritation, etc. The concerns above have led to restrictions on the permissible levels of formaldehyde in consumer products. Today, most clothing contains relatively low levels of formaldehyde, which have decreased from as high as 3,000 parts per million in the 1960s to less than 500 parts per million by the 1980s. Various certification standards have been developed to assist firms in guaranteeing that the textiles they use are of appropriate quality and safe for use [13].

6.4.2 Azo Dyes

Azo dyes, which comprise 60–80% of all colorants, are responsible for the vibrant hues in many textiles. Azo dyes break down into compounds called aromatic amines, some of which have been linked to cancer. In 2003, the European Union prohibited using some of the azo dyes that could potentially lead to the production of around 22 of the most hazardous aromatic amines. However, this prohibition is only applicable in the state of California in the United States [29].

6.4.3 Heavy Metals

In dyeing operations, heavy metals such as antimony, cadmium, lead, mercury, and chromium VI are employed. Lead is used as a catalyst in synthetic textiles and as a flame-retardant synergist. Silver is used as an antibacterial, and titanium and zinc are water-repellents and odor-preventive agents. Allergenic metals such as cobalt, copper, chromium, and lead are used as metal complex dyes. After being absorbed by the body, these metals have the potential to build up in the liver or kidney, leading to severe health complications. The European Union has long prohibited the use of cadmium, lead, and chromium VI, as all are classified as carcinogenic [48].

6.4.4 Organotin Compounds

An organotin is a chemical compound that contains at least one tin-carbon bond. In addition to serving as a catalyst or stabilizer in PVC, polyester, and polyurethane foams, organotins are also used as biocides in sports textiles, cotton, and polyester products to inhibit smells. Tin is the primary component of organotin compounds, which is known for its antibacterial properties and odor inhibitor in garments. Much like heavy metals, these tin compounds can build up in the body and impact the immune and reproductive systems. Dioctyltin (DOT) and tributyltin (TBT) are two categories of organotin compounds that are prohibited by the European Union (EU) when present in amounts exceeding 0.1%. Even at extremely low concentrations, organotins exhibit high toxicity not only toward humans but also toward freshwater and marine organisms.

6.4.5 Chlorobenzene

In the dyeing process, chlorobenzenes serve as transporters or leveling agents and are predominantly employed in polyester and polyester blend materials. Their accumulation in the body has the potential to cause injury to the thyroid, liver, and central nervous system over time. Additionally, they may be toxic when inhaled or in contact with the epidermis. The European Union establishes a limit of no more than 1.0 mg/kg for the use of chlorobenzenes [15].

6.5 Enzymes in Textiles

Enzymes have become a popular choice for processing in the textile industry due to their non-toxic and environmentally favorable properties. Enzymes, which are biological catalysts, enable the acceleration of chemical reactions by converting the substrates into products. Enzymes can lower activation energy by creating an intermediate enzyme–substrate complex and finishing the reaction without being consumed in the process. Highly efficient, safe, and simple to regulate, they are also biodegradable and substrate-specific. Primarily, they are sourced from microbes, animal tissue, and plants. Microbial enzymes are typically preferred for industrial applications due to their availability in sufficient quantities. *Aspergillus, Trichoderma, Streptomyces, and Bacillus*, etc. are a few of the microorganisms that produce a large number of industrially valuable enzymes. It is common for textile processing to utilize abrasive chemicals, such as urea, caustic soda, acids, bases, salts, bleach, and others, that are environmentally harmful. Consequently, enzymes offer a significantly more environmentally favorable and cleaner alternative (Table 6.1) (Mojsov).

Table 6.1 Enzymes and their applications in textile processing

Enzymes	Microbial source	Substrate	Biotechnological application
Catalase	*Aspergillus niger*, *Micrococcus luteus*, *Serratia marcescens*	Hydrogen peroxidases	Bleach clean-up, improved dyeing, degradation of reactive dyes
Amylase	*Bacillus subtilis*, *B. amyloliquefacines*, *B. licheniformis*, and *Aspergillus oiyzae*	Starch	Desizing and laundry detergent formulation
Pectinases	*B. sonorensis*, *Chryseobacterium indologenes*, *Pseudomonas*, *Streptomyces*, *Lactobacillus*	Pectin	Bio-scouring
Proteases	*Aspergillus oryzae*, *Trichoderma*, *Streptomyces*, *B. licheniformis*	Protein	Anti-shrinkage, increased dyeing affinity, remove stiff and dull gum layer of sericin from the raw silk
Esterase	*Trichoderma*, *Aspergillus*, *Cryphonectria*, *Rhizophus*, *Pseudomonas*, *Lactobacillus*	Ester bond	Polyster modification, reduce pilling, bio-bleaching
Lipase	*Pseudomonas*, *Burkholderia*, *Penicillium*, *Rhizopus*, *Fusarium*, *Geotrichum*, *Candida antarctica*, *Cryptococcus*, *Trichosporon*, *Aureobasidium*, *Rhodotorula*	Fats and oils	Degreasing, dyeability, desizing, surface modification, improve wool quality
Glucose oxidase	*Aspergillus*, *penicillium* and *saaccharomyce*	Pigments	Enzymatic bleaching
Peroxidases	*B. sphaericus*, *B. subtilis*, *Pseudomonas* sp., *Citrobacter* sp., *Anabaena* sp., *Candida krusei*, *Coprinopsis cinerea*, *Phanerochaete chrysosporium*	Chromophores	Decolorizing, improve enzyme stability, remove hydrogen peroxide residues, bioremediation

(continued)

Table 6.1 (continued)

Enzymes	Microbial source	Substrate	Biotechnological application
Laccases	*Phanerochaete chrysosporium, Theiophora terrestris, and Lenzites, betulina, Phlebia radiate, Pleurotus ostreatus* and *Trametes versicolour*	Chromophores	Bio-bleaching, dyeing, rove scouring, finishing, neps removal, printing, washing off treatment, dye synthesis and effluent treatment
Transglutaminase	*Streptomyces mobaraensis, Candida albicans*	Protein	Strengthen wool, reduce shrinkage, improve color fastness, smooth wool, improve wettability

In the textile industry, hydrolases and oxidoreductases are the two primary categories of enzymes that are employed for a variety of enzymatic applications. The majority of enzyme applications in textiles are restricted to cotton processing, including desizing, scouring, and bleaching; biofinishing to improve the appearance; biostoning, or stone washing, jeans to give them a stylish aged look; and bleaching clean-up to eliminate any leftover H_2O_2 before dying [34].

6.5.1 Hydrolases

The hydrolase enzyme class catalyzes the hydrolysis of numerous bonds, including glycosidic, ester, peptide, and carbon–carbon bonds in the presence of water, resulting in the breakdown of bigger molecules into small ones. Given the challenging nature of textile processing, it is crucial to isolate microbial strains that can produce robust enzymes capable of withstanding high temperatures and pH levels, etc. Additionally, optimizing the industrial growth conditions through the choice of fermentation method and substrate is essential for large-scale enzyme production [18].

6.5.1.1 Amylase

Amylase is a crucial enzyme that has extensive applications in the textile industry. These enzymes hydrolyze starch molecules into monosaccharides such as glucose and maltose. Amylases can be classified into two types: α-amylases and β-amylases where α-amylases randomly target starch molecules, and β-amylases selectively remove maltose from the nonreducing end of the starch chain. Rhozyme DX, Rhozyme GC, and Diastafor LCD are all desizing α-amylase enzymes of bacterial origin [45].

6.5.1.2 Cellulase

Cellulases are enzymes that break down cellulose into glucose and oligosaccharides. Cellulases are widely used in the textile and laundry industries. They are primarily used in biopolishing and softening cellulosic fibers. Furthermore, they also improve the appearance, sheen, color, and texture of fabrics. The elimination of hair (biopolishing) in jute-cotton mixed textiles is optimized using Denimax L cellulase. They are also effective on cotton and lyocell fabric, particularly for knits, rayon, and linen. Commercial cellulase, such as Gempil 4 L, eliminates pill and fuzz from colored and knitted fabrics. Similarly, the G-ZYME VGB ST is highly effective in facilitating the activity of cellulases on reactively dyed cotton and is particularly suitable for bio-polishing various types of spun yarn knitted garments [16].

6.5.1.3 Pectinase

Pectinases are a set of enzymes that hydrolyze complex polysaccharides found in plant cell walls called pectic compounds. They are derived from bacterial and fungal origins. They are mostly used to pretreat cotton fabric, which prepares it for dyeing by making it more absorbent [49]. Moreover, it enhances the brightness of the textile without causing harm to the cellulose structure. Using pectinase in retting can result in stronger, softer, and more consistently sized fibers for textiles. The efficacy of commercially available pectinolytic enzymes, namely Pectinex ultra Sp-L and Flaxzyme, in retting jute ribbons. Furthermore, the application of commercially available pectinase (0.01%) and xylanase (0.1%) successfully decreases the rigidity of jute cotton fibers [12].

6.5.1.4 Protease

The enzyme protease catalyzes the hydrolysis of peptide bonds within proteins. Proteolytic enzymes comprise endo- and exopeptidases where endopeptidases recognize specific amino acids in the middle of the peptide, whereas exopeptidases recognize one or two terminal amino acids. Proteases from plant sources are papain and ficin, and animal proteases are mostly pepsin and trypsin. Likewise, microbial proteolytic enzymes are derived from various bacteria and fungi. Protease-containing commercial enzyme mixtures are utilized for bioscouring. Proteases (*Bacillus*) and cellulases enable successful scouring, coloring, and water absorption [7]. Proteases are used to remove stiff and dull sericin gum layers from raw silk to achieve improved shine and softness. Protease treatments can also modify the surface of wool and silk fibers to provide unique finishes [14].

6.5.1.5 Esterase/Lipase

Esterases/lipases are a group of biochemical enzymes that catalyze the hydrolysis of fats, oils, and ester linkages. They are derived from microorganisms, plant, and animal sources. Lipases improve the dye-ability of the fabric by removing lubricants. It can be used with amylases to eliminate tallow compounds and natural fats during the desizing and scouring process. Lipases process and modify synthetic fibers and can also be employed to obtain a high level of whiteness in bio-bleaching and bio-scouring procedures [10].

6.5.1.6 Cutinase

The enzyme cutinase is employed to remove wax or bioscouring cotton fiber at low temperatures. Even when significant concentrations of detergents are present, cutinases can effectively eliminate waxes from cotton. Furthermore, they enhance the pectinase activity, which helps in the removal of pectin from cloth fibers [20].

6.5.2 Oxidoreductases

An oxidoreductase is an enzyme that facilitates oxidative reduction processes and transfers electrons across substrates such as cellulose. Oxidoreductases are classified according to the donor–acceptor groups they contain. The molecular oxygen (O_2) found in the enzyme functions as an oxidase, that is responsible for accepting electrons from substrates [53]. Oxidoreductases often employed in the textile industry include laccase, catalase, and peroxidase.

6.5.2.1 Glucose Oxidase/Peroxidase

Glucose oxidase, or peroxidase, catalyzes the decomposition of glucose into hydrogen peroxide and gluconic acid upon exposure to oxygen. Glucose-oxidase is a biobleaching agent that helps in desizing cotton, linen material at high temperatures and pH conditions, producing peroxide. The biozyme combination of pullulanase and glucose-oxidase synergistically generates H_2O_2, which is adequate for bleaching and results in the highest level of whiteness at alkaline pH compared to neutral and acidic pH [6].

6.5.2.2 Catalase

Hydroperoxides, or catalases (CATs), are enzymes that facilitate the conversion of H_2O_2 into H_2O and organic oxygen (O_2). The majority of catalases are ubiquitous oxidoreductases present in plants, bacteria, fungi, and archaea. Catalase is primarily employed in the textile industry to remove the residual hydrogen peroxide from fabrics that remain after the bleaching process. Bleach like hydrogen peroxide is often used to make clothes whiter, but if it stays on the fabric, it can weaken, discolor, and deteriorate it. Catalase enzyme keeps fabric from getting damaged and improves product quality by efficiently eliminating hydrogen peroxide. Likewise, alkaline catalase can also be employed to generate a sulphide native redox system that is beneficial for the reduction of sulfur pigments in cellulosic coloring [3].

6.5.2.3 Laccase

Laccases are members of the blue-multicopper oxidases family. It oxidizes various aromatic and non-aromatic phenolic compounds and depolymerizes the substrate through a radical-catalyzed reaction process. The complicated enzymes Laccase and

Peroxidase are known to efficiently degrade and remove lignin from flax fiber to generate whiteness. The bleaching linen with ecolite II is a commercial laccase, and H_2O_2 allows for better dye absorption for both reactive and cationic dyes. Laccases can also improve the color of cotton textiles in many ways. It is possible to create colored pigments in situ on cotton textiles to color cellulosic fibers by oxidizing flavonoids with the laccase enzyme [8].

6.6 Applications of Enzymes in Textile Processing

6.6.1 De-pilling

This process is also known as "biopolishing," "biofinishing," and "reformation." It relates to the depilling of completed garments or unfinished cotton fabrics using cellulase enzymes, especially *Trichoderma* cellulase enzyme mixtures that are primarily composed of at least 80% cellulases. With cellulase-enriched enzyme combinations, it is possible to reduce fabric degradation by approximately 88% during enzyme treatment and achieve more effective pill removal. The fabric's quality and value are enhanced by the cellulase treatment, which smooths the fabric's surface during depilling to improve its softness and appearance. Cellulase enzymes hydrolyze the fibrils (fabric's most exposed area) beta-1, 4 links in cellulose and are consequently eliminated. The removal of fibrils is said to improve the color and cleanliness of the clothing, as well as softness, by removing soil that has adhered to the fibers and facilitating the penetration of other cleaning agents. Preventing the formation of new fibrils also benefits from the early elimination of existing fibrils [2].

Cellulase enzymes give cotton fibers several benefits in terms of smoothness and gloss as compared to conventional fabric softeners. Conventional softeners, primarily clay or cationic surfactants, are applied to the fabric, giving it an unwelcome greasy feel. On the other hand, the effects become more obvious and economical if the proportion of cotton is more than 60% by weight; the best results are obtained if the percentage of cotton is more than 75% by weight. Using cellulase component mixes that are superior to the natural combination has been one method of minimizing the loss of fabric strength during the depilling process [31].

6.6.2 De-sizing

"Desizing" refers to the process of removing size (a starch-based protective material) from the cotton fiber to reduce hairiness and boost strength. Warp cotton yarns are coated with a sizing agent before weaving. So, insufficient removal of size formulation after the weaving process may result in uneven dyeing, printing, or finishing.

Desizing can be accomplished by heating for an extended period or utilizing strong chemicals such as oxidizing agents, bases, acids, or enzymes [23].

In recent times, amylopectic enzymes have been widely employed in industrial applications. Commercial enzymes, such as amylase and maltase, are frequently used to degrade starch, which is subsequently converted into maltose, dextrose, and short-chain carbohydrates which can readily removed by washing or dissolving in hot water [4]. Additionally, as amylase is selective to starch, it does not compromise the cellulosic fabric and enhances the preservation of strength. Amylase desizing works best at temperatures between 30 and 60 ° and pH levels between 5.5 and 6.5. Furthermore, it aids in water conservation as there is no necessity to repeatedly rinse the material in order to eliminate residual chemicals [50].

6.6.3 Bleaching

The purpose of bleaching cotton fibers is to remove the natural pigmentation provided by flavonoids and to give the fibers a white appearance, enabling them to be dyed in a variety of colors at a later time. Traditional bleaching employs hydrogen peroxide (H_2O_2), chlorine, and sodium hypochlorite. Excessive amounts of these chemicals can cause fiber reactions that slow down polymerization and damage the fibers severely [39]. Enzymes usage in the bleaching process instead of these chemicals reduces fiber damage, uses significantly less water to remove H_2O_2, and produces a less hazardous effluent. Pectinases, amyloglucosidases, glucose oxidases, and laccases are among the enzymes used in combination to produce the white color that results from the oxidation of flavonoids. The benefit of laccase is that it specifically targets the indigo dyes in denim, preserving the natural raw white of the weft yarn and resulting in a darker shade of woven fabric, which is not possible with hypochlorite bleaching. Furthermore, the effectiveness of laccase in bleaching cotton textiles was improved by the moderate ultrasonic energy (7 W) input [8].

6.6.4 Bioscouring

Fabrics made from cellulosic fibers often contain non-cellulosic contaminants in varying concentrations, such as proteins, pectins, waxes, lipids, minerals, natural colorants, etc. These factors restrict the effectiveness of dyeing and finishing processes, leading to poor quality and appearance of the fabric [40]. Scouring refers to eliminating these contaminants from the cotton surface, normally accomplished by using a potent alkali like sodium hydroxide [31]. However, removing these contaminants can be accomplished by using enzymes referred to as bioscouring. Consequently, the fabric's quality is preserved while attaining an exceptionally soft finish. Furthermore, bioscouring has no negative impact on the environment.

Microbiological enzymes, including cellulases, pectinases, proteases, and cutinases, are employed in bioscouring. Cellulases enzymatically break down the cellulosic materials in the main cell wall beneath the cuticle. Pectinases catalyze the breakdown of the cuticle of cotton fibers to digest and eliminate pectin. Similarly, Proteases facilitate the separation of proteinaceous impurities from cellulose fibers [1], and cutinases are responsible for the degradation of cutin and wax [20].

6.6.5 Enzymatic Degumming of Silk

Silk degumming refers to the procedure of eliminating sericin, the adhesive sheath that envelops silk strands. This method makes silk suitable for textile application by improving its color, softness, and shine. Silk is composed of two protein complexes, namely fibroin and ceresin. Cocoon silk fibroins are naturally gummed together with contaminants such as wax, protein, sericin, pectin, and colors. Their removal using traditional methods, including detergent washing, alkali, and oxidizing chemical treatment, could reduce the quality of silk. However, Enzymatic treatment can facilitate the hydrolysis of peptide bonds of ceresin protein, enabling its removal from silk fibers without causing harm. This process also enhances the softness as well as the increased dye absorption of the silk fibers [39].

6.6.6 Anti-shrinkable Wool

The anti-shrink treatment's primary objective is to preserve the fabric's texture and prevent unintended shrinkage. Wool's shrink resistance is significant for its use in clothing, comforters, curtains, cushions, and bedding. Various treatments have been developed over the years to make wool supple and shrinkage-resistant. The chlorination treatment, also known as the chlorine-Hercosett treatment, is the most cost-effective method of shrink-resistant treatment which is followed by a polyamide-epichlorohydrin coating [22]. However, a more recent development is the use of enzymes in antishrinkage treatments. Proteases are used to enhance the resilience and anti-shrinkage of wool. Pronase, a bacterial protease, is recognized for its higher degree of selectivity in targeting the highly swellable cell membrane complex (CMC) present underneath the cuticular scales. Papain a plant protease, is widely used for its high efficacy in reducing shrinkage. When applied at 50 °C in a 1% sodium bisulfite solution, this treatment achieved optimal results on coarser fibers as well [9]. Furthermore, esperase protease is responsible for the enhanced dye-ability, whiteness, and shrink-resistance of wool. Transglutaminase (TGase), a protein-crosslinking enzyme, is also a popular choice, which assists in the reduction of wool shrinkage, preservation, or enhancement of fabric strength. Wool fibers treated with protease may benefit more from subsequent TGase treatments, presumably because the fibers will be more vulnerable to protein crosslinking, and the wool's structure may become more porous [46].

6.6.7 Denim Finishing/Biostoning

Denim washing or biostoning is a cosmetic process used on denim textiles to enhance their durability and visual appeal and to achieve the desired faded appearance. Stone washing using chemicals such as potassium permanganate and sodium hypochlorite, commonly referred to as pumice stones, was traditionally employed to achieve the desired appearance and velvety texture [39]. While washing denim, the pumice stone removes dye specks from the yarn's surface. However, these techniques had certain drawbacks, such as fabric wear and tear and back staining. Biostoning of denim using cellulase enzyme started in the late 80s. Cellulase can remove the indigo color embedded in the denim fabric, resulting in worn-out, uneven shade fading which is a very effective and eco-friendly technique. Furthermore, laccase enzyme can also be employed in the denim biostoning process. Their mechanism of action involves the oxidation of phenolic molecules, which degrades the dye and thus leads to a lighter color of the fabric. The method can be paired with additional enzymes such as cellulase and amylase to generate the intended aged or distressed effects on denim, thereby improving the visual appeal of the fabric without inducing major damage [38].

6.6.8 Modification of Synthetic Fiber by Enzymes

The use of enzymes enhances the hydrophilicity, as well as its anti-pilling, dyeability, and antistatic charge generation, even though the majority of synthetic fibers are hydrophobic. The dye molecules can more readily penetrate the synthetic fibers as the amorphous and paracrystalline regions of the fabric are more accessible to enzymatic hydrolysis. Various types of enzymes, including cutinases, lipases, and esterases, can modify polyester [34]. Esterase exhibited potential for polyester modification by surface hydrolysis of polylactic acid (PLA) and polyethylene terephthalate (PET) fabric [44]. Similarly, lipases were employed to enhance the digitally printed polyester cloth [41]. The hydrophilicity of nylon 66 fabrics was improved by laccases. Nylon 66 fabric dyed with reactive and acid dyes was treated with proteases, amidases, and cutinases to increase dye bath fatigue [47]. Proteases are known to improve the hydrophilicity and cationic dye affinity of nylon fabric without compromising its mechanical properties. Likewise, Polyacrylonitrile (PAN) fabrics can undergo hydrolysis of their vinyl acetate constituents through lipases and cutinases [37].

6.6.9 Biological Polymer Synthesis and Functionalization

Enzyme-based synthesis and functionalization of textile polymers are novel topics of technological advancement. Enzyme specificity could allow for the exact modification of polymer structures without the toxicity risk associated with solvents and metal catalyst residues [31]. Several industrial-scale studies have investigated the enzymatic synthesis of various polymers, including polyesters, polyamides, polycarbonates, polysaccharides, polypeptides, polyanilines, polyphenols, polystyrene, polymethyl methacrylate, and cellulose [33].

The enzyme lipase, which breaks down fatty esters, is employed as a catalyst for polymerization, particularly in the production of polyester and polyamides. In addition, they are commonly used in polymer transesterification, polycondensation, ring-opening polymerization, and polymer modifications [32]. Enzymes such as xylanase, chitinase, and cellulase also show a high propensity toward natural polymer modification. Peroxidases and laccases are known to help generate polyaromatic compounds from polyanilines and their derivatives (phenol and aniline). In addition to hydrolases, the catalytic properties of oxidoreductase enzymes, such as laccase and peroxidase, have been studied in connection to their ability to catalyze significant polymerization processes. Horseradish peroxidase (HRP) is one of the several systems of radical-forming enzymes used for iterative polymerization of vinyl monomers, including acrylamides and styrene. HRP is also used as a biocatalyst in the manufacture of water-soluble, conducting poly(N-methylaniline) (PNMA), poly(N-ethylaniline) (PNEA), poly(N-butylaniline) (PNBA), and poly(Phenylethanolamine) (PNPEA) materials.

Enzymatic polymerization was implemented to generate functional water-soluble polyesters by employing polyethylene glycol and dimethyl itaconate as building elements. Enzyme-catalyzed ring opening polymerization of cyclic esters in ionic liquids has recently been proposed as a more environmentally friendly option. Significant potential exists for immobilizing biocatalysts in enzymatic polymerization [28].

6.6.10 Textile Effluent Treatment

The uncontrollable discharge of dye-contaminated effluents into water bodies has contributed to a surge in water pollution. In addition to posing health risks to textile industry workers, these dyes also have significant detrimental effects on the health of aquatic species, as they are mostly carcinogenic, mutagenic, and poisonous. Besides dyes, other contaminants such as surfactants, bases, acids, salts, detergents, etc., are also present in the textile effluents, making it imperative to treat it before it is discharged into the environment [54]. Current dye decolorization processes are unsuitable for large-scale use due to high costs and sludge generation, which requires safe disposal. Enzymatic procedures are recommended because they are eco-friendly, cost-effective, and non-toxic. Ligninases, including laccases and peroxidases, can decolorize certain synthetic colors found in textile effluents [51]. In addition, Catalases convert H_2O_2 to oxygen and water and protect fibers against reactive oxygen species. Enzymes like these reduce the need for water, energy, and time while making effluents cleaner [35].

6.7 Current Advancements to Improve Enzyme Stability and Efficiency

6.7.1 Genetic Engineering

The prospects for enzyme innovations in textiles appear to be highly encouraging. Researchers are continuously investigating new enzymes and processing techniques to enhance textile production. The enzymes and commercial strains used in bioprocessing are often sourced from their native habitats, which makes them vulnerable to severe industrial conditions that might reduce their efficiency and, in turn, their cost-effectiveness. The issue can be resolved in two primary ways: identifying new strains with distinctive or robust characteristics or improving the productivity and efficiency of the wild-type strains. Approaches such as strain improvement through genetic engineering can be implemented to increase enzyme productivity and efficiency. The technique is highly efficient and potent, as it introduces point mutations to modify the genetic makeup of wild-type strains. This frequently leads to overproduction, enhanced temperature or pH tolerance, selectivity, and other benefits for the targeted enzymes (Table 6.2).

Novel enzymes have been synthesized to tackle the challenges encountered by the textile industry. Enzymes specifically designed to operate in harsh conditions, such as higher temperatures and pH, have been developed to enhance the effectiveness of textile processing. Furthermore, multi-enzyme cocktails have been formulated to help

Table 6.2 A few examples of genetically modified enzymes used in the textile industry

Enzyme	Modifications	Application
Amylases	Independent amino acid substitutions—His(133) and Ala(209) using random mutagenesis	Increases thermostability
Pectinases	Point mutations at Ala-118His, tyr-190leu, ala197gly, ser-208lys, ser263lys, Asn275Tyr	Improved thermostability
Cellulases	5 Point mutations at Glu223ser, Ala224His, Leu25Val, Thr226Ala, Asp262GLy	Increased alkaline optimum pH
Proteases	Single substitution of Ser for Asn at position 218	Stability at higher temperature
Lipases/ esterases	Exchange of Met by Leu at position 72	Increased stability toward oxidation

remove stains and contaminants, leading to effective and environmentally friendly textile manufacturing.

6.7.2 Enzyme Immobilization

The second innovation in this sector is the use of immobilized enzymes. Prolonged exposure to severe processing conditions frequently results in the denaturation of free enzymes, which ultimately slows down the production process. Enzyme immobilization was developed to address these limitations, which not only aids in enzyme reuse but also enhances their stability and activity. Furthermore, the technique of adsorbing the catalyst onto a solid substrate streamlines the purification procedure, thus improving the profitability of manufacturing. Adsorption, entrapment, encapsulation, or covalent binding/cross-linking are commonly employed techniques for immobilizing enzymes on carrier materials that are biocompatible, thermostable, and non-toxic. Natural biopolymers and nanomaterials are increasingly becoming recognized as preferred options in this regard [17].

The immobilization of enzymes directly on textiles results in the acquisition of distinct characteristics, including improved moisture-wicking capabilities, antimicrobial properties, and better durability. Recent studies have demonstrated that certain naturally existing enzymes possess the capacity to serve as biological protective coatings. Enzymes like lysozyme have been immobilized onto textiles to generate smart fabrics with characteristics like self-detoxification and antibacterial activity. Lysozyme was covalently immobilized on cotton fibers through the use of glycine or glycine dipeptide and its antibacterial activity was proven. Immobilization of various biomolecules on cotton-based wound dressings has been shown to improve their functionalization. Similarly, functional molecules are linked by peroxidase and laccase to generate textiles with improved technical properties. For e.g. laccase is employed to add lauryl gallate, a water-insoluble phenolic molecule, onto wool, thereby conferring it with antioxidant, water-repellent, and antibacterial properties. Deposition of NDGA onto wool by laccase yields textiles with UV protection, tensile strength, and shrink resistance. Likewise, flame retardancy can be achieved in natural materials by the in-situ synthesis of phenolic resin utilizing laccase and enzymatic phosphorylation with kinases [24]. Enzyme immobilization is also a promising method used for altering synthetic fibers to decrease their surface hydrolysis activity, as demonstrated in denim fading, wool anti-shrinkage, and bio-polishing. Furthermore, this technology can also be employed in treatment of textile effluent [26]. The incorporation of enzymes into the textile production process allows manufacturers to provide consumers with an array of functional and eco-friendly textile goods.

6.8 Conclusion

Enzymes provide an economically viable and renewable substitute for chemical processes in the textile sector. Their characteristics, such as high catalytic rate, strong substrate specificity, and eco-friendly nature, render them ideal for substituting hazardous chemicals at different phases of textile processing. Furthermore, enzymes aid in lowering environmental impact and improve effluent quality by decreasing the demand for toxic chemicals, water, and energy. In order to guarantee a more sustainable future for the textile industry, it is imperative to conduct additional research to completely integrate enzyme-based processes in every aspect of textile bioprocessing.

Declaration The authors declare that they have no known competing financial interests or personal relationships that could have appeared to influence the work reported in this chapter. Generative AI and AI-assisted technologies have been used in the writing process to improve the readability and language of the work only. After using this tool, the authors carefully reviewed and revised the content to ensure accuracy and quality. The authors take full responsibility for the final version of this publication.

References

1. Abdullah Al-Dhabi N, Ali Esmail G, Mohammed Ghilan A-K et al (2020) Characterization and fermentation optimization of novel thermo stable alkaline protease from *Streptomyces* sp. Al-Dhabi-82 from the Saudi Arabian environment for eco-friendly and industrial applications. J King Saud Univ - Sci 32:1258–1264. https://doi.org/10.1016/j.jksus.2019.11.011
2. Acharya S, Chaudhary A (2012) Bioprospecting thermophiles for cellulase production: a review. Braz J Microbiol 43:844–856. https://doi.org/10.1590/S1517-83822012000300001
3. Andrés CMC, Pérez de la Lastra JM, Juan CA et al (2022) Chemistry of hydrogen peroxide formation and elimination in mammalian cells, and its role in various pathologies. Stresses 2:256–274. https://doi.org/10.3390/stresses2030019
4. Aydemir D, Gecili F, Özdemir N, Nuray Ulusu N (2020) Synthesis and characterization of a triple enzyme-inorganic hybrid nanoflower (TrpE@ihNF) as a combination of three pancreatic digestive enzymes amylase, protease and lipase. J Biosci Bioeng 129:679–686. https://doi.org/10.1016/j.jbiosc.2020.01.008
5. Bajpai P (2012) Enzymatic pre-bleaching. In: Bajpai P (ed) Environmentally benign approaches for pulp bleaching, 2nd edn. Elsevier, Boston, pp 189–223
6. Bauer JA, Zámocká M, Majtán J, Bauerová-Hlinková V (2022) Glucose oxidase, an enzyme "Ferrari": its structure, function, production and properties in the light of various industrial and biotechnological applications. Biomolecules 12:472. https://doi.org/10.3390/biom12030472
7. Beckhorn EJ, Labbee MD, Underkofler LA et al (1965) Food additives, production and use of microbial enzymes for food processing. J Agric Food Chem 13:30–34. https://doi.org/10.1021/jf60137a010
8. Brugnari T, Braga DM, dos Santos CSA et al (2021) Laccases as green and versatile biocatalysts: from lab to enzyme market—an overview. Bioresour Bioprocess 8:131. https://doi.org/10.1186/s40643-021-00484-1
9. Chakraborty J, Jaruhar P (2014) Dyeing of cotton with sulphur dyes using alkaline catalase as reduction catalyst. Indian J Fibre Text Res 39:303–309

10. Chandra P, Enespa SR, Arora PK (2020) Microbial lipases and their industrial applications: a comprehensive review. Microb Cell Factories 19:169. https://doi.org/10.1186/s12934-020-01428-8
11. Chequer F, Olivera GAR, Ferraz E et al (2013) Textile dyes: dyeing process and environmental impact. Eco-Friendly Text Dye Finish 6:151–176. https://doi.org/10.5772/53659
12. Chiliveri SR, Koti S, Linga VR (2016) Retting and degumming of natural fibers by pectinolytic enzymes produced from Bacillus tequilensis SV11-UV37 using solid state fermentation. Springerplus 5:559. https://doi.org/10.1186/s40064-016-2173-x
13. David E, Niculescu V-C (2021) Volatile organic compounds (VOCs) as environmental pollutants: occurrence and mitigation using nanomaterials. Int J Environ Res Public Health 18:13147. https://doi.org/10.3390/ijerph182413147
14. Garg G, Singh A, Kaur A et al (2016) Microbial pectinases: an ecofriendly tool of nature for industries. 3 Biotech 6:47. https://doi.org/10.1007/s13205-016-0371-4
15. Ghaly A, Ananthashankar R, Alhattab M, vasudevan ramakrishnan V (2014) Production, characterization and treatment of textile effluents: a critical review. J Chem Eng Process Technol 5
16. Ghosh S, Godoy L, Yongabi K et al (2021) Fungal cellulases: current research and future challenges, pp 263–298
17. Guisan JM, Fernandez-Lorente G, Rocha-Martin J, Moreno-Gamero D (2022) Enzyme immobilization strategies for the design of robust and efficient biocatalysts. Curr Opin Green Sustain Chem 35:100593. https://doi.org/10.1016/j.cogsc.2022.100593
18. Gulzar T, Farooq T, Kiran S et al (2019) 1—Green chemistry in the wet processing of textiles. In: Shahid-ul-Islam, Butola BS (eds) The impact and prospects of green chemistry for textile technology. Woodhead Publishing, pp 1–20
19. Guru R, Singh S, Rani J, Grewal D (2024) Dyeing and printing techniques used in the handloom sector. Asian Dyer 20:48–52
20. Gururaj P, Khushbu S, Monisha B et al (2021) Production, purification and application of cutinase in enzymatic scouring of cotton fabric isolated from Acinetobacter baumannii AU10. Prep Biochem Biotechnol 51:550–561. https://doi.org/10.1080/10826068.2020.1836655
21. Hasan R, Aktar N, Kabir SMdT et al (2020) Pectinolytic bacterial consortia reduce jute retting period and improve fibre quality. Sci Rep 10:5174. https://doi.org/10.1038/s41598-020-61898-z
22. Hassan M, Carr C (2019) A review of the sustainable methods in imparting shrink resistance to wool fabrics. J Adv Res 18. https://doi.org/10.1016/j.jare.2019.01.014
23. Imran MA, Hussain T, Memon M, Shafai MM (2015) Sustainable and economical one-step desizing, scouring and bleaching method for industrial scale pretreatment of woven fabrics. J Clean Prod 108. https://doi.org/10.1016/j.jclepro.2015.08.073
24. Jesionowski T, Zdarta J, Krajewska B (2014) Enzyme immobilization by adsorption: a review. Adsorption 20:801–821. https://doi.org/10.1007/s10450-014-9623-y
25. Judd S, Jefferson B (eds) (2003) 3.3—The textile industry. In: Membranes for industrial wastewater recovery and re-use. Elsevier Science, Amsterdam, pp 132–158
26. Khan MR (2021) Immobilized enzymes: a comprehensive review. Bull Natl Res Cent 45:207. https://doi.org/10.1186/s42269-021-00649-0
27. Kiron MI (2015) Difference between natural fiber and man made fiber. In: Text. Learn. https://textilelearner.net/difference-between-natural-fiber-and-man-made-fiber/. Accessed 17 Sep 2024
28. Kundys A, Białecka-Florjańczyk E, Fabiszewska A, Małajowicz J (2018) Candida antarctica lipase B as catalyst for cyclic esters synthesis, their polymerization and degradation of aliphatic polyesters. J Polym Environ 26:396–407. https://doi.org/10.1007/s10924-017-0945-1
29. Lacasse K, Baumann W (2012) Textile chemicals: environmental data and facts. Springer Science & Business Media
30. Lin L, Jiang T, Liang Y et al (2022) Combination of pre- and post-mercerization processes for cotton fabric. Materials 15:2092. https://doi.org/10.3390/ma15062092

31. Madhu A, Chakraborty JN (2017) Developments in application of enzymes for textile processing. J Clean Prod 145:114–133. https://doi.org/10.1016/j.jclepro.2017.01.013
32. Maghraby YR, El-Shabasy RM, Ibrahim AH, Azzazy HME-S (2023) Enzyme immobilization technologies and industrial applications. ACS Omega 8:5184–5196. https://doi.org/10.1021/acsomega.2c07560
33. Martins M, Silva C, Cavaco-Paulo A (2019) 6—Enzyme stabilization for biotechnological applications. In: Cavaco-Paulo A, Nierstrasz VA, Wang Q (eds) Advances in textile biotechnology, 2nd edn. Woodhead Publishing, pp 107–131
34. Mendes KF, Sousa RD, Mielke KC (2022) Biodegradation Technology of organic and inorganic pollutants. BoD—Books on Demand
35. Mojsov K (2011) Application of enzymes in the textile industry: a review
36. Ömeroğulları Basyigit Z (2017) Functional finishing for textiles
37. Parvinzadeh Gashti M (2012) Surface modification of synthetic fibers to improve performance: recent approaches. Glob J Phys Chem 3:1–10
38. Periyasamy AP, Periyasami S (2023) Critical review on sustainability in Denim: a step toward sustainable production and consumption of Denim. ACS Omega 8:4472–4490. https://doi.org/10.1021/acsomega.2c06374
39. Rahman M, Hack-Polay D, Billah MM, Nabi MNU (2020) Bio-based textile processing through the application of enzymes for environmental sustainability. Int J Technol Manag Sustain Dev 19:87–106. https://doi.org/10.1386/tmsd_00017_1
40. Rajulapati V, Dhillon A, kumar Gali K et al (2020) Green bioprocess of degumming of jute fibers and bioscouring of cotton fabric by recombinant pectin methylesterase and pectate lyases from *Clostridium thermocellum*. Process Biochem 92:93–104. https://doi.org/10.1016/j.procbio.2020.02.024
41. Rathore B (2023) Textile industry 4.0: a review of sustainability in manufacturing. Int J New Media Technol 10:38–43. https://doi.org/10.58972/eiprmj.v10i1y23.41
42. Raveendran S, Parameswaran B, Ummalyma SB et al (2018) Applications of microbial enzymes in food industry. Food Technol Biotechnol 56:16–30. https://doi.org/10.17113/ftb.56.01.18.5491
43. Rehman A, Saeed A, Asad W et al (2023) Eco-friendly textile desizing with indigenously produced amylase from Bacillus cereus AS2. Sci Rep 13:11991. https://doi.org/10.1038/s41598-023-38956-3
44. Ribitsch D, Herrero Acero E, Greimel K et al (2012) A new esterase from thermobifida halotolerans hydrolyses polyethylene terephthalate (PET) and polylactic acid (PLA). Polymers 4:617–629. https://doi.org/10.3390/polym4010617
45. Saha P, Khan MF, Patra S (2018) Potential application of α-amylase for desizing of fabrics in textile industries
46. Sen A, Kapila R, Chaudhary S, Nigam D (2021) Biotechnological applications of microbial enzymes to replace chemicals in the textile industry—a review 1 2 2 2. 82:68–73
47. Sh Abd ElSalam DI (2012) Enzymatic treatment of polyester fabrics digitally printed. J Text Sci Eng 02. https://doi.org/10.4172/2165-8064.1000113
48. Sima MF (2022) Determination of some heavy metals and their health risk in T-shirts printed for a special program. PLoS ONE 17. https://doi.org/10.1371/journal.pone.0274952
49. Singh R, Kumar M, Mittal A, Mehta PK (2016) Microbial enzymes: industrial progress in 21st century. 3 Biotech 6:174. https://doi.org/10.1007/s13205-016-0485-8
50. Strade E, Kalnina D, Kulczycka J (2020) Water efficiency and safe re-use of different grades of water—topical issues for the pharmaceutical industry. Water Resour Ind 24:100132. https://doi.org/10.1016/j.wri.2020.100132
51. Tapia-Tussell R, Pereira-Patrón A, Alzate-Gaviria L et al (2020) Decolorization of textile effluent by Trametes hirsuta Bm-2 and lac-T as possible main laccase-contributing gene. Curr Microbiol 77:3953–3961. https://doi.org/10.1007/s00284-020-02188-9
52. Textile-Engineering (2023) Raw materials used in the textile industry. In: Text. Eng. https://textileengineering.net/raw-materials-used-in-the-textile-industry/. Accessed 3 Sept 2024

53. Van Noorden CJF (2014) Metabolic mapping by (quantitative) enzyme histochemistry. In: McManus LM, Mitchell RN (eds) Pathobiology of human disease. Academic Press, San Diego, pp 3760–3774
54. Yuan H, Chen L, Cao Z, Hong FF (2020) Enhanced decolourization efficiency of textile dye Reactive Blue 19 in a horizontal rotating reactor using strips of BNC-immobilized laccase: optimization of conditions and comparison of decolourization efficiency. Biochem Eng J 156:107501. https://doi.org/10.1016/j.bej.2020.107501

Chapter 7
Enzymatic Textile Processing for Biotechnological Cleaning of Cellulosic Textiles

Sri Sneha Jeyakumar, Suryapratap Ray, Rahul Vashishth, Bairabathina Venkatesh, and Abilash Valsala Gopalakrishnan

Abstract The textile industry has witnessed an upsurge in the use of enzymatic treatments since they are energy and environmentally efficient alternatives. The focus of biotechnology and enzyme modification advancements has been on several uses in textile processes. Key enzymes include amylases, used in both bleaching and desizing processes. Peroxidase, which also plays a role in these procedures, and cellulase, activated for denim finishing and bio-polishing. These enzymes provide environmentally acceptable and long-lasting substitutes for conventional chemical treatments. The textile industry can cut back on energy, water, and chemical waste by using enzymes. To accomplish environmental goals and achieve cleaner production processes, enzymes are essential. A sizeable share of the world's textile industry is made up of cellulosic textiles, and cleaning them frequently requires the use of harsh chemicals and a lot of energy, which raises environmental concerns. By efficiently using the catalytic power of enzymes to break down stains and dirt, enzymatic methods provide a sustainable substitute. This paper examines current developments in the creation and use of unique enzymes for cellulosic textile biotechnology cleaning. It includes new oxidative enzymes, hemicellulases, and cellulases from various fungal and microbial origins. These enzymes are highly suitable for industrial-scale textile cleaning procedures because of their increased stability,

S. S. Jeyakumar · S. Ray · R. Vashishth (✉)
Department of Biosciences, School of Biosciences and Technology, Vellore Institute of Technology, Vellore, India
e-mail: rahul.vashishth@vit.ac.in

S. Ray
e-mail: suryapratap1504@gmail.com

B. Venkatesh
Department of Textile Technology, National Institute of Fashion Technology, Hyderabad, Telangana, India
e-mail: bairabathina.venkatesh@nift.ac.in

A. V. Gopalakrishnan
Department of Biomedical Sciences, School of Biosciences and Technology, Vellore Institute of Technology (VIT), Vellore, Tamil Nadu, India
e-mail: abilash.vg@vit.ac.in

S. Biranje et al. (eds.), *Sustainable Finishing Techniques in Textiles*,
https://doi.org/10.1007/978-981-96-4860-3_7

activity, and specificity. This study also focuses on the possible synergistic effects of different enzyme combinations for enhanced cleaning efficacy in cellulosic textile cleaning.

Keywords Cellulose textile · Enzymatic process · Sustainable development

7.1 Introduction

Enzymes function as biological catalysts that expedite chemical reactions while remaining unaltered during the process. Their significance is particularly pronounced in biotechnological applications, encompassing textile processing. The attributes of enzymes, including high specificity, efficiency, and environmental sustainability, render them appealing substitutes for conventional chemical methodologies. Within biotechnological frameworks, enzymes facilitate a multitude of reactions, such as the degradation or modification of particular compounds. These biomolecules can be sourced from diverse origins, including microorganisms (such as bacteria and fungi), as well as from flora and fauna. Enzymes utilized in textile processing are predominantly extracted from microorganisms via fermentation methodologies. The role of enzymes in textile processing encompasses various dimensions. They may be utilized in pre-treatment phases to eliminate sizing agents, waxes, and other contaminants from cellulosic fabrics, which bestow desirable characteristics such as softness and a faded appearance upon fabrics. The incorporation of enzymes in textile processing yields numerous benefits, including enhanced efficiency, diminished water and energy usage, and a reduced ecological footprint when juxtaposed with traditional chemical processes. Biotechnological operations advance the evolution of sustainable and environmentally friendly textile processing methodologies by leveraging the specificity and catalytic efficacy of enzymes. Enzymes, serving as biological catalysts, have become essential elements in a variety of biotechnological operations, including textile production [1]. These protein entities accelerate chemical reactions without experiencing irreversible alterations, providing a remarkably specific and effective alternative to traditional chemical approaches [2]. The utilization of enzymes in textile processing encompasses several phases, ranging from pre-treatment to finishing. In the pre-treatment stage, enzymes aid in the elimination of sizing agents, waxes, and various impurities from fabrics [3]. The procurement of enzymes for textile applications predominantly hinges on microbial fermentation processes, employing bacteria and fungi as production hosts [4]. This method provides scalability and consistency in enzyme production, which are pivotal elements for industrial applications. The implementation of enzymatic processes in textile manufacturing offers numerous advantages over traditional chemical techniques. These advantages encompass improved process efficiency, lower water and energy consumption, and a diminished environmental impact [5]. The specificity of enzymes enables precise modifications of textile materials, thereby minimizing unintended consequences on fabric characteristics [6]. As the textile sector increasingly emphasizes sustainability, the incorporation of enzyme-based biotechnological processes signifies a promising pathway toward the establishment of eco-friendly processing practices [7]. Through the utilization of the catalytic capabilities inherent in enzymes, producers within the textile sector can potentially diminish their dependence on aggressive chemical agents while attaining results in fabric quality and performance that are either comparable or superior. Enzymes, which are defined as biocatalysts that augment reaction velocities without undergoing consumption or alteration, play a pivotal role in a multitude of biotechnological applications, particularly in the realm of textile processing [8]. Their remarkable specificity, efficiency, and environmentally benign characteristics render them attractive alternatives to traditional chemical methodologies [9]. Within biotechnological frameworks, enzymes facilitate a wide spectrum of reactions, including the degradation or modification of targeted compounds [10]. These biocatalysts can be obtained from a variety of organisms, encompassing microbes (such as bacteria and fungi), flora, and fauna [11]. In the textile processing domain, applied enzymes are primarily extracted from microbial origins through fermentation techniques [12]. The implementation of enzymes in textile processing is multifaceted, playing a crucial role during pretreatment stages where they assist in the elimination of sizing agents, waxes, and other impurities from cellulosic fabric surfaces [13].

7.2 Historical Development and Advancements in Enzymatic Textile Processing

The use of enzymes in textile processing has a rich history dating back to the early twentieth century (Fig. 7.1). The need for sustainable and eco-friendly alternatives to traditional chemical processes has driven the historical development and advancements in enzymatic textile processing. Enzymes have become integral to textile processing due to their ability to operate under mild conditions, reduce environmental impact, and improve product quality. This transition has been facilitated by advancements in biotechnology and enzyme engineering, which have expanded the applications and efficiency of enzymes in the textile industry. The evolution of enzymatic textile processing can be broadly categorized into several key phases.

7.3 Role of Enzymes in the Textile Industry

Enzymes play a pivotal role in the textile industry by offering sustainable, eco-friendly alternatives to traditional chemical processes. Cellulases and laccases are used in denim finishing processes to create various effects, including localized abrasion, stone-washed looks, and bleached or faded patterns. Their application spans

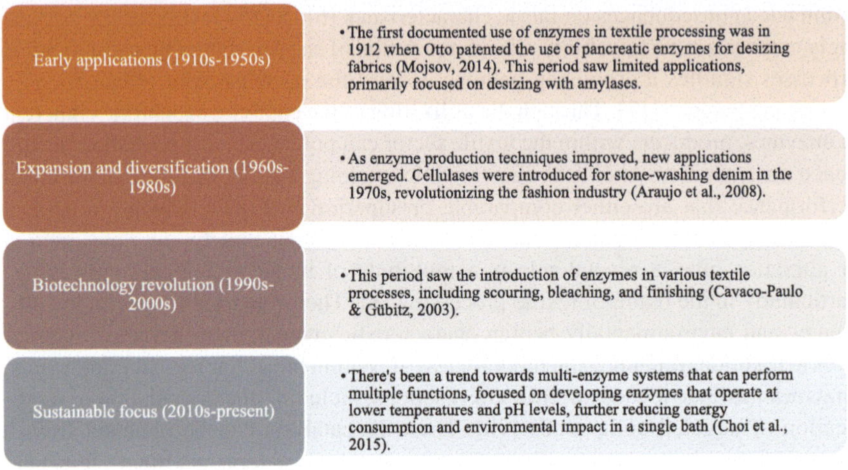

Fig. 7.1 Historical evolution of textile processing enzymes

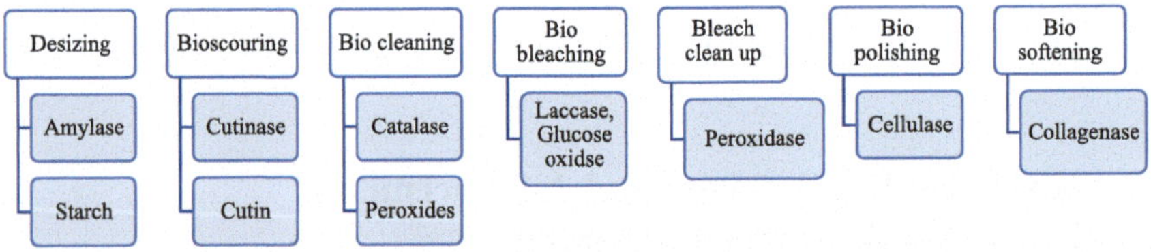

Fig. 7.2 Role of different enzymes in the textile industry

various stages of textile processing, contributing to environmental conservation, cost reduction, and improved product quality (Fig. 7.2).

7.3.1 *Desizing*

Amylases are commonly used for desizing, which is the process of removing sizing agents (such as starch) from woven fabrics before bleaching and dyeing. Alpha-amylases break down the long-chain starch molecules into smaller dextrin units, facilitating their removal [14]. This enzymatic process is more environmentally friendly compared to chemical methods, as it operates under mild conditions and reduces water and energy consumption [15].

7.3.2 Scouring and Bleaching

Pectinases and lipases are employed in scouring processes to remove non-cellulosic impurities like pectins, waxes, and oils from cotton and other cellulosic fibers [16]. Catalases and glucose oxidases are used in bleaching processes to remove residual hydrogen peroxide and improve whiteness. These enzymes enhance the effectiveness of scouring and bleaching, leading to improved absorbency, dyeability, and brightness of the fabrics.

7.3.3 Bio-polishing and Stone Washing

Cellulases are widely used for the bio-polishing of cellulosic fabrics, particularly cotton. These enzymes selectively remove the fuzzy fibers from the fabric surface, creating a smooth and lustrous appearance [17]. Cellulases, along with other enzymes like pectinases and amylases, are also used in stonewashing processes to create a worn or faded look on denim fabrics. Studies have shown that factors such as enzyme concentration, treatment time, and temperature significantly affect the bio-polishing outcome. For instance, a low-temperature and time enzymatic process can minimize fabric strength loss while achieving desired polishing effects [18]. Bio-polishing with cellulases improves the tactile features of fabrics, such as tensile resilience and surface smoothness, without compromising other properties. This process enhances comfort and luster, making fabrics more appealing [19].

7.3.4 Anti-pilling and Anti-felting

Cellulases and proteases are used in anti-pilling treatments for fabrics such as cotton, wool, and synthetic blends. These enzymes selectively target and remove protruding fibers on the fabric surface, which are responsible for forming pills. Pills are small, unwanted fiber balls that develop during wear and washing due to friction and abrasion, diminishing the fabric's appearance and quality. Cellulases act on cellulose-based fibers like cotton, breaking the protruding fibers into smaller fragments that can be easily washed away. Similarly, proteases target protein-based fibers such as wool, effectively removing loose fibers without compromising the fabric's structural integrity. These enzymatic treatments result in a smoother fabric surface, enhancing the fabric's aesthetic appeal and texture. In addition to improving the appearance of fabrics, enzymatic anti-pilling treatments contribute to prolonging the lifespan of textiles by preventing further damage caused by pilling. This enhances the durability and usability of the fabric, making enzymatic treatments an eco-friendly alternative to mechanical or chemical methods traditionally used for anti-pilling [20].

7.3.5 Enzymatic Bioscouring

A combination of pectinases, cellulases, and hemicellulases is used in enzymatic bioscouring processes for cotton and other cellulosic fibers. This process effectively removes non-cellulosic impurities, improves absorbency and dyeability, and reduces the consumption of harsh chemicals.

7.4 Properties of Enzymes Used in Textiles

Firstly, the enzyme accelerates the reaction by lowering the activation energy, which remains intact at the end of the reaction by acting as a catalyst. Secondly, enzymes operate under milder conditions. Enzymes can be used in catalytic concentrations at low temperatures and pH values near to neutral. Thirdly, enzymes are the best alternative to toxic, hazardous, and polluting chemicals. Fourthly, enzymes act only on specific substrates, for example, enzymes used in desizing do not affect cellulose hence there is no loss of strength of cotton. Fifthly, enzymes are easy to control because their activity depends upon optimum conditions. Sixthly, enzymes are biodegradable. At the end of the reaction enzymes are used can drain the remaining solution because enzymes are biodegradable and do not produce toxic waste on degradation hence there is no pollution. The properties of enzymes and their significance are as follows.

7.4.1 Catalytic Acceleration of Reactions

Enzymes dramatically increase reaction rates by lowering activation energy barriers. As biological catalysts, they remain unchanged at the reaction's conclusion, allowing for repeated use [21]. This catalytic efficiency enables industrial processes to occur under milder conditions and shorter timeframes.

7.4.2 Mild Operating Conditions

Each enzyme exhibits optimal activity within specific temperature and pH ranges. Outside these optima, enzyme activity typically decreases. This characteristic allows for energy-efficient processing, as extreme temperatures or harsh chemical environments are often unnecessary [22].

7.4.3 Environmentally Friendly Alternatives

Enzymes offer a sustainable substitute for many toxic, hazardous, and potentially carcinogenic chemicals traditionally used in textile processing. Their use significantly reduces environmental pollution and workplace hazards [23].

7.4.4 Substrate Pecificity

Most enzymes demonstrate high specificity, catalyzing reactions with only one or a few substrates. This selectivity is particularly valuable in textile processing. For instance, amylases used in desizing specifically target starch without degrading cellulose, preserving fabric strength [12].

7.4.5 Controllability

Enzyme activity can be precisely controlled by adjusting environmental conditions such as temperature, pH, and substrate concentration. This controllability allows for fine-tuning industrial processes to achieve desired outcomes [24].

7.4.6 Biodegradability

Enzymes are naturally biodegradable, breaking down into harmless components after use. This property eliminates concerns about persistent environmental pollutants and simplifies wastewater treatment in textile processing. These characteristics make enzymes particularly well-suited for textile processing applications, offering numerous advantages over traditional chemical methods in terms of efficiency, product quality, and environmental sustainability.

7.5 Cellulosic Cleaning

Cellulosic cleaning refers to the process of removing impurities and preparing cellulosic fibers, such as cotton, for further processing in textile manufacturing. This step is crucial for ensuring the quality and performance of the final textile product. Cellulase enzymes have become increasingly significant in various industrial applications due to their ability to efficiently break down cellulose into glucose monomers. The cellulose derivatives used in the study have a weight average molecular weight ranging from 50,000 to 200,000 g/mol, showing their effectiveness in enhancing cleaning performance [25]. This capability is crucial for industries seeking cost-effective and environmentally sustainable solutions. The demand for cellulases has surged, driven by their diverse applications in sectors such as biofuels, textiles, paper, and detergents. The commercial viability of cellulases hinges on their efficiency in cellulose hydrolysis, which is influenced by several factors including enzyme structure, production methods, and technological advancements. Cellulases are widely used for biopolishing, which enhances the smoothness and appearance of cotton fabrics by removing protruding fibers. Cellulosic cleaning in textiles primarily involves the use of cellulase enzymes to treat fabrics, particularly those made from cotton, which is a cellulosic fiber. Recent developments in biotechnology have introduced enzymes that specifically target and break down impurities in cellulosic fibers. Conditioning woven cotton fabrics with a cellulase enzyme during wash-dry cycles leads to topographical changes on the yarn micro-surfaces. These changes are quantified using mathematical methods to show how the cleanability of the cotton fabrics is improved [26]. The conceptual model explains how the topographical effects resulting from washing and conditioning with cellulase enzyme contribute to enhancing the cleanability and soil release of plain-woven cotton fabrics. Cleaning cellulase enzyme selectively attacks amorphous cellulose regions on cotton fabrics, consisting of small hills in a matrix of flat crystalline regions, which enhances soil release performance. The topographical changes caused by treatment with cleaning cellulase enzyme and carboxymethyl cellulose on cotton fabrics were analyzed on three different length scales to interpret their cooperation on water and oil absorption mechanisms, ultimately affecting the cleanability of the fabrics stained with liquid–solid, liquid, and solid soils [27]. A novel test model for monitoring the growth and removal of biofilms on textiles using Pseudomonas fluorescens and *Pseudomonas aeruginosa* as model organisms was developed and research findings indicate that a combined formulation of detergent and enzymes (F1/2) efficiently removed robust biofilms of Pseudomonas spp. on woven cellulose fabrics, leading to a significant reduction in viable bacteria within the biofilms [28].

Cellulase enzymes are used in the textile industry to perform biopolishing, which is a process that cleans and enhances the fabric's surface by removing protruding fibers. A study demonstrated the use of cellulase enzymes derived from sugarcane bagasse, achieving significant fabric weight and strength reduction, comparable to commercial enzymes [17]. This method exemplifies the potential for sustainable enzyme sourcing and application in textile processing. This results in a smoother texture and improved appearance of the fabric. The use of enzymes for cellulosic cleaning is considered environmentally friendly. It reduces the need for harsh chemicals traditionally used in textile processing, thereby minimizing the environmental impact. The cellulase enzymes were extracted from wasted sugarcane bagasse, which is a byproduct of sugar production. This not only provides a sustainable source of enzymes but also adds value to agricultural waste. The research identified optimal conditions for the enzymatic treatment, which include a pH of 4.8, a temperature of 55 °C, and a treatment duration of 1 h. These conditions were found to be effective in achieving significant fabric weight and strength reduction, indicating successful biopolishing. The study demonstrated that the enzymes extracted from sugarcane bagasse were as effective as commercial enzymes used in the industry. This suggests that locally sourced, waste-derived enzymes can be a viable alternative for textile processing. Enzymes are also employed in functional finishing processes, such as imparting antibacterial properties, UV protection, and flame resistance to textiles. These applications utilize biomaterials and nanobiotechnology, offering eco-friendly alternatives to traditional chemical finishes. The cleaning process helps in removing natural impurities like waxes, pectins, and other non-cellulosic materials from the fibers. This enhances the absorbency and dye affinity of the textiles, which is essential for producing vibrant and uniform colors in dyed fabrics. This not only improves the efficiency of the cleaning process but also minimizes the use of harsh chemicals, aligning with sustainable textile production goals. Proper cleaning of cellulosic fibers can significantly affect the mechanical properties of textiles, such as strength and softness. By ensuring that fibers are free from impurities, the cleaning process contributes to the durability and comfort of the final product. Cleaned cellulosic textiles are better prepared for subsequent finishing processes, such as bleaching, dyeing, and printing [29]. This preparation is essential for achieving high-quality finishes and ensuring that the textiles meet industry standards. Efficient cellulosic cleaning can lead to cost savings in the textile industry by reducing the need for additional processing steps and minimizing waste. This efficiency can also translate into lower production costs and more competitive pricing for textile products [21, 30, 31].

7.6 Applications of Cellulases

In textile companies, enzymes including cellulases, catalase, and laccase are frequently utilized. These enzymes are used to break down lignin, remove starch, and bleach fabrics, and break down excess hydrogen peroxide. The use of enzymes in the textile industry is expanding quickly because of their highly specialized, effective, non-toxic, and environmentally benign qualities. The most recent commercial developments include the use of cellulases for denim finishing and lactases for bleaching and decolorizing textile effluents. Enzymes also lead to shorter processing times, water and energy savings, better product quality, and possible process integration. Cellulases play a pivotal role in converting lignocellulosic biomass into fermentable sugars, which are essential for bioethanol production. This process is crucial for the biofuel industry, as it provides a sustainable alternative to fossil fuels [32]. Cellulases are employed in the paper industry to improve pulp processing and paper quality. They help in reducing the need for harsh chemicals, thus promoting eco-friendly manufacturing practices [33]. The inclusion of cellulases in detergents aids in the removal of microfibrils from cotton fabrics, enhancing cleaning efficiency and maintaining fabric integrity [34]. The exploration of novel cellulolytic microorganisms and the optimization of microbial fermentation processes have been crucial in enhancing cellulase production. Submerged and solid-state fermentation techniques are commonly used to meet industrial demands [35]. The high production cost of cellulases is a significant barrier to their widespread industrial use. Advances in genetic engineering and fermentation technologies have been pivotal in reducing these costs, making cellulases more commercially viable [36]. The use of cellulases in textile processing aligns with green chemistry principles, offering a sustainable alternative to toxic chemicals traditionally used in the industry. Enzymatic processes are biodegradable, work at lower temperatures, and reduce the overall environmental footprint of textile production [37, 38].

7.7 Environmental and Economic Impact

Enzymatic textile processing offers a promising alternative to traditional chemical methods, significantly reducing the environmental footprint and providing economic benefits. Enzymatic processes operate under milder conditions, such as lower temperatures and neutral pH, significantly reducing water and energy consumption. This reduces water usage by up to 50% and energy savings between 20–30%. Enzymes are biodegradable, making them a sustainable alternative to harsh chemicals traditionally used in textile processing [39]. The adoption of enzymatic processing contributes to a decrease in the carbon footprint of the textile industry by minimizing greenhouse gas emissions and reducing the release of toxic effluents into water bodies [40]. Although the initial cost of enzymes can be higher than that of traditional chemicals, the overall economic benefits outweigh this expense. Savings from reduced water and energy consumption and lower wastewater treatment costs make enzymatic processing a cost-effective choice in the long term. Large-scale operations, in particular, can capitalize on these savings, ensuring the economic viability of enzymatic processing on an industrial level [41]. The adoption of enzymatic methods not only improves sustainability but also enhances product quality. By operating under gentle conditions, enzymatic processes preserve the integrity of fabrics, ensuring superior durability and finish. This dual benefit of environmental and economic advantages makes enzymatic textile processing a valuable tool in promoting greener and more responsible practices across the textile industry.

7.8 Conclusion

One of the primary challenges in enzymatic textile processing is the limited stability of enzymes under industrial conditions. High temperatures, extreme pH levels, and long processing times can reduce enzyme activity and efficiency. Innovations like enzyme immobilization and nano-immobilization have been developed to address these limitations. These techniques enhance enzyme stability, reusability, and catalytic efficiency, making them better suited for industrial applications [42]. Immobilized enzymes also provide the added benefit of being recoverable and reusable, reducing overall costs and improving the feasibility of enzymatic processes on a commercial scale. Despite significant research efforts, many enzymatic processes remain confined to the laboratory, with limited industrial adoption. However, some enzymatic methods have achieved commercial success, such as enzymatic desizing with amylase and biopolishing with cellulase. These processes have been implemented at larger scales and have demonstrated their effectiveness in improving fabric quality while reducing chemical use and wastewater pollution [43]. Recent enzyme production and engineering advancements have further improved enzyme stability and reproducibility. Researchers have created enzymes that can withstand industrial conditions and perform consistently by optimizing enzyme structures and production methods. These advancements have made enzymes a viable and environmentally friendly alternative to conventional chemical reagents in processing cellulosic textiles. The adoption of enzymatic processes has led to significant reductions in water consumption, energy use, and harmful chemical discharge, contributing to more sustainable textile manufacturing practices [4]. Biotechnology and enzyme modification have been critical in expanding the applications of enzymes in various textile processes, including desizing, bleaching, biopolishing, and denim finishing. For example, enzymatic bleaching offers a gentler alternative to traditional methods, preserving fabric strength while minimizing environmental impact. In denim finishing, enzymes such as laccases and cellulases achieve desired aesthetic effects without abrasive treatments or harsh chemicals. These advancements align closely with the goals of the fourth industrial revolution, which emphasizes the integration of green technology and sustainable practices in industrial processes [15]. In conclusion, while enzymatic textile processing faces challenges, ongoing enzyme technology, and production innovations are steadily overcoming these obstacles. By aligning with sustainability goals and leveraging biotechnological advancements, the textile industry is moving toward greener, more efficient processing methods that reduce environmental impact and improve product quality.

Declaration The authors declare that they have no known competing financial interests or personal relationships that could have appeared to influence the work reported in this chapter. Generative AI and AI-assisted technologies have been used in the writing process to improve the readability and language of the work only. After using this tool, the authors carefully reviewed and revised the content to ensure accuracy and quality. The authors take full responsibility for the final version of this publication.

References

1. Liu Z, Smith SR (2021) Enzyme recovery from biological wastewater treatment. Waste Biomass Valorization 12(8):4185–4211. https://doi.org/10.1007/s12649-020-01251-7
2. Johnson BJ, Russ Algar W, Malanoski AP, Ancona MG, Medintz IL (2014) Understanding enzymatic acceleration at nanoparticle interfaces: approaches and challenges. Nano Today 9(1):102–131. https://doi.org/10.1016/j.nantod.2014.02.005
3. Cavaco-Paulo A, Nierstrasz VA, Wang Q (eds) (2019) Advances in textile biotechnology. Woodhead Publishing

4. Luo Z, Yan Y, Du S, Zhu Y, Pan F, Wang R, Xu Z, Xu X, Li S, Xu H (2023) Recent advances and prospects of Bacillus amyloliquefaciens as microbial cell factories: from rational design to industrial applications. Crit Rev Biotechnol 43(7):1073–1091. https://doi.org/10.1080/07388551.2022.2095499
5. Rodríguez-Couto S (2019) Green nanotechnology for biofuel production. In: Sustainable approaches for biofuels production technologies: from current status to practical implementation, pp73–82
6. Zhang Y, Rather LJ, Li Q (2022) Recent advances in the surface modification strategies to improve functional finishing of cotton with natural colourants-a review. J Clean Prod 335
7. Bhatt P (2023) Industrial applications of microbial enzymes. CRC Press
8. Chatha SAS, Asgher M, Iqbal HMN (2017) Enzyme-based solutions for textile processing and dye contaminant biodegradation-a review. Environ Sci Pollut Res Int 24(16):14005–14018. https://doi.org/10.1007/s11356-017-8998-1
9. Kirk O, Borchert TV, Fuglsang CC (2002) Industrial enzyme applications. Curr Opin Biotechnol 13(4):345–351. https://doi.org/10.1016/s0958-1669(02)00328-2
10. Pandey A, Tripathi AH, Tripathi PH (2022) Production, purification, and application of the microbial enzymes. In: Industrial applications of microbial enzymes. CRC Press, Boca Raton, pp 19–39. https://doi.org/10.1201/9781003202998-2
11. Bharathiraja B, Chakravarthy M, Kumar RR, Yuvaraj D, Jayamuthunagai J, Kumar RP, Palani S (2014) Biodiesel production using chemical and biological methods-a review of process, catalyst, acyl acceptor, source and process variables. Renew Sustain Energy Rev 38:368–382
12. Araújo R, Casal M, Cavaco-Paulo A (2008) Application of enzymes for textile fibres processing. Biocatal Biotransformation 26(5):332–349. https://doi.org/10.1080/10242420802390457
13. Shahid M, Mohammad F, Chen G, Tang R-C, Xing T (2016) ChemInform abstract: enzymatic processing of natural fibres: white biotechnology for sustainable development. ChemInform 47(23). https://doi.org/10.1002/chin.201623273
14. Niyonzima FN et al (2023) Microbial enzymes used in textile industry. In: Biotechnology of microbial enzymes. Academic Press, pp 649–684
15. Kabir SMM, Koh J (2021) Sustainable textile processing by enzyme applications. In: Biodegradation technology of organic and inorganic pollutants, pp 339–366
16. Tochetto GA, Aragão AMI, de Oliveira D, Immich APS (2022) Can enzymatic processes transform textile processes? A critical analysis of the industrial application. Process Biochem 123:27–35. https://doi.org/10.1016/j.procbio.2022.10.030
17. Ikbal MS, Tisha FA, Asheque AI, Hasnat E, Uddin MA (2024) Eco-friendly biopolishing of cotton fabric through wasted sugarcane bagasse-derived enzymes. Heliyon 10(4)
18. Marzoug IB, Cheriaa R (2023) Effects optimization of bio-polishing industrial process parameters. J Text Sci Technol 9(1):30–51
19. Kumar M, Shukla SR, Arputharaj A, Saxena S, Patil S, Patil PG, Varghese E, Amarowicz R (2021) Biopolishing of cellulosic fabrics: a study on low-stress mechanical properties, microstructure, and dye uptake. Fiber Polym 22(10):2803–2814. https://doi.org/10.1007/s12221-021-0356-8
20. Bahtiyari Mİ, Ekmekçi Körlü A, Bilisik K (2021) Bioprocessing of natural textile fibres and clothes. In: Fundamentals of natural fibres and textiles. Elsevier, pp 221–262. https://doi.org/10.1016/b978-0-12-821483-1.00007-3
21. Bischof Vukusic S, Katović D, Soljačić I (2004) Enzymatic treatment of cotton: reactive dyeing characteristics. Color Technol 120(3):141–145
22. Cavaco-Paulo A, Gubitz G (eds) (2003) Textile processing with enzymes. Woodhead publishing, vol 29
23. Shahid M, Mohammad F (2013) Perspectives for natural product based agents derived from industrial plants in textile applications-a review. J Clean Prod 57:2–18
24. Mojsov K (2011) The textile industry and enzymes. Tech Text Cloth 10
25. Hellmuth H, Schulz A, Kropf C, Schmiedel P (2016) Cellulose ether improving cleaning performance

26. Calvimontes A, Lant NJ, Dutschk V (2012) Cleanability improvement of cotton fabrics through their topographical changes due to the conditioning with cellulase enzyme. J Surfactants Deterg 15(2):131–137. https://doi.org/10.1007/s11743-011-1293-0
27. Calvimontes A, Lant NJ, Dutschk V (2011) Cooperative action of cellulase enzyme and carboxymethyl cellulose on cotton fabric cleanability from a topographical standpoint. J Surfactants Deterg 14(3):307–316. https://doi.org/10.1007/s11743-011-1248-5
28. Agustín MDR, Stengel P, Kellermeier M, Tücking K-S, Müller M (2023) Monitoring growth and removal of Pseudomonas biofilms on cellulose-based fabrics. Microorganisms 11(4):892. https://doi.org/10.3390/microorganisms11040892
29. Eid BM, Ibrahim NA (2021) Recent developments in sustainable finishing of cellulosic textiles employing biotechnology. J Clean Prod 284:124701. https://doi.org/10.1016/j.jclepro.2020.124701
30. Shahidi S, Wiener J, Ghoranneviss M (2012) The use of enzymes in the textile industry. Eur J Exp Biol 2(1):271–276
31. Slifka MK, Whitton JL (2000) Clinical implications of dysregulated cytokine production. J Mol Med 78:74–80. https://doi.org/10.1007/s001090000086
32. Ilić N, Milić M, Beluhan S, Dimitrijević-Branković S (2023) Cellulases: from lignocellulosic biomass to improved production. Energies 16(8):3598
33. Sarwan J, Uddin N (2024) Enhanced production of microbial cellulases as an industrial enzyme-a short review. J Multidiscip Res Adv 2(1):59–70
34. Singh A, Bajar S, Devi A, Pant D (2021) An overview on the recent developments in fungal cellulase production and their industrial applications. Bioresour Technol Rep 14100652. https://doi.org/10.1016/j.biteb.2021.100652
35. Korsa G, Konwarh R, Masi C, Ayele A, Haile S (2023) Microbial cellulase production and its potential application for textile industries. Ann Microbiol 73(1)
36. Ranjan R, Rai R, Bhatt SB, Dhar P (2023) Technological road map of Cellulase: A comprehensive outlook to structural, computational, and industrial applications. Biochem Eng J 198:109020. https://doi.org/10.1016/j.bej.2023.109020
37. Jajpura L (2018) Enzyme: A bio catalyst for cleaning up textile and apparel sector. In: Textile science and clothing technology. Springer, Singapore, pp 95–137. https://doi.org/10.1007/978-981-10-4876-0_5
38. Yusuf M (2021) Current and future perspectives of enzyme treatments for cellulosic fibers: a review. In: Green chemistry for sustainable textiles, pp 69–76
39. Molina-Espeja P et al (2023) Metagenomics and new enzymes for the bioeconomy to 2030. In: Biotechnology of microbial enzymes. Academic Press, pp 165–178
40. Leal Filho W, Dinis MAP, Liakh O, Paço A, Dennis K, Shollo F, Sidsaph H (2024) Reducing the carbon footprint of the textile sector: an overview of impacts and solutions. Text Res J
41. Estévez S, Mosca Angelucci D, Moreira MT, Tomei MC (2024) Techno-environmental and economic assessment of color removal strategies from textile wastewater. Sci Total Environ 913:169721. https://doi.org/10.1016/j.scitotenv.2023.169721
42. Kundu D, Thakur MS, Patra S (2021) Textile fabric processing and their sustainable effluent treatment using enzymes—insights and challenges. In: Immobilization strategies. Springer, Singapore, pp 645–666. https://doi.org/10.1007/978-981-15-7998-1_19.
43. Andreaus J et al (2019) Processing of cotton and man-made cellulosic fibers. In: Advances in textile biotechnology. Elsevier, pp 185–238. https://doi.org/10.1016/b978-0-08-102632-8.00009-8

Chapter 8
Polymeric Materials in Textile Finishing

Melkie Getnet Tadesse, Molla Tadesse Abate, and Jörn Felix Lübben

Abstract Polymeric materials have emerged as pivotal components in textile finishing, offering a diverse array of applications to enhance the properties and functionalities of textiles. This review explores the multifaceted role of polymeric materials in textile finishing processes, delving into their significance in achieving desired attributes such as durability, comfort, aesthetics, and functionality. We started the exploration of various polymeric materials employed in textile finishing, encompassing polymers like silicones, fluoropolymers, polyurethanes, and beyond. The discussion unfolds the intrinsic properties of these polymers that contribute to their effectiveness in textile applications. The review extends to the techniques employed in applying polymeric finishes, encompassing methods such as padding, coating, and spray applications. The challenges associated with ensuring uniform application and durability are addressed, providing insights into optimizing these processes for enhanced performance. Furthermore, the environmental implications of polymeric finishes are discussed, with a focus on sustainable and eco-friendly practices. Innovations in developing biodegradable and recyclable polymeric finishes highlight the industry's efforts toward mitigating environmental impact. The integration of polymeric materials in textile finishing represents a dynamic and evolving field. This review serves as a comprehensive guide for researchers, textile engineers, and industry professionals seeking to understand and harness the potential of polymeric materials in advancing the performance and sustainability of finished textiles.

Keywords Textile finishing · Polymeric materials · Coating · Spray coating · Biodegradable

M. G. Tadesse (✉) · J. F. Lübben
Sustainable Engineering (STE), Albstadt-Sigmaringen University, Albstadt, Germany
e-mail: melkiegetnet23@gmail.com

M. G. Tadesse · M. T. Abate
Ethiopian Institute of Textile and Fashion Technology, Bahir Dar University, Bahir Dar, Ethiopia

M. T. Abate
Engineering Department, Division of Sustainable Textiles, Hof University of Applied Sciences, Münchberg, Germany

S. Biranje et al. (eds.), *Sustainable Finishing Techniques in Textiles*,
https://doi.org/10.1007/978-981-96-4860-3_8

8.1 Introduction

Textile finishing is essential to enhancing the practical and visual characteristics of textiles and their appeal and performance [1]. Polymeric materials are among the many materials used in textile finishing that are highly effective and indispensable [2–4]. These polymeric materials add various functionalities to textile materials, such as electroluminescence, conductivity, and other functional properties. Polymeric materials not only provide functional properties but also have acceptable comfort properties [5–11]. These materials, made up of extended chains of repeating units, have a wide range of properties that can be amended to suit specific needs. Since polymeric materials have special qualities, including flexibility, durability, and adaptability, they are used widely in textile finishing. They can impart various functionalities such as improved strength, water repellency, flame resistance, and color retention by forming films, coatings, and finishes. Because of their multiplicity, textiles can be tailored to various needs, from high-performance technical textiles to stylish and cozy daily wear applications. Considering these improvements, the use of polymeric materials in textile finishing processes has changed to meet contemporary needs and problems. The search for environmentally friendly and sustainable methods has become more important, leading to research into recyclable and biodegradable polymer alternatives. Advances in nanotechnology have also made it easier to develop nanocomposite polymer coatings, giving fabrics better microscopic properties. This introduction provides a foundation for exploring the different roles polymers play in textile finishing. The ongoing connection between polymers and textiles continues to shape the industry, driving improvements in both sustainability and performance, from traditional uses to modern innovations.

8.1.1 Importance of Polymeric Materials in Textile Finishing

The impact of polymeric materials in textile finishing is significant, as they greatly influence the functionality, performance, and appearance of textiles. Their use in many applications in textile finishing offers several key benefits due to their unique properties, as outlined below.

- *Enhanced durability*—Textiles treated with polymers are stronger and more durable. Polymers are made of long chains of repeating units that can create a protective coating on textiles, improving their resistance to abrasion tears, and wear. This is especially important for textiles used in harsh environments or for those under frequent use, to extend the life span.
- *Water and stain resistance*—Fabrics with special coatings made from polymers can resist water and stains [12]. These coatings create a layer that repels liquids like water and oil, preventing them from penetrating the fabric. This is useful for outdoor and performance fabrics, as well as everyday clothing, to protect them from spills.

- *Improving Comfort*—Fabrics can be made more comfortable by using certain polymers [9]. These materials improve breathability, moisture control, and flexibility, making clothes feel better on the skin [13]. This is especially important for sportswear, underwear, and activewear.
- *Flame Retardancy*—Fabrics can be made safer by adding flame-retardant properties using polymeric materials [14]. This is important in industries like automotive, aerospace, and protective gear for workers in hazardous environments.
- *Color Retention*—Polymeric treatments help fabrics keep their color vibrant over time by preventing fading. This is crucial for clothing, home furnishings, and other products where appearance matters.
- *Customization for Specific Applications*—Polymers are versatile and can be adjusted for different uses. For example, they can create fabrics with conductive coatings for smart textiles, UV protection for outdoor fabrics, or antimicrobial properties for medical textiles [15].
- *Sustainability*—There is a growing focus on creating eco-friendly and sustainable polymer coatings. Research is being done on recyclable and biodegradable polymer systems to reduce the environmental impact of fabric-finishing processes [16].

Polymeric materials play an important role in improving the performance and usefulness of fabrics. Ongoing research is helping make textiles more durable, comfortable, safe, and eco-friendly.

8.1.2 Overview of Polymeric Materials

Polymeric materials are widely used in the textile industry to improve the functionality, durability, and aesthetic appeal of fabrics. These materials, made up of long chains of repeating units (monomers), can be either synthetic or natural, and they serve a variety of purposes in textile finishing, dyeing, printing, and coating processes. This review provides an overview of key polymeric materials commonly used in textile finishing, highlighting their applications, benefits, and contributions to the textile industry. Several important polymeric materials are discussed in this section.

- *Polyurethanes (PU)*—Polyurethanes are highly adaptable polymers that find widespread use in textile finishing because of their excellent abrasion resistance, flexibility, and durability [17]. They are commonly used in textile applications such as coatings, laminations, and elastic fibers, making it particularly popular in outdoor textiles, industrial fabrics, sportswear, and upholstery.
- *Fluoropolymers*—Fluoropolymers are synthetic polymers known for their strong resistance to heat, chemicals, oil, and water. They are widely used to make fabrics water, oil, and stain-resistant. These polymers are often applied as surface treatments, forming a protective layer that is ideal for outdoor clothing, performance fabrics, and upholstery. A well-known example is polytetrafluoroethylene (PTFE),

which is highly effective at repelling water and oil [18]. Fluoropolymer-treated textiles are easier to clean, last longer, and perform well under tough conditions, making them essential for high-performance fabrics.

- *Silicones*—Silicones are widely used in the textile industry due to their unique properties, which enhance both the comfort and functionality of fabrics. Comfort is very important, especially for functional fabrics, because these materials are made for specific activities or environments where movement and wearability are paramount [6, 7]. Silicon treatments improve softness, provide water repellency while maintaining the breathability of the fabric, and improve durability by protecting them from water, heat, and chemicals. These properties make them more comfortable and durable ideal for items like activewear and medical textiles, where both flexibility and durability are essential [19].
- *Acrylics*—Acrylic-based polymers are commonly used as binders in coatings, providing durability and flexibility. They offer excellent resistance to UV radiation, weathering, and chemicals [20]. Textile finishes that incorporate acrylics help maintain colour and protect fabrics from environmental damage. Acrylic finishes are commonly used in outdoor fabrics, awnings, and marine textiles, where durability and appearance are crucial.
- *Polyolefins*—Commonly known polyolefins such as polyethylene (PE) and polypropylene (PP) are lightweight polymers with good chemical resistance for non-woven fabrics [21]. Coatings made from these polymers offer several advantages that enhance the performance and durability of fabrics. Polyolefins serve as effective polymer coatings in textiles, providing numerous benefits such as water resistance, durability, and flexibility. Their applications span outdoor, home, and industrial textiles, medical textiles, disposable protective clothing, and breathable membranes for outdoor and sportswear.
- *Biodegradable Polymers*—In light of environmental concerns, biodegradable polymers such as chitosan, alginate, and cellulose derivatives are becoming increasingly popular in textile finishing [22]. These polymers break down naturally, offering a sustainable option. Biodegradable polymers are investigated for eco-friendly textile finishes, specifically in industries where sustainability is a key focus.

The textile industry is dynamic and continually evolving. The use of polymeric materials in textile finishing processes is expanding and getting integrated into most parts of textile finishing processes. Each polymeric material possesses distinct properties and functionalities, enabling textile specialists to tailor finishes to meet specific requirements. As technological advancements progress and sustainability emerge as a critical focus, exploring innovative polymeric materials is expected to intensify. This synergy between chemistry and textiles highlights the pivotal role of polymeric materials in enhancing the performance and sustainability of textiles, significantly impacting everyday applications.

8.2 Polymeric Materials in Textile Finishing

In the field of textile finishing, polymeric materials are now critical elements that prominently improve the effectiveness and graphic appeal of textiles. Polymeric materials possess a variety of features that make them useful instruments for researchers and textile technologists who want to customize fabrics for particular uses. The main polymeric materials used in textile finishing are examined in this overview, which also sheds light on their uses and characteristics that have influenced the development of contemporary textiles.

8.2.1 Types of Polymeric Materials

Various types of polymeric materials are used in textile finishing, each offering distinctive properties and functionalities that enhance the performance and aesthetics of textile fabrics. Polyurethanes (PU), fluoropolymers, silicones, acrylics, polyethylene (PE), polypropylene (PP), polyesters, biodegradable polymers (e.g., poly lactic acid—PLA), vinyl polymers, polyamides (nylon), polyvinyl chloride (PVC), polyvinylidene fluoride (PVDF), styrene butadiene rubber (SBR), and ethylene vinyl acetate (EVA) are some of the polymeric materials that have been widely used in the finishing of textile materials. In addition to the conventional polymers, polymers with special properties including conductive polymers such as polyacetylene (PA), polyaniline (PAni), polypyrrole (PPy), polythiophenes, and poly(3,4-ethylenedioxythiophene) (PEDOT)–polystyrene sulfonates (PSS) are among them which have been used in textile functional finishing [23]. In this section, we will discuss some of the polymeric materials that have been extensively employed in the textile finishing sector.

8.2.1.1 Polyurethanes

In textile finishing, polyurethane (PU) is a flexible and regularly used polymeric material with several qualities that improve the functionality, toughness, and visual appeal of fabrics. The utilization of polyurethane in textile finishing requires a number of procedures, each designed to ensure certain needs. Polyurethane is frequently used in textile finishing for the following reasons (Fig. 8.1) [17, 24, 25].

Due to the hydrophobic properties of PU [26], fabric treated with it is suitable for outdoor apparel, sportswear, and technical textiles. Industrial fabrics, workwear, and other garments exposed to high levels of wear and tear can be benefited when finished with PU as it enhances abrasion resistance. PU also has excellent elastic recovery, making it ideal for use in sportswear, swimwear, and activewear. Its strong adhesion properties allow for vibrant and long-lasting prints on fabrics. Furthermore, jackets with insulating properties can be achieved by finishing with PU because PU

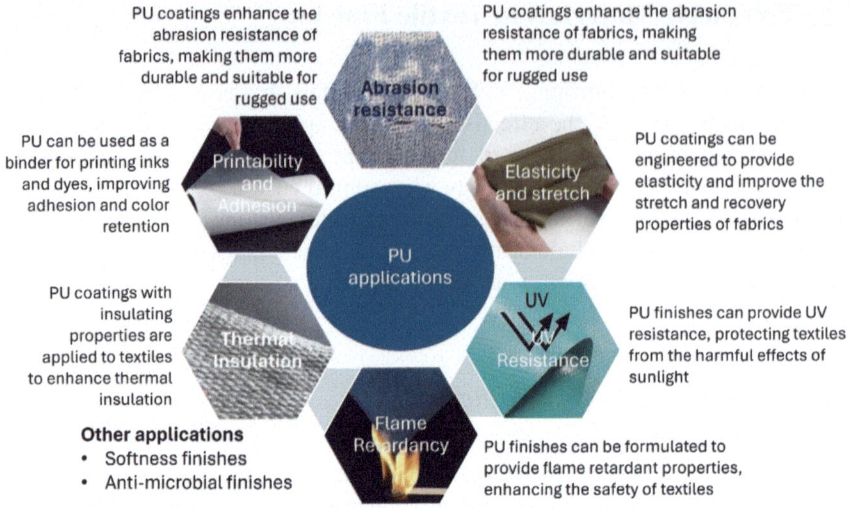

Fig. 8.1 Various applications of polyurethane in textile finishing

is an excellent insulator. Additionally, PU-treated fabrics provide UV protection due to their natural UV resistance.

Because of its versatility, PU is a valuable material in textile finishing, offering durability, improved comfort, and enhanced performance across a wide range of applications. It can be tailored to meet specific performance requirements, making it essential to modern textile technology.

8.2.1.2 Silicon

Because of their special qualities, silicones are frequently employed in textile finishing, which enhances the functionality, comfort, and visual appeal of fabrics. The use of silicones in textile finishing entails a number of procedures designed to provide certain results. Silicones are frequently used in textile finishing as shown in Fig. 8.2 [27, 28].

Luxurious and silky touch finished fabrics can be used for lingerie, activewear, and bed linens. Friction can be reduced using silicon finishing which provides comfort to the wearer an hence hosiery and intimate apparels fabrics can be beneficial [29]. Silicon can prevent the penetration of water for fabric used for outdoor activities and rainwear fabrics. Stain resistance of upholstery fabrics, table linens, and garments can be enhanced by silicon finishes due to their inherent nature. On the other hand, the water absorption of bath towels and sportswear can be improved by silicon treatment. Uniforms and work wear exposed to build up of static electricity can be benefited from silicon treatment as it can reduce static electricity generation.

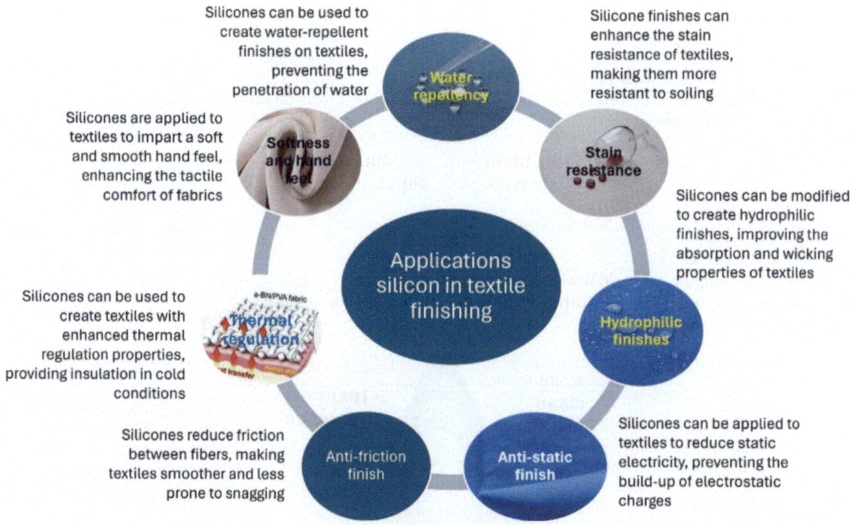

Fig. 8.2 Applications of silicon in textile finishing

The wide range of uses for silicones in textile finishing demonstrates how important it is to fulfill the comfort and performance needs of materials. Silicones play a major role in the production of high-performance and informal textiles by improving the hand feel and offering practical qualities like breathability and water repellency.

8.2.1.3 Fluoropolymers

Fluoropolymers are important components of textile finishing because they offer useful qualities that improve the robustness and use of textiles. Because of their special qualities, such as their low surface energy and good chemical resistance, fluoropolymers are especially well-suited for a wide range of textile industry applications. Fluoropolymers are used in textile finishing for the following noteworthy applications [30] (Fig. 8.3).

Fluoropolymers have excellent water-resistance properties, ensuring rainwear, outdoor garments, and high-performance textiles for hydrophobic barriers, which help to prevent the penetration of water [31]. In addition, it contributes to the stain resistance of upholstery fabric and carpets to possess good resistance to stain for the specified application. Furthermore, fluoropolymer possesses outstanding chemical and UV-resistant properties and is used for such finishing the respective use.

Although there are many benefits to using fluoropolymers in textile finishing, it's crucial to take sustainability and the environment into account. There is a push to create environmentally acceptable substitutes and procedures that mimic or preserve the advantageous aspects of fluoropolymers with the least number of negative effects on the environment.

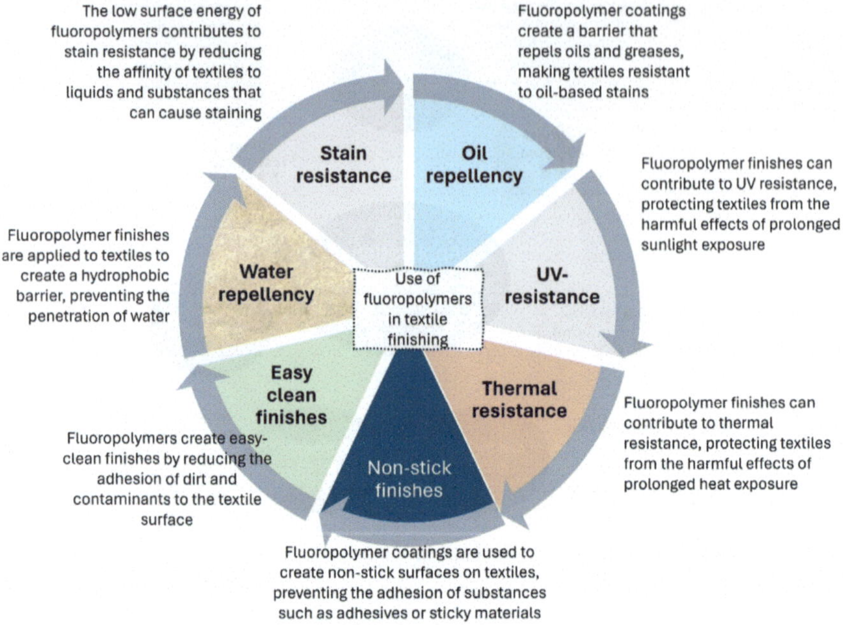

Fig. 8.3 Various applications of fluoropolymers in textile finishing

8.2.1.4 PEDOT-PSS

PEDOT-PSS, also known as poly(3,4-ethylenedioxythiophene) polystyrene sulfonate, is a conductive polymer that is frequently used in textile finishing procedures [32, 33]. PEDOT-PSS is used in textile and leather applications to give materials functionality and electrical conductivity [34]. The polymer sticks to a variety of textile surfaces well, enhancing their conductivity without compromising their comfort or flexibility. This novel approach to textile finishing creates opportunities for the incorporation of electronic elements into textiles, such as flexible displays or wearable sensors. They have very good electromechanical properties [4]. Because of their remarkable washability and durability, textiles treated with PEDOT-PSS are well-suited for use in wearable technology, smart textiles, and other cutting-edge electronic textile applications [35] (Fig. 8.4).

8.2.1.5 Other Polymeric Materials for Textile Finishing

Acrylic polymers are often used to improve the durability and colorfastness of textiles. They can be applied as coatings to enhance resistance to sunlight, abrasion, and chemicals. Acrylic polymers have been used to increase the thickness of printing paste [36] and to improve the physicochemical properties [37]. Another

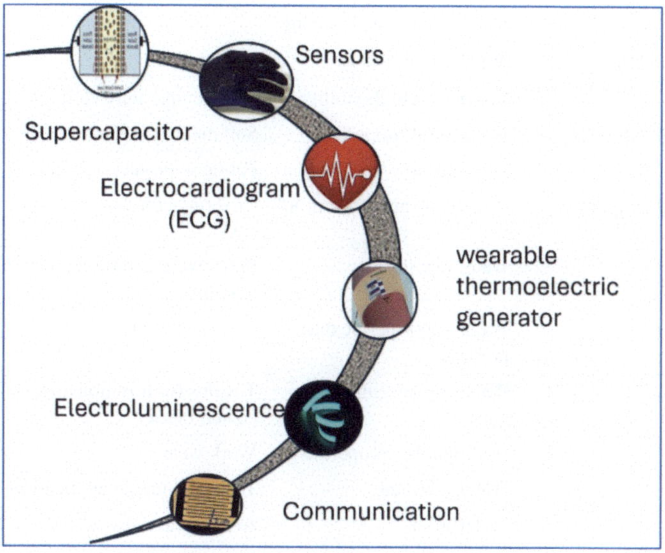

Fig. 8.4 Applications of PEDOT-PSS in textile finishing

important polymeric material that has been investigated and used in textile finishing is Polyvinyl Chloride (PVC). PVC coatings are employed to make textiles waterproof [38] and resistant to environmental elements. They are commonly used in rainwear and outdoor gear. Table 8.1 shows the summary of applications of PVC in textile finishing.

Derived from chitin, a natural polymer found in crustacean shells, chitosan is used for antimicrobial and antibacterial finishes in textiles [48]. It finds applications in medical textiles and sportswear [49, 50] (Fig. 8.5).

The use of chitosan in textile finishing aligns with the growing demand for functional and sustainable textiles. As research continues, new applications and formulations of chitosan in textile finishing are likely to emerge, contributing to the development of innovative and high-performance textile products.

Finally, cellulosic polymers (e.g., cellulose acetate, viscose) have various applications in textile finishing. Derived from plant-based sources, these polymers are used in textiles like rayon and acetate. They offer a natural feel and are used in a variety of clothing and home textile applications.

8.2.2 *Applications of Polymeric Materials in Textile Finishing*

In textile finishing, polymeric materials are essential for improving the functional and visual qualities of textiles. These applications use the special properties of polymers

Table 8.1 Use of polyvinyl chloride in textile finishing

Types of finishing	Improved properties	Applications	Refs.
Coating	Viscosity and flexibility	High temperature cloth	[39]
Coating and laminating	Waterproof properties	Sail making industries	[38]
Coating	Waterproof	Fabric to protect weather	[40]
Coating and laminating	Breathable and waterproof	Coverings for convertible cars	[41]
Coating	Water repellent	Sportswear and protective clothing	[42]
Impregnation	Water vapor-permeable fabrics	Outer garments	[42]
Coating	Waterproof weldable fabric	Textiles for filtration and coated fabrics	[43]
Coating	Hydrophobic properties	Work wear	[44]
Coating	Water resistant	Technical textile application	[45]
Coating	Waterproof	Architectural membrane	[46]
Coating	Waterproof	Protective clothing	[47]

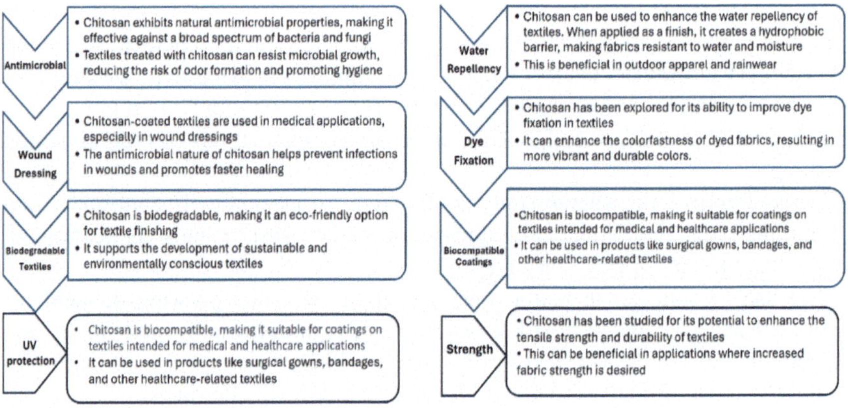

Fig. 8.5 Various applications of chitosan in textile finishing

to accomplish a variety of goals. Polymeric materials are frequently used in textile finishing for the following purposes.

8.2.2.1 Water Repellency and Hydrophobic Finishes

In textile finishing, water repellency is achieved through different methods, usually by applying chemicals or coatings that change the surface properties of the fabrics. The main mechanisms to impart water repellency are:

- *Hydrophobic Coatings*—Textiles are coated with hydrophobic materials that repel water [51]. These coatings are made of polymeric materials such as paraffin, silicone, or fluorocarbons. Since these materials repel water, they cause water droplets to roll off the fabric and form beads. Such treated fabrics are useful in situations where protection from rain is needed, helping to keep water from getting through.
- *Micro-structuring*—Some water-repellent treatments create microstructures on the surface of the fabric, that look like uneven surfaces or raised fibers [52]. These microstructures improve water repellency by reducing the ability of water to be drawn into the fabric, due to changes in the surface texture.
- *Chemical Modifications*—The molecular structure of textile fibers can be changed by incorporating hydrophobic groups [53]. This modification lowers the surface energy of the fibers, making them less likely to attract water. Often, highly water-repellent perfluoroalkyl groups are added to achieve this effect.
- *Cross-Linking Agents*—Cross-linking agents can be used to connect polymer chains in the fabric or coating, creating a stronger and more water-resistant structure [54]. This increased stability helps prevent water from penetrating the material.
- *Nano-coatings*—Nanotechnology is increasingly used in textile finishing. Nano-sized particles, like hydrophobic nanoparticles, can be applied to the fabric [55]. These particles create a protective layer that repels water while still allowing the fabric to remain flexible and breathable.
- *Fluorocarbon Finishes*—Compounds based on fluorocarbons are frequently used in textile finishing to provide water repellency [56]. These compounds form a durable coating on the fabric's surface. The strong bond between carbon and fluorine gives fluorocarbons excellent water-repellent properties.

By using these methods, either individually or combined, textiles can be made as water-repellent as needed. The choice of method depends on factors like fabric type, intended use, and specific performance requirements.

Types of Water-Repellent Polymeric Finishes

Water-repellent polymeric finishes play a crucial role in enhancing the performance of textiles by providing resistance against water penetration. Various types of water-repellent polymeric finishes are employed in textile industries, each with its unique properties and applications. Here are some common types [57–59] (Fig. 8.6).

8.2.2.2 Anti-microbial Finishes Using Polymers

Textiles are treated with anti-microbial finishes made of polymers to prevent the growth and spread of microbes such as bacteria, fungi, and algae. These finishes

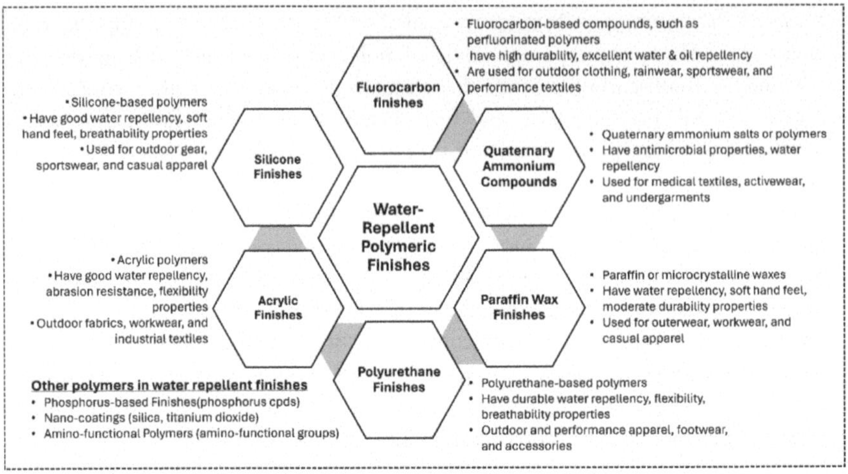

Fig. 8.6 Various applications of polymers in water repellent finish

are crucial for improving hygiene, preventing odors, and making fabrics generally cleaner in various applications. The following are some typical techniques and polymer kinds utilized in anti-microbial finishes (Table 8.2).

These antimicrobial treatments help maintain cleaner and more hygienic environments, particularly in places like hospitals or in situations where textiles contact the

Table 8.2 Applications of various polymeric materials for anti-microbial finishes

Polymers	Chemistry	Mechanism	Applications	Refs.
Quaternary ammonium (QA) compounds	Polymers with QA groups	Positively charged QA ions disrupt the microbial cell membranes, leading to cell lysis	Medical textiles, sportswear, undergarments	[60]
Polyhexamethylene biguanide (PHMB)	Polymer with biguanide groups	Interferes with microbial cell membranes and inhibits cell metabolism	Bedding, medical textiles, sportswear	[61]
Silver (Ag) nanoparticles	Nanoscale Ag particles incorporated into polymers	Silver ions released from nanoparticles disrupt microbial cell membranes and inhibit growth	Medical textiles, socks, sportswear	[62]
Chitosan	Derived from chitin, a natural polymer found in crustacean shells	Positively charged chitosan disrupts microbial cell membranes and binds to bacterial cell walls	Medical textiles, wound dressings, activewear	[63]
Zinc pyrithione	Polymer with zinc pyrithione groups	Releases zinc ions, which interfere with microbial cell metabolism	Socks, sportswear, undergarments	[64]

human body. When applying anti-microbial is finished, safety and regulatory criteria must be followed with consideration for potential effects on human health and the environment. Polymeric finishes for antimicrobial properties took the next step in functional finishing of textiles.

8.2.2.3 Polymeric Materials for Stain Resistance

Textile stain-resistance polymeric compounds are made to repel liquids and stop stains from leaching through. These substances coat the fabric's surface to provide protection and ease of cleaning and upkeep. These are a few typical varieties of polymeric materials that are used to repel stains. The most common polymeric materials for stain-resistance finish is fluorocarbon-based compounds, such as polytetrafluoroethylene (PTFE) or perfluorinated polymers [65]. Fluoropolymers have excellent water and oil repellency that can be able to provide effective stain resistance. Fabrics that are used for upholstery and carpets are finished with fluoropolymers. Silane-based materials are also other important polymer used for treating of textile materials as can be formulated to create elastomers with stain-resistant properties [12]. Furthermore, there are lots of polymeric materials that have been involved in finishing of textiles materials for stain-resistant properties including silicon [66] and acrylic-based [67].

It is crucial to consider the specific application, the desired performance qualities, and environmental factors when choosing polymeric materials for stain resistance. Additionally, optimizing the efficacy of stain-resistant polymeric materials requires careful surface preparation and application methods.

8.2.2.4 Flame Retardant Polymeric Finishes

In several sectors, flame-retardant polymeric coatings are essential for improving the fire safety of materials—particularly textiles. These coatings are intended to lessen polymers' flammability and stop or slow the spread of fire. For various uses, flame-retardant polymeric coatings come in a variety of forms. In addition, different kinds of polymeric materials have been employed in flame retardant finishes. The first polymeric material that comes to mind when thinking of polymeric materials for flame retardant finish is phosphorus-based flame retardants [68]. Among the phosphorus compounds phosphate and phosphonates are effective flame-retardant in reducing the flame spread and releasing water or other water-inhibiting gases. The basic drawback of this compound is some are extremely toxic to humans. Halogenated flame-retardants are also the basic ingredients in this application [69]. Most recently, graphene-based flame retardants have been researched widely due to various reasons [70]. Graphene and its derivatives have the potential to be effective flame retardants; in particular, using graphene as an adjuvant in conjunction with inorganic nanomaterials may make graphene a viable flame-retardant material. There are lots of polymeric materials that have been widely used in the flame-retardant finishes including silicon [71], boron [72], melamine-based flame-retardants [73], nitrogen-based [74], and aluminum trihydrate and thermoplastic polyurethane-based flame-retardants [75] are among the many polymeric materials. The type of material required flame resistance, and particular application requirements all play a role in the flame-retardant polymeric finish selection. It's crucial to remember that laws and guidelines pertaining to flame retardant use are in place to guarantee public health and environmental sustainability.

8.2.2.5 Polymeric Materials for UV Blocking

UV-blocking polymeric materials are essential in a variety of applications to guard against the damaging effects of UV radiation. These substances are made to scatter, absorb, or reflect UV rays in order to keep them from entering and causing harm. The following are a few typical polymeric materials for UV blocking (Table 8.3).

Numerous industries, including building, automotive, textiles, packaging, and outdoor equipment, use UV-blocking polymeric materials. The intended use, the amount of durability necessary, and the level of UV protection required all influence

Table 8.3 Applications of polymeric materials for UV protective finish

Types of polymers	Properties	Applications	Refs.
Polyethylene (PE)	PE can be modified to include UV stabilizers, which absorb and block UV radiation	Agricultural films, outdoor furniture, packaging materials	[76]
Polypropylene (PP)	PP with UV stabilizers provide UV resistance	Outdoor furniture, automotive parts, ropes, and twines	[77]
Polyvinyl chloride (PVC)	PVC formulations with UV absorbers and stabilizers offer UV protection	Construction materials, window frames, outdoor signage	[78]
Polycarbonate (PC)	PC inherently absorbs UV radiation and can be enhanced with additional UV stabilizers	Eyewear lenses, safety shields, greenhouse panels	[79]
Acrylic (PMMA)	Acrylic inherently has good UV resistance and optical clarity	Outdoor signage, lenses, windows, and skylights	[80]
Polyurethane (PU)	PU coatings can provide UV protection to various substrates	Outdoor fabrics, coatings for furniture, footwear	[81]
Ethylene vinyl acetate (EVA)	EVA with UV stabilizers is used in outdoor applications	Footwear, outdoor equipment, solar cell encapsulation	[82]
Fluoropolymers (e.g., PTFE)	Some fluoropolymers provide inherent UV resistance	Architectural fabrics, high-performance coatings	[18]
Silicone rubber	Silicone rubber can be formulated to resist UV degradation	Outdoor seals, gaskets, and components	[83]

the choice of material. During the production process, UV stabilizers and absorbers are frequently added to improve the UV-blocking properties of these polymers.

8.3 Techniques for Applying Polymeric Finishes

Polymeric finishes can be applied to textiles using several methods. The choice of techniques depends on the material, the intended result, and the textile surface. The following are some standard methods for applying polymeric finishes.

8.3.1 Padding and Immersion Techniques

In the padding method, a polymer solution is applied to the fabric using a machine called a padder [84, 85]. The padder has two rollers positioned horizontally, which press together as they rotate, creating a line of contact where the fabric is treated with the solution. Polymeric materials/solution are kept in the bath while the fabric slowly passes between the two rollers, dipped/immersed into the solution, and then squeezed using squeeze rollers and drying for fixation. Immersion is another technique for polymeric materials application onto the textile surface by applying chemical solutions using a chemical container and keeping them for some time followed by drying and fixation. The schematic diagram for padding and immersion technique is illustrated in Fig. 8.7a, b, respectively.

To guarantee adhesion and a long-lasting finish, adequate surface preparation including cleaning, sanding, and priming is necessary prior to applying polymeric coatings. The type of substrate, the intended finish qualities, and the project's size are some of the variables that influence the application method selection.

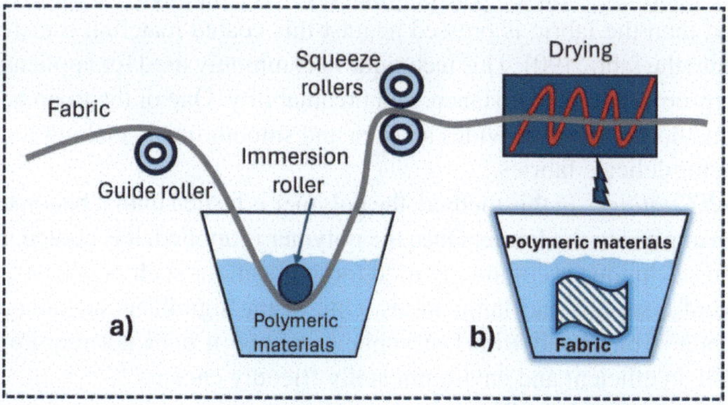

Fig. 8.7 Schematic view of padding and immersion methods in fabric finishing

8.3.2 Coating Methods

Coating is a process where polymer materials are applied to the surface of textiles using techniques such as a knife, roller, or brush. The polymers must be in resin or paste form to ensure they are spread evenly on the fabric. To achieve this, the polymer mixture needs to have the right thickness or viscosity. Different coating methods are used depending on the type of polymer, the equipment, and the level of precision needed. Common techniques to apply the polymeric materials onto the fabric surface include doctor blade coating, spray coating, and roller coating.

- *Doctor Blade Coating*—This is a more precise method, and it uses a blade or knife to spread the polymer finish onto the fabric surface. It is often used to create films with controlled thickness [86].
- *Spray coating*—In this technique, the polymer finish is sprayed onto the fabric using high-pressure equipment, creating a fine mist. This method covers large areas smoothly and evenly and is preferred for applications that require thick, dense coatings [87].
- *Roller coating*—This method is used for applying polymer coatings to flat surfaces. It creates a thicker coating than spraying, and the texture of the finish can be influenced by the type of roller used [88].
- *Dip Coating*—Dip coating is a common method used to apply polymer coatings onto textiles, providing a uniform and protective layer [89]. In this process, the textile is immersed into a polymer solution, allowing the fibers to absorb or become coated with the polymer. After dipping, the material is carefully removed from the solution and the excess coating is drained off. The textile is then dried and cured, typically through heating, to ensure the polymer adheres well to the fibers and forms a durable coating. This method is commonly used for applications such as waterproofing, oil repellency, and stain resistance. One of the key advantages of this technique is its simplicity and effectiveness, making it particularly suitable for achieving full coverage on the fabric.
- *Transfer Coating/printing*—The polymer is initially applied to a transfer paper or film, and then the fabric is pressed against this coated material, transferring the finish onto the fabric [90]. This technique is commonly used for applications such as waterproofing, stain resistance, and breathability. One of the main advantages of this method is that it provides an even and smooth finish, making it especially suitable for delicate fabrics.
- *Hot-Melt Coating*—In this method, the polymer is heated until it becomes a liquid and then applied to the fabric. Once the polymer is applied, it is cooled to solidify [91]. This technique is commonly used for applications such as adhesive finishes, waterproof coatings, and laminations. One of the significant advantages is that it creates a strong bond with fast application, and it does not require solvents, making it an efficient and environmentally friendly choice.
- *Solvent-Based Coating*—The polymer is dissolved in a solvent and then applied to the fabric [92]. As the solvent evaporates, the polymer forms a coating on the

fabric's surface. This method is commonly used for water resistance, oil repellency, and stain resistance applications. While it produces strong and even finishes, it is less eco-friendly due to using solvents.

8.3.3 Lamination

This process involves applying a thin layer of polymer, such as polyurethane (PU), polyethylene (PE), or polytetrafluoroethylene (PTFE), onto textiles, either by heat or adhesive bonding. This process involves applying a thin polymer layer to the fabric through heat or adhesive bonding, creating a composite material that combines the properties of both the fabric and the polymer. Laminated textiles are widely used for applications requiring water resistance, breathability, or protection from environmental factors. Key benefits include waterproofing, breathability, and increased durability, making the fabric resistant to abrasion and chemicals. Additionally, lamination allows for the customization of textiles with properties like flame resistance, UV protection, or antimicrobial effects. This technique is crucial in industries such as fashion, outdoor gear, protective clothing, and medical textiles, where high performance and specific functionality are essential.

8.3.4 Screen and Digital Printings

In screen printing, polymeric finishes can be applied in patterns or specific areas using a screen-printing technique. This method is particularly useful for applications such as conductive finishes for smart textiles, decorative coatings, and self-cleaning coatings. One of the main advantages of screen printing is that it allows for precise application, enabling the creation of detailed patterns and targeted finishes on the fabric. Using digital printing, polymeric finishes can be applied using a digital printer, which allows for the creation of patterns or functional designs on the fabric. This technique is used for applications such as conductive finishes, functional coatings, and aesthetic designs. The advantages of using digital printing include high precision, customization options, and minimal waste.

8.4 Environmental Considerations in Polymeric Finishes in Textile

Most polymeric materials are toxic and they cause health and environmental problems [93]. For instance, fluorine is the most frequently used polymeric material in anti-stain resistance finishes. This section covers the sustainability of polymeric finishes, the development of biodegradable options, and eco-friendly practices in textile finishing.

Key factors to consider when using polymeric materials in textile finishing include selecting sustainable materials with biodegradable properties, using low volatile organic compounds (VOCs), ensuring recyclability, choosing biodegradable options, and applying water-based formulations. It's important to assess the environmental impact of these materials before use. Figure 8.8 shows the steps to follow for sustainable production of polymer-finished textiles.

Adopting sustainable practices during production and use of polymeric finishes helps reduce the environmental impact of manufacturing processes and promotes a more sustainable and ethical industry. Recycling textiles with polymeric finishes can be challenging because they are complex materials made up of polymer coatings and synthetic fibers. However, several strategies can be explored to improve their recycling as discussed below:

- **Separation Technologies**—Invest in advanced separation technologies that can identify the different fiber and coating components in polymeric-finished textiles [94]. Methods like heat degradation, solvent dissolution, and mechanical separation can help break down the materials into their individual parts.
- **Chemical Recycling**—Explore chemical recycling methods that produce valuable chemicals or the original building blocks (monomers) of polymer coatings [95]. Techniques such as hydrolysis, glycolysis, and pyrolysis can convert polymeric materials into feedstocks for making new polymers or other useful products.
- **Mechanical Recycling**—Use mechanical recycling methods to recover synthetic fibers from polymer-finished textiles [96]. Techniques such as melt spinning or extrusion can reprocess the fibers into yarns or other textile products without significantly losing their quality.

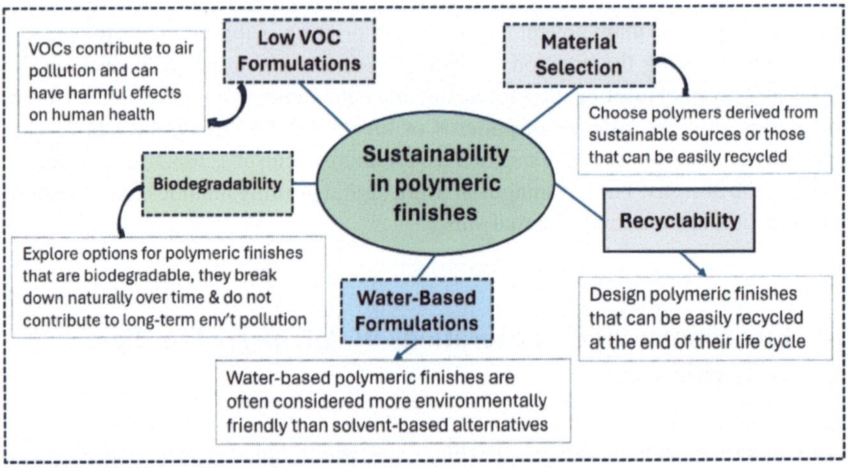

Fig. 8.8 Schematic representation of factors that should be considered for the sustainability of polymeric finishes in textile applications

- **Biological Recycling**—Examine biological recycling strategies that use enzymes or microbes to break down polymeric coatings or speed up the breakdown of blended textile materials [97]. These biological processes can be particularly effective for blends containing natural fibers and provide a sustainable alternative to traditional recycling methods.

By taking a comprehensive approach that includes technological innovation, collaboration, and consumer engagement, the recycling of polymeric-finished textiles can be significantly improved. This will help the textile industry transition to a more sustainable and circular model.

8.5 Future Trends and Developments

This section explores the future of polymeric materials in textile finishing, including innovations, prospects, and recommendations.

8.5.1 Innovations in Polymeric Materials

Innovations in polymeric materials for textile finishing are continuously advancing to improve performance, sustainability, and functionality. Here are several notable innovations in this field:

- *Nanotechnology*—One of the key developments is *nanotechnology* [98], where nanomaterials such as silica, titanium dioxide, and silver nanoparticles are incorporated into polymeric finishes to provide antimicrobial properties, water repellency, stain resistance, and UV protection, enhancing fabric performance [99].
- *Smart Textiles*—Another innovation is the creation of smart textiles using polymeric materials with embedded sensors and conductive elements [100]. These allow for advanced functions like temperature control, moisture management, health monitoring, and interactive clothing.
- *Biodegradable Polymers*—Using biodegradable polymers derived from renewable resources or microbial fermentation to develop eco-friendly textile finishes. Biodegradable polymers offer a sustainable alternative to conventional synthetic polymers and facilitate the development of biodegradable and compostable textiles [101]. Environmental sustainability/eco-friendly polymeric materials are becoming a fundamental requirement for a green future.
- *Self-cleaning Coatings*—Self-cleaning coatings are emerging. It uses hydrophobic or superhydrophobic materials to repel dirt, water, and oil [102]. This reduces the need for frequent washing and enhances the fabric's durability.
- *Phase Change Materials (PCMs)*—Polymeric finishes with phase change materials (PCMs) are used to improve temperature control in textiles [103]. PCMs

absorb and release heat during phase transitions, enabling textiles to maintain a comfortable temperature in response to environmental changes. For example, clothing with PCM microcapsules can reduce body heat loss by about 6.5–13.2 W, depending on the number of clothing layers, during changes in the environment [104]. In this regard, natural-based materials have been getting a lot of attention.
- *Responsive Polymers*—Design of polymeric materials that respond to external stimuli such as temperature, pH, light, or mechanical stress to alter their properties or functionalities [105]. Responsive polymers enable the development of adaptive textiles with dynamic performance characteristics tailored to specific applications.
- *Biobased Polyurethanes*—Utilization of biobased polyurethanes derived from renewable resources such as vegetable oils, soybeans, or castor oil for textile finishing applications. Biobased polyurethanes offer comparable performance to petroleum-based counterparts while reducing reliance on fossil fuels and minimizing environmental impact [106].
- *Breathable Membranes*—Development of breathable polymeric membranes with controlled permeability to water vapor and air, enhancing comfort and moisture management in textiles [107]. These membranes are commonly used in outdoor apparel, sportswear, and protective clothing to maintain a dry and comfortable microclimate next to the skin.
- *Recyclable and Recycled Polymers*—Exploration of recyclable and recycled polymeric materials for textile finishing to promote circularity and reduce environmental impact. Recyclable polymers enable the recovery and reuse of materials at the end of their life, while recycled polymers utilize post-consumer or post-industrial waste as feedstock for new textile products [108].

These innovations demonstrate the diverse range of polymeric materials and applications in textile finishing, contributing to the advancement of sustainable, functional, and high-performance textiles for various industries.

8.5.2 Prospects and Recommendations

The prospects for polymeric materials in textile finishing are promising, with ongoing advancements to enhance performance, sustainability, and functionality. A key area of development is sustainability, where efforts are directed toward creating polymeric materials from renewable resources, biodegradable polymers, and recyclable formulations. Collaborative efforts across the textile industry can accelerate the adoption of eco-friendly practices and facilitate the transition to a circular economy. Another exciting avenue is the development of functional polymeric coatings. Researchers are designing coatings with multiple features, such as self-cleaning, antimicrobial properties, flame resistance, and conductivity. The incorporation of functional additives and nanomaterials into these coatings enhances textile performance for various applications. Simultaneously, the integration of polymeric materials with smart textile technologies, such as sensors and embedded electronics, is enabling the creation of responsive and adaptive textiles. These advancements open new opportunities in wearable technology, healthcare, and interactive clothing. Polymeric materials are also tailored to improve performance attributes like durability, weather resistance, breathability, and comfort. Advanced techniques in polymer chemistry and processing contribute to enhanced performance in challenging conditions. Additionally, these materials enable customization and personalization of textile finishes to meet individual preferences and specific applications. Innovations like digital printing and additive manufacturing simplify the production of personalized textile products with unique designs and features. Cross-industry collaboration is essential for driving innovation in polymeric materials for textile finishing. Partnerships among textile producers, polymer suppliers, universities, and research institutions foster knowledge sharing, technology transfer, and solutions to complex challenges. Furthermore, regulatory compliance plays a vital role in ensuring the safety and sustainability of polymeric materials. By adhering to regulations governing hazardous substances, VOC emissions, and chemical safety, the industry can build trust and credibility in sustainable textile products.

8.6 Conclusions

In conclusion, polymeric materials play a key role in textile finishing by providing enhanced durability, improved functionality, and better aesthetics. Continuous research and innovation have allowed these materials to evolve in response to industry needs while addressing environmental challenges. Polymeric materials boost textile performance by increasing durability, weather resistance, breathability, and comfort through advanced chemistry and processing techniques. They also add functional properties like water repellency, stain resistance, flame retardancy, antimicrobial protection, and conductivity, expanding their use in various sectors. With a growing focus on sustainability, efforts are being made to use renewable resources, biodegradable polymers, recyclable formulations, and eco-friendly processes to reduce environmental impact and promote circular production. Collaboration between textile manufacturers, polymer suppliers, researchers, and regulators are essential to driving innovation in polymeric materials, ensuring the development of sustainable solutions that meet industry standards and consumer expectations. Looking ahead, the future of polymeric materials in textiles is promising, with further advancements in performance, smart textiles, personalization, and sustainability. By embracing sustainable practices, fostering collaboration, and continuing research, the textile industry can fully harness the potential of polymeric materials to create high-performance, eco-friendly products that meet the needs of both consumers and industries.

Declaration The authors confirm that they have no competing financial interests or personal relationships that could have influenced the work presented in this chapter. AI tools were used exclusively to improve the text's readability and language. The authors have carefully reviewed and revised the content as needed and accept full responsibility for the final published version.

References

1. Schindler WD, Hauser PJ (2004) Chemical finishing of textiles. Elsevier
2. Tadesse MG, Dumitrescu D, Loghin C, Chen Y, Wang L, Nierstrasz V (2018) 3D printing of NinjaFlex filament onto PEDOT: PSS-coated textile fabrics for electroluminescence applications. J Electron Mater 47:2082–2092
3. Tadesse MG, Loghin C, Chen Y, Wang L, Catalin D, Nierstrasz V (2017) Effect of liquid immersion of PEDOT: PSS-coated polyester fabric on surface resistance and wettability. Smart Mater Struct 26(6):065016
4. Tadesse MG, Mengistie DA, Chen Y, Wang L, Loghin C, Nierstrasz V (2019) Electrically conductive highly elastic polyamide/lycra fabric treated with PEDOT: PSS and polyurethane. J Mater Sci 54(13):9591–9602
5. Tadesse M, Nagy L, Nierstrasz V, Loghin C, Chen Y, Wang L (2018) Low-stress mechanical property study of various functional fabrics for tactile property evaluation. Materials 11(12) 10:2466
6. Tadesse M, Nierstrasz V, Loghin E-C, Loghin M-C (2020) Quality inspection and prediction of the comfort of fabrics finished with functional polymers. Industria Textila 71(4):340–349
7. Tadesse MG (2019) Quality inspection and evaluation of smart or functional textile fabric surface by skin contact mechanics. Högskolan i Borås
8. Tadesse MG, Chen Y, Wang L, Nierstrasz V, Loghin C (2019) Tactile comfort prediction of functional fabrics from instrumental data using intelligence systems. Fibers Polym 20:199–209
9. Tadesse MG, Harpa R, Chen Y, Wang L, Nierstrasz V, Loghin C (2019) Assessing the comfort of functional fabrics for smart clothing using subjective evaluation. J Ind Text 48(8):1310–1326
10. Tadesse MG, Loghin C, Dulgheriu I, Loghin E (2021) Comfort evaluation of wearable functional textiles. Materials 14(21):6466
11. Tadesse MG, Loghin E, Pislaru M, Wang L, Chen Y, Nierstrasz V, Loghin C (2019) Prediction of the tactile comfort of fabrics from functional finishing parameters using fuzzy logic and artificial neural network models. Text Res J 89(19–20):4083–4094
12. Sfameni S, Lawnick T, Rando G, Visco A, Textor T, Plutino MR (2022) Functional silane-based nanohybrid materials for the development of hydrophobic and water-based stain resistant cotton fabrics coatings. Nanomaterials 12(19):3404
13. Atalie D, Tesinova P, Tadesse MG, Ferede E, Dulgheriu I, Loghin E (2021) Thermo-physiological comfort properties of sportswear with different combination of inner and outer layers. Materials 14(22):6863
14. Rosace G, Migani V, Guido E, Colleoni C (2015) Flame retardant finishing for textiles. In: Flame retardants: polymer blends, composites and nanocomposites, pp 209–246
15. Attia NF, Osama R, Elashery SE, Kalam A, Al-Sehemi AG, Algarni H (2022) Recent advances of sustainable textile fabric coatings for UV protection properties. Coatings 12(10):1597
16. Abdellatif FHH, Abdellatif MM (2021) Utilization of sustainable biopolymers in textile processing. Green chemistry for sustainable textiles. Elsevier, pp 453–469
17. Hassabo AG, Mohamed NA, Gouda NZ, Khaleed N, Shaker S, El-Salam A, Neaama A, El-Aziz A (2024) Polyurethane (PU) in textile finishing process. J Text, Color Polym Sci 21(1):173–186
18. Sonnendecker A, Viljoen D, Ameduri B, Crouse P (2020) Fluoropolymer-based architectural textiles: production, processing, and characterization. Fascinating fluoropolymers and their applications. Elsevier, pp 337–399
19. Fortess F (1954) Silicone resins in textiles. Ind Eng Chem 46(11):2325–2331
20. Özgenç Ö, Durmaz S, Şahin S, Boyaci IH (2020) Evaluation of the weathering resistance of waterborne acrylic-and alkyd-based coatings containing HALS, UV absorber, and bark extracts on wood surfaces. J Coat Technol Res 17:461–475
21. Wang Z, Wang D, Fang Y, Shen J (2018) Preparation and properties of antistatic polyethylene film coated polypropylene non-woven fabrics. Fibers Polym 19:934–940

22. Osman E (2018) Fabrication and characterization of biodegradable polymer as multifunction finishing agent for natural fabrics. Egypt J Chem 61(4):679–689
23. Namsheer K, Rout CS (2021) Conducting polymers: a comprehensive review on recent advances in synthesis, properties and applications. RSC Adv 11(10):5659–5697
24. Choi C-N, Kim J-B, Hwang T-Y, Lee M-S, Kim R, Jung I-B (2001) Preparation of polyurethane derivative microcapsule and its application to textile finishing. Text Sci Eng 38(4):174–181
25. Das A, Mahanwar P (2020) A brief discussion on advances in polyurethane applications. Adv Ind Eng Polym Res 3(3):93–101
26. Pulat M, Akdoğan A (2002) The diffusion and bulk properties of polyurethane (PU)-based hydrophilic and hydrophobic membranes. J Appl Polym Sci 85(1):193–198
27. Teli MD (2015) Softening finishes for textiles and clothing. In: Functional finishes for textiles: improving comfort, performance and protection, pp 123–152
28. Chruściel JJ (2022) Modifications of textile materials with functional silanes, liquid silicone softeners, and silicone rubbers—a review. Polymers 14(20):4382
29. Özdil N, Anand S (2014) Recent developments in textile materials and products used for activewear and sportswear. Electron J Veh Technol/Tasit Teknolojileri Elektronik Dergisi 8(3)
30. Ebnesajjad S, Khaladkar PR (2017) Fluoropolymer applications in the chemical processing industries: the definitive user's guide and handbook. William Andrew
31. Hunter L, Fan J (2009) Waterproofing and breathability of fabrics and garments. In: Engineering apparel fabrics and garments, pp 283–308
32. Kasaw E, Haile A, Getnet M (2020) Conductive coatings of cotton fabric consisting of carbonized charcoal for E-textile. Coatings 10(6):579
33. Tadesse MG, Nierstrasz V (2016) Effects of process parameters on electrical properties of PEDOT: PSS coated polyester fabrics. In: Aachen-Dresden-Denkendorf international textile conference, Dresden, November 24–25
34. Kebede ZT, Tadesse MG, Chane TE, Mengistie DA (2023) Application of PEDOT: PSS conductive polymer to enhance the conductivity of natural leather: retanning process. J Nanomater 2023
35. Tseghai GB, Mengistie DA, Malengier B, Fante KA, Van Langenhove L (2020) PEDOT: PSS-based conductive textiles and their applications. Sensors 20(7):1881
36. Nuessle A, Kine B (1953) Acrylic resins in textile processing. Ind Eng Chem 45(6):1287–1293
37. Inna S, Ya SO (2020) Investigation of physicochemical properties of styrene-acrylic and urethane polymers used in finishing of textile materials. Известия высших учебных заведений. Химия и химическая технология 63(3):88–93
38. Fung W (2002) Coated and laminated textiles, vol 23. Woodhead Publishing
39. Sen AK (2008) Coated textiles. Principles and applications, 2nd edn, pp 72–73
40. Holmes DA (2000) Waterproof breathable fabrics. Handbook of technical textiles 12:282
41. Moolla A (2011) Breathable and waterproof coated textiles. Technická Univerzita v Liberci
42. Loghin C, Ciobanu L, Ionesi D, Loghin E, Cristian I (2018) Introduction to waterproof and water repellent textiles. Waterproof and water repellent textiles and clothing. Elsevier, pp 3–24
43. Bajaj P, Sengupta A (1985) Industrial applications of textiles: textiles for filtration and coated fabrics. Text Prog 14(1):1–30
44. Sacevičienė V, Jucienė M, Šveikauskaitė L (2016) The influence of mechanical effects on coated textile hydrophobic properties. In: Development of polyester/cellulosic blend woven fabric for better comfort
45. Patel JP, Patel PR, Mahera HK, Patel P (2015) Effect of PU and PVC coating on different fabrics for technical textile application. Int J Sci Technol Eng 1(11):279–284
46. Ghosh S, Kundu B, Ghosh D, Saha K, Roy P (2022) A review on mechanical characterization of PVC coated fabrics. In: AIP conference proceedings. AIP Publishing
47. Akovali G (2012) Advances in polymer coated textiles. Smithers Rapra
48. Lim S-H, Hudson SM (2003) Review of chitosan and its derivatives as antimicrobial agents and their uses as textile chemicals. J Macromol Sci, Part C: Polym Rev 43(2):223–269

49. Rajendran S, Anand S (2002) Developments in medical textiles. Text Prog 32(4):1–42
50. El-Ghazali S, Khatri M, Kobayashi S, Kim IS (2022) An overview of medical textile materials. Med Text Nat Resour 3–42
51. Ivanova N, Zaretskaya A (2010) Simple treatment of cotton textile to impart high water repellent properties. Appl Surf Sci 257(5):1800–1803
52. Zhang W, Ding W, Fernandino M, Dorao CA (2019) Water-repellent surfaces consisting of nanowires on micropyramidal structures. ACS Appl Nano Mater 2(12):7696–7704
53. Ali A, Shaker K, Nawab Y, Jabbar M, Hussain T, Militky J, Baheti V (2018) Hydrophobic treatment of natural fibers and their composites—a review. J Ind Text 47(8):2153–2183
54. Tillet G, Boutevin B, Ameduri B (2011) Chemical reactions of polymer crosslinking and post-crosslinking at room and medium temperature. Prog Polym Sci 36(2):191–217
55. Hsieh C-T, Wu F-L, Chen W-Y (2009) Super water- and oil-repellencies from silica-based nanocoatings. Surf Coat Technol 203(22):3377–3384
56. Sayed U, Dabhi P (2014) Finishing of textiles with fluorocarbons. Waterproof and water repellent textiles and clothing. Elsevier, pp 139–152
57. Ozcan G (2007) Performance evaluation of water repellent finishes on woven fabric properties. Text Res J 77(4):265–270
58. Zahid M, Heredia-Guerrero JA, Athanassiou A, Bayer IS (2017) Robust water repellent treatment for woven cotton fabrics with eco-friendly polymers. Chem Eng J 319:321–332
59. Özek HZ (2018) Silicone-based water repellents. Waterproof and water repellent textiles and clothing. Elsevier, pp 153–189
60. Kamel MY, Hassabo AG (2021) Anti-microbial finishing for natural textile fabrics. J Text, Color Polym Sci 18(2):83–95
61. Ibrahim A, Laquerre J-É, Forcier P, Deregnaucourt V, Decaens J, Vermeersch O (2021) Antimicrobial agents for textiles: types, mechanisms and analysis standards. Text Funct Appl 13(2):261–293
62. Khoddami A, Shokohi S, Sebdani Z (2012) A facile method for anti-bacterial finishing of cotton fabrics using silver nanoparticles. Industria Textilă 63(1):20–26
63. Yang H-C, Wang W-H, Huang K-S, Hon M-H (2010) Preparation and application of nanochitosan to finishing treatment with anti-microbial and anti-shrinking properties. Carbohydr Polym 79(1):176–179
64. Simončič B, Tomšič B (2017) Recent concepts of antimicrobial textile finishes. In: Textile finishing: recent developments and future trends, pp 1–68
65. Churchley D, Rees GD, Barbu E, Nevell TG, Tsibouklis J (2008) Fluoropolymers as low-surface-energy tooth coatings for oral care. Int J Pharm 352(1–2):44–49
66. Hassan MM, Leighs SJ (2017) Effect of surface treatments on physicomechanical, stain-resist, and UV protection properties of wool fabrics. Appl Surf Sci 419:348–356
67. Kim H-J, Hong J-P, Kwak D-S, Kwon D-J (2021) Acrylic polymer treatment on Nylon fibers improves color strength and durability in thermal transfer printing. Colloids Surf, A 615:126268
68. Levchik SV, Weil ED (2006) A review of recent progress in phosphorus-based flame retardants. J Fire Sci 24(5):345–364
69. Bocchini S, Camino G (2010) Halogen-containing flame retardants. In: Fire retardancy of polymeric materials, vol 2, pp 75–106
70. Sang B, Li Z-W, Li X-H, Yu L-G, Zhang Z-J (2016) Graphene-based flame retardants: a review. J Mater Sci 51:8271–8295
71. Kashiwagi T, Gilman JW (2000) Silicon-based flame retardants. In: Flame retardancy of polymeric materials. Marcel Dekker, New York, NY, pp 353–389
72. Shen KK (2014) Review of recent advances on the use of boron-based flame retardants. In: Polymer green flame retardants, pp 367–388
73. Wang X, Wang B, Xing W, Tang G, Zhan J, Yang W, Song L, Hu Y (2014) Flame retardancy and thermal property of novel UV-curable epoxy acrylate coatings modified by melamine-based hyperbranched polyphosphonate acrylate. Prog Org Coat 77(1):94–100

74. Klatt M (2014) Nitrogen-based flame retardants. In: Morgan AB, Wilkie CA (eds) Non-halogenated flame retardant handbook, pp 143–168
75. Pinto UA, Visconte LLY, Gallo J, Nunes RCR (2000) Flame retardancy in thermoplastic polyurethane elastomers (TPU) with mica and aluminum trihydrate (ATH). Polym Degrad Stab 69(3):257–260
76. Won J, Said MA, Seyam A-FM (2013) Development of UV protective sheath for high performance fibers for high altitude applications. Fibers Polym 14:647–652
77. Lai M-F, Huang C-H, Lou C-W, Chuang Y-C, Wei C-Y, Lin J-H (2023) Using melt extraction technique to produce polypropylene/TiO_2-coated stainless steel wrapped yarns and functional fabrics: evaluations of UV resistances and electromagnetic shielding. Fibers Polym 24(5):1661–1670
78. Krehula LK, Papić A, Krehula S, Gilja V, Foglar L, Hrnjak-Murgić Z (2017) Properties of UV protective films of poly(vinyl-chloride)/TiO_2 nanocomposites for food packaging. Polym Bull 74:1387–1404
79. Alongi J, Di Blasio A, Carosio F, Malucelli G (2014) UV-cured hybrid organic–inorganic layer by layer assemblies: effect on the flame retardancy of polycarbonate films. Polym Degrad Stab 107:74–81
80. Rawat A, Soni U, Malik RS, Pandey SC (2018) Facile synthesis of UV blocking nano-sized zinc oxide and polymethyl-methacrylate polymer nanocomposite coating material. Nano-Struct Nano-Objects 16:371–380
81. Muzaffar S, Abbas M, Siddiqua U, Arshad M, Tufail A, Ahsan M, Alissa S, Abubshait SA, Abubshait H, Iqbal M (2021) Enhanced mechanical, UV protection and antimicrobial properties of cotton fabric employing nanochitosan and polyurethane based finishing. J Market Res 11:946–956
82. Malucelli G, Barbalini M (2019) UV-curable acrylic coatings containing biomacromolecules: a new fire retardant strategy for ethylene-vinyl acetate copolymers. Prog Org Coat 127:330–337
83. Refaat E, Mohamed M, Youssif A, Ali S, Fathallah A, Maamoun D, Ahmed HM, Hassabo AG, Khattab TA (2024) UV protection and water repellent finishing of polyester fabric printed with photochromic dyes. J Text, Color Polym Sci 21(1):87–90
84. Singh N, Sheikh J (2020) Microencapsulation and its application in production of functional textiles
85. Pasichnyk M, Gaálová J, Minarik P, Václavíková M, Melnyk I (2022) Development of polyester filters with polymer nanocomposite active layer for effective dye filtration. Sci Rep 12(1):973
86. Yang H, Jiang P (2010) Self-cleaning diffractive macroporous films by doctor blade coating. Langmuir 26(15):12598–12604
87. Ganesan A, Yamada M, Fukumoto M (2013) Cold spray coating deposition mechanism on the thermoplastic and thermosetting polymer substrates. J Therm Spray Technol 22:1275–1282
88. Barroso G, Li Q, Bordia RK, Motz G (2019) Polymeric and ceramic silicon-based coatings—a review. J Mater Chem A 7(5):1936–1963
89. Tang X, Yan X (2017) Dip-coating for fibrous materials: mechanism, methods and applications. J Sol-Gel Sci Technol 81:378–404
90. Chen L, Degenaar P, Bradley DD (2008) Polymer transfer printing: application to layer coating, pattern definition, and diode dark current blocking. Adv Mater 20(9):1679–1683
91. Achanta A, Adusumilli P, James K, Rhodes C (1997) Development of hot melt coating methods. Drug Dev Ind Pharm 23(5):441–449
92. Gutoff EB, Cohen ED (2016) Water- and solvent-based coating technology. Multilayer flexible packaging. Elsevier, pp 205–234
93. Lithner D, Larsson Å, Dave G (2011) Environmental and health hazard ranking and assessment of plastic polymers based on chemical composition. Sci Total Environ 409(18):3309–3324
94. Pandey P, Chauhan R (2001) Membranes for gas separation. Prog Polym Sci 26(6):853–893
95. Ghosal K, Nayak C (2022) Recent advances in chemical recycling of polyethylene terephthalate waste into value added products for sustainable coating solutions—hope vs. hype. Mater Adv 3(4):1974–1992

96. Ribul M, Lanot A, Pisapia CT, Purnell P, McQueen-Mason SJ, Baurley S (2021) Mechanical, chemical, biological: moving towards closed-loop bio-based recycling in a circular economy of sustainable textiles. J Clean Prod 326:129325
97. Koshti R, Mehta L, Samarth N (2018) Biological recycling of polyethylene terephthalate: a mini-review. J Polym Environ 26:3520–3529
98. Zille A, Almeida L, Amorim T, Carneiro N, Esteves MF, Silva CJ, Souto AP (2014) Application of nanotechnology in antimicrobial finishing of biomedical textiles. Mater Res Express 1(3):032003
99. Hassan B, Islam G, Haque A (2019) Applications of nanotechnology in textiles: a review. Adv Res Text Eng 4(2):1038
100. Carpi F, De Rossi D (2005) Electroactive polymer-based devices for e-textiles in biomedicine. IEEE Trans Inf Technol Biomed 9(3):295–318
101. Chandra R, Rustgi R (1998) Biodegradable polymers. Prog Polym Sci 23(7):1273–1335
102. Parkin IP, Palgrave RG (2005) Self-cleaning coatings. J Mater Chem 15(17):1689–1695
103. Huang J, Luo Y, Weng M, Yu J, Sun L, Zeng H, Liu Y, Zeng W, Min Y, Guo Z (2021) Advances and applications of phase change materials (PCMs) and PCMs-based technologies. ES Mater Manuf 13:23–39
104. Shim H, McCullough E, Jones B (2001) Using phase change materials in clothing. Text Res J 71(6):495–502
105. Liu F, Urban MW (2010) Recent advances and challenges in designing stimuli-responsive polymers. Prog Polym Sci 35(1–2):3–23
106. Schneiderman DK, Vanderlaan ME, Mannion AM, Panthani TR, Batiste DC, Wang JZ, Bates FS, Macosko CW, Hillmyer MA (2016) Chemically recyclable biobased polyurethanes. ACS Macro Lett 5(4):515–518
107. Chang Y, Liu F (2023) Review of waterproof breathable membranes: preparation, performance and applications in the textile field. Materials 16(15):5339
108. Ignatyev IA, Thielemans W, Vander Beke B (2014) Recycling of polymers: a review. ChemSusChem 7(6):1579–1593

Chapter 9
Textile Finishing with Plasma Technology: Fundamentals, Advancements, and Sustainable Practices

Nadim Ibn Sayed, A Z M Mofasser, Musarrat Jahan Mridu, Niger Nahid, and Rony Mia

Abstract Plasma, the fourth phase of matter, can confer significant attributes to various textile materials. Plasma-induced surface modification of textiles especially grabs the attention of researchers as it is a potential process to impart multi-functional properties to the substrates. One of the major advantages of using plasma technology is that several chemical and physical properties are altered in the outer layer of the substrate, keeping the bulk material intact. Among the thermal and non-thermal plasmas, the second type is appropriate for treating textiles despite some limitations in industrial-scale processing. Several functionalities such as antimicrobial, antistatic, hydrophilic, and hydrophobic properties can be endowed to textile substrates using plasma. Additionally, flame retardancy, UV protection, and vintage denim look can be achieved by plasma treatment. As plasma technology is a dry, non-polluting, energy-efficient process for treating materials, this can replace the traditional wet chemical processes in textile finishing.

9.1 Introduction

The treatment of textiles using plasma technology is gaining attention around the world due to its eco-friendliness. When a gas or vapor is excited through electric or electromagnetic fields, it forms plasma, which comprises equal numbers of positive ions and electrons and free radicals, UV radiation, and neutral species [1]. Plasma can be categorized into two main types: thermal and nonthermal. In thermal plasmas, electrons and heavy particles (ions and neutrals) are at the same temperature, achieving thermal equilibrium. On the other hand, nonthermal plasmas have

N. Ibn Sayed · A Z M Mofasser · M. J. Mridu · N. Nahid
Department of Textile Engineering, European University of Bangladesh, Mirpur, Dhaka, Bangladesh

R. Mia (✉)
Wilson College of Textiles, North Carolina State University, Raleigh, NC, USA
e-mail: mroni_mia@yahoo.com

S. Biranje et al. (eds.), *Sustainable Finishing Techniques in Textiles*,
https://doi.org/10.1007/978-981-96-4860-3_9

high electron temperatures, while the heavy particles remain at much lower temperatures, often close to room temperature. Non-thermal plasmas are appropriate for heat-sensitive polymeric and textile materials [2]. Traditional textile wet processing involves abundant use of water and toxic chemicals, while plasma technology facilitates dry and non-polluting processing of textiles [3]. Chemical finishing of textile materials using plasma can be achieved through two approaches: surface modification via discharges and surface grafting of compounds onto the substrates. Plasma alters only the outermost atomic layer of the material without affecting its bulk properties [4, 5].

Different types of plasma treatments on textile materials can be carried out to obtain some superior properties like antimicrobial, UV protection, flame retardant, and antistatic properties. Plasma can modify the surfaces of different substrates to impart better adhesion properties. Consequently, different nano-particles can be easily incorporated in the substrate surfaces pre-treated with plasma. For instance, O_2-plasma increases the hydroxyl groups on the fiber surface, which enhances the water molecule adsorption of the substrate [6, 7]. In addition, low-pressure O_2-plasma can improve the adhesion property of cotton fiber, facilitating the addition of ZnO nano-particles to the substrate surface, resulting in the introduction of ultraviolet protection properties to cotton substrates [8]. Flame retardant fishing can be done on cotton fiber surfaces using argon plasma-induced grafting and polymerization of flame retardants [9]. Moreover, antistatic properties can be imparted to several textiles, especially, polyester fabrics by means of non-thermal plasma at an optimized condition [10]. The color strength of dyed textiles can also be improved using plasma treatment [11–13].

In this chapter, the fundamentals of plasma, advancements in textile finishing using plasma technology, and the sustainability aspects of plasma technology have been demonstrated. Different plasma generation techniques have been elucidated here as well. Furthermore, the applications of plasma technology to add important functionalities to textile materials have been demonstrated. The importance of plasma technology in textile finishing to attain sustainable approach like circular economy has been also depicted in this chapter mentioning some limitations and future trends.

9.2 Fundamentals of Plasma Technology

9.2.1 What is Plasma?

Plasma, often referred to as the fourth state of matter, is an ionized gas containing nearly equal amounts of positively charged ions and negatively charged electrons. When energy is supplied to a substance, it can transition from a solid to a liquid and then to a gaseous state. If the supply of energy continues, the atomic shells of the matter disintegrate, and then positively charged ions and negatively charged electrons are created. Plasma can also contain non-ionized neutral particles [14].

Plasma can also be defined as the quasi-neutral discharge of charged and neutral particles showing collective behavior [15, 16].

9.2.2 Types of Plasma

Plasmas are categorized based on their electrical and physical parameters such as temperature, energy levels, and pressure. Primarily, plasmas can be classified into two types: thermal plasma and non-thermal plasma. Thermal plasmas, also known as hot plasma, are characterized by the local thermodynamic equilibrium where the temperature of the electrons, ions, and gas are approximately equal, and the temperature is typically less than 10^4 K [17]. On the other hand, non-thermal plasmas, known as cold plasmas, exhibit low energy density and significant inequality between the temperature of the electrons and other particles like ions and gas. Considering the parameter 'pressure', plasmas can be categorized into two types such as low-pressure plasma and atmospheric-pressure plasma. Low-pressure plasmas are mainly non-thermal plasma generated under vacuum or reduced pressure, whereas atmospheric-pressure plasmas are generated near atmospheric pressure (100 kPa) [17].

9.2.3 Plasma Generation Techniques

Generally, plasma is generated by high-frequency electromagnetic oscillation, electrical discharge, and intense radiation. Electrical discharge is widely utilized in industrial applications to generate plasma. The density of plasma is primarily influenced by the temperature of gas and electron energy, making the discharge type less significant if similar energies and temperatures are achieved. The selection of a specific method depends on factors such as cost, process uniformity, flexibility, and processing rate [17].

9.2.3.1 DC Glow Discharge Method

With this method, plasma is generated by applying a DC voltage across two conductive electrodes placed in a low-pressure gas where the electric field is generated using a power supply with high resistance [18]. A small number of free electrons exist in gases due to natural radioactivity, cosmic rays, or processes like photoionization and field emission. As voltage increases in a discharge tube, these electrons gain energy in the electric field and collide with gas atoms or molecules. Initially, collisions are elastic, with minimal energy loss. As electrons gain enough energy, inelastic collisions occur, efficiently transferring energy and causing ionization. This process produces more free electrons, sustaining a chain reaction. When ionization and recombination reach equilibrium, the discharge becomes self-sustaining, resulting in a stable glow discharge [19].

9.2.3.2 Radio Frequency (RF) Discharge Method

In RF discharge, plasma is generated by employing a relatively high frequency within the radiofrequency range, typically between 3 kHz and several MHz, to the gas targeted for discharge [20]. Capacitive and inductive coupling are widely utilized for generating RF discharge in textile surface plasma engineering, both relying on RF voltage as the primary energy source to create plasma. In capacitive coupling, two electrodes within the treatment chamber generate an electric field that ionizes the gas to generate plasma. On the other hand, inductive coupling employs an external coil to produce a magnetic field, which induces electric currents that ionize the gas molecules to create plasma, eliminating the need for electrodes inside the treatment chamber [21]. RF discharge offers greater control over the energy distribution of plasma species compared to many other discharge methods, owing to its broader and adjustable frequency range [22].

9.2.3.3 Plasma Generation by Microwave

A power source operating at a frequency of 2.45 GHz is used to generate microwave plasma. The induction of plasma by microwave energy resembles that achieved with radio frequency (RF) methods, with the primary distinction arising from their differing frequency ranges. Nonetheless, sustaining a microwave discharge at low pressures (below 1 torr) is more challenging compared to direct current (DC) or RF discharges [23]. While RF glow discharges can extend nearly throughout the entire reactor, their spatial dimensions are significantly smaller than the wavelength of the RF field (approximately 22 m at 13.56 MHz) [24]. In contrast, microwave plasma exhibits its highest glow intensity within the coupling microwave cavity, with rapid attenuation beyond this region due to the much shorter wavelength of microwaves (roughly 12.24 cm at a frequency of 2.45 GHz). The electric field strength may fluctuate within the reactor, with dimensions comparable to the microwave wavelength. As a result, active species from the plasma discharge can persist in areas beyond the visible plasma glow, known as the afterglow region [19].

9.2.3.4 Dielectric Barrier Discharge Method

Dielectric-barrier discharges (DBDs) typically occur between two dielectric boundaries that are positioned between metal electrodes. These discharges are usually initiated within the gap between two parallel electrodes, with one electrode coated in dielectric material. The presence of the dielectric material ensures quasi-non-equilibrium plasma conditions, as it prevents the discharge from evolving into electrical arcs [25, 26]. The gap between the electrodes is kept minimal, typically ranging from 0.1 mm to a few centimeters [27]. The gas used for plasma generation flows through the gap between the electrodes and operates within a pressure range of 10^4 to 10^6 Pa. Plasma is produced by applying either a sinusoidal AC voltage with a frequency between 50 Hz and 1 MHz or a pulsed power source across the gap [28, 29]. The nature of the discharge, whether filamentary or glow, depends on factors such as the composition of the working gases, the applied voltage, and frequency.

9.2.3.5 Corona Discharge

This type of plasma is non-thermal and characterized by low current density [30]. A corona discharge process consists of an anode and cathode wire, with pulsed direct current power supplied to the system. The plasma generates a luminous 'corona' around the wire forming a crown-like emergence. In the corona discharge mechanism, an electric field that is non-uniform creates a glowing effect around a sharp electrode. Its applications in material processing is limited because the discharge is confined to a compact area. With a view to overcoming this limitation, 2D arrays of electrodes are employed. This plasma typically operates at a potential difference of 2 to 5 kV, producing a small current range of 10^{-1} A to 10^{-5} A. When the current exceeds 10^{-5} A, the voltage increases without bound, leading to the formation of micro-arcs [31]. In devices, coronal discharges typically carry currents below the threshold that would cause arcing, such as currents around 5×10^{-4} A [32, 33].

9.2.3.6 Atmospheric Pressure Plasma Jet

The approach for cost reduction and solution of the limitations in production paved the way for developing an atmospheric pressure plasma generation system [34]. Several researchers worked on this and found that under atmospheric pressure stable plasma could also be generated [35–37]. Cold plasmas are generated by most of the atmospheric pressure discharge within the inter-electrode gap, driven by numerous transient micro-discharges, which are streamers (current filaments) of very short duration (in the microsecond or nanosecond range). Consequently, the plasma's lifetime is typically short and it is essential to either ensure rapid transfer of energy from the plasma region to the surface or generate active species directly at the surface being treated [38].

9.3 Plasmas Used in the Processing of Textiles

As most of the textile fibers are thermosensitive, mainly low-temperature plasmas are used for the processing of textiles [39–42]. Plasmas can be generated in a closed chamber at low pressure or at atmospheric pressure [43]. Low-pressure plasmas are generated using microwave energy or radio frequency and this can contribute a more uniform effect compared to atmospheric pressure plasma on textile materials. Modification of polymers by polymerization, etching, and formation of free radicals on the textile substrates has been done using low-pressure plasmas [44]. It is arduous to implement the large-scale processing of textiles using low-temperature plasma in a closed chamber as the size of the closed chamber is limited [45, 46]. Therefore, using plasma under atmospheric pressure conditions is more feasible for the bulk processing of textile materials.

9.4 Recent Advances in Plasma-Based Textile Finishing

With a view to achieving some superior attributes like wettability, dyeability, adhesion, resistance to shrinkage, and softness; various reactive species are introduced to the surfaces of textile materials by treating them with plasma. At present, plasma technology is of priority over conventional wet chemical processing for surface modification on experimental and commercial scales as it is a simple, efficient, and eco-friendly process [17]. In addition, the surface modification methods based on plasma technology offer minimal degradation of the built-in properties of the materials as the reactions that occur by plasma are confined only within the outer layers of the material [47].

During plasma processing two opposite phenomena are possible such as polymer formation and ablation. Plasma polymerization may occur due to the high content of carbon and hydrogen atoms. Plasmas generated from ethanol, ethylene, and methane have the propensity to cause plasma polymerization and hence create very thin, insoluble, highly cross-linked, and pinhole-free plasma polymer films. There are two approaches for the plasma-induced ablation of materials namely chemical etching and physical sputtering [48].

The process that involves the removal of materials from a surface by producing volatile products through chemical reaction followed by physical sputtering is known as plasma sputtering and cleaning process. The sputtering process involves a negative voltage which is implemented on the substrate and in the meantime, argon plasma is produced by RF glow discharge or electron cyclotron resonance method. The argon ions are propelled toward the substrate through the electric field. Instead of getting into the depth of the material, the argon plasmas convey a large portion of energy to the atoms of the surface of the material with the help of inelastic and elastic collisions with the material as the plasmas do not have significantly high amounts of energy. Thus, the energized surface atoms escape into the vacuum, revealing layers beneath for continued etching. If enough time is given for sputtering, the contaminants are eliminated from the surface of the material. The plasma sputtering and cleaning process is applicable for further implantation and deposition [49].

The plasma reaction with a polymer results in two conflicting outcomes: degradation and modification. When the degradation effect is the primary influence, the polymer surface will undergo an etching process. In this case, the loss of weight due to etching essentially depends on the energy of plasma and the properties of the polymer. Polymers like polyolefins having no oxygen functionalities are capable of retaining their original properties like degree of polymerization, chemical structure, elemental composition, and crystallinity, as the weight loss occurs only in the top layer of the polymeric materials. In contrast, polymers having ester, carboxylic acid, and ether groups are vulnerable to degradation to a larger extent when subjected to plasmas. When the modification effect is the primary influence, because of plasma polymerization, plasma-graft copolymerization, and ion beam interaction, the characteristics of the polymer alter [50–52].

9.4.1 Applications of Plasma Technology in Textile Finishing Processes

Plasma technology can be implemented in antimicrobial [53], ultraviolet protection [54], fire retardant [9], and anti-static finishing [10] of textiles.

9.4.1.1 Use of Plasma to Enhance Hydrophilic Properties

Enhancement of surface hydroxyl (OH) groups and molecular water adsorption can be gained with ZnO exposed to flowing O_2 gas in the microwave plasma system at a power of 300 W for 300 s. In this method of microwave plasma treatment, the frequency was maintained at 2.45 GHz. Spectroscopy of the ZnO surface showed OH radicals and molecular water enhancement because of oxygen plasma treatment. Moreover, treatment with O_2-plasma increased the photocatalytic activity of ZnO nanoparticles by around 60%, as indicated by the diminished absorption peak of the degraded methylene blue organic dye solution. While O_2-treated ZnO decolorized 50% of the dye in 15 min, the untreated version took upwards of 40 min to reach the same level. The investigation confirms that the O_2 plasma treatment offers increased OH radicals and molecular water on the surface of ZnO, and minimizes the oxygen and zinc vacancies on the ZnO surface [6].

9.4.1.2 Use of Plasma in Antimicrobial Finishing

The physical and chemical incorporation of functional agents can enhance the antimicrobial properties of textile materials [53]. Antimicrobial properties can be added to cellulosic fibers like cotton, viscose, lyocell, and linen by surface modification with oxygen and nitrogen plasma which can introduce active sites and amine groups to the fabric. Silver nanoparticles produced using biological methods were then combined with certain antibiotics and applied to the fabrics [55]. Plasma-treated fabrics with AgNPs exhibit improved antibacterial activity even after 15 washes, resulting from stronger bonding facilitated by plasma-induced carboxylic acid groups [56–59].

Research demonstrated a plasma-induced polymerization process by pre-treating the cellulose substrate surface with O_2 plasma and acrylic acid. The study found that after being treated with acrylic acid, the microfibrils of the cellulosic fibers became aligned with the fiber axis and the fibers appeared smoother and cleaner [60]. Senthilkumar et al. [61] studied the impact of the variables in argon plasma therapy on the antimicrobial efficacy and comfort of polyester fabrics. Polyesters treated with plasma exhibited enhanced antibacterial properties along with coagulability, and water–vapor permeability compared to the untreated polyesters. The treatment conditions (power condition of 600 W, processing time of 30 s, and distance between the electrodes of 2.8 mm) were optimized by the Program Design Expert. The effectiveness of the process was dependent on the treatment's intensity, as the intensity of the corona discharge and atmospheric pressure depends on both the time of exposure and the discharge power.

DC glow discharge plasma improves the surface adhesion properties of silk fabrics. Prior to plasma treatment, the incorporation of ZnO nano-particles can provide protection to the silk substrates from microbes [62]. Silk substrate treated with Low pressure plasma and subsequently coated with ZnO nano-particles exhibits not only antibacterial and antifungal properties but also anti-static and ultraviolet protection behavior [63]. With a view to endow desirable antibacterial properties to wool substrates, enzyme/plasma treatment can be applied [64].

9.4.1.3 Use of Plasma in Ultraviolet Protection Finishing

The exposure of cotton fabrics to low-pressure non-equilibrium gaseous plasma can uplift the dye uptake and offer a considerable ultraviolet protection factor (UPF). Gorjanc et al. [54] generated plasma inside a glass tube through an electrodeless RF discharge using oxygen or ammonia as the medium maintained at a pressure of 50 Pa to facilitate the creation of oxygen and nitrogen groups. Substantially high color yield was achieved in the case of ammonia plasma-treated fabric as the fabric surface enriched in amino groups strengthened the bonding between dye (curcumin) and fiber. Furthermore, the treated fabrics exhibited excellent protection against ultraviolet rays [65]. Shahidi and Moazzenchi [66] treated wool fabrics with plasma sputtering using copper and iron in order to enhance their UPF value and improve their affinity for natural dyes. The naturally dyed plasma sputtered wool fabrics achieved UPF values of greater than 25 which indicates very good UV protection properties.

To improve the ZnO nano-particle adsorption property of the cotton fabric, low-pressure oxygen plasma produced using an electrode-less radiofrequency (RF) discharge can be implemented [8].

9.4.1.4 Use of Plasma in Flame Retardant Finishing

To produce fireproof textiles from natural fibers is a challenge as only a surface treatment can possibly be applied to them. Strong fixation of flame retardants with the fabric surface is important with a view to conferring reasonable resistance to washing and weathering conditions. Tsafack and Levalois-Grützmacher [9] observed the grafting and polymerization of flame-retardant monomers using argon plasma on cotton substrates. The presence of nitrogen in the diethyl(acryloyloxyethyl)phosphoramidate and acryloyloxy-1,3-bis(diethylphosphoramidate)propan monomers enhances the effectiveness of phosphorus-based flame retardants which results in excellent fire-resistant properties. Thi et al. [67] processed cotton fabric with plasma atmospheric-pressure dielectric barrier discharge (APDBD) and subsequently treated with flame retardants following the pad-dry-cure method. This study revealed that to obtain a limiting oxygen index of 25, the fabrics finished by the process without the use of plasma required a temperature of 180 °C and 114 s. On the contrary, the fabric processed by plasma needed 160 °C temperature and 90 s time. Ömeroğulları and Kut [68] developed polyester fabrics with enhanced hydrophilicity by padding them using alkyl-phosphonate-based flame retardants and subsequently carried out a low-frequency oxygen plasma treatment at a low temperature.

Palaskar et al. [69] demonstrated that the deposition of nano-TiO_2/SiO_2 on the surface of cotton fabric by means of atmospheric pressure He-O_2 plasma can elevate flame retardant properties. Besides, the treatment can also confer some important functionality to cotton substrates such as anti-microbial properties, and ultraviolet protection.

9.4.1.5 Use of Plasma in Anti-Static Finishing

Kan and Yuen [10] applied low-temperature plasma under variable conditions to improve the antistatic property of polyester fabrics so that the fabrics can achieve the capacity to hold moisture and thus the static electricity can be dissipated. The relationship between the half-life decay time of static charges and moisture content was analyzed. The result revealed that increased moisture shortens the dissipation period.

Low-temperature plasma processing increases the concentration of oxygen-based polar groups on polyester fibers. These polar groups interact with moisture via hydrogen bonding, enhancing its ability to penetrate and adhere to the fiber surface.

The ionization of polar groups by water molecules creates a conductive layer on the fiber surface which enhances the dissipation of static charges, thereby reducing the fiber's half-life. Plasma treatment at low temperatures can be used to manage this process [70].

9.4.1.6 Use of Plasma in Denim Finishing

To achieve a worn-out effect on denim garments mainly stone and enzyme washing are the popular methods [71]. In the stone-washing process, there are some significant drawbacks such as the use of pumice stones can damage the fabrics and produce huge amounts of dust of pumice stones causing harm to the environment. Enzyme washing is considered as the alternative for stone-washing of garments. However, this process consumes a lot of water and chemicals [72, 73]. A similar vintage effect on denim fabrics can be achieved by plasma treatment without changing the mechanical properties of the fabrics, which is a more eco-friendly and efficient process [72].

9.4.1.7 Effect of Plasma on Mechanical Properties of Textiles

The use of plasma technology can increase C-O bonds and reduce C–C bonds. When a substrate is exposed to plasma, depending on the treatment time, the surface of the substrate gets etched and grooved resulting in the roughness of the surface. This physical modification of the substrate has an impact on the mechanical properties of it. If plasma etching occurs both in the outer layer and the bulk of a material, it can be concluded that the treated material has poor mechanical properties [8].

9.4.2 Application of Plasma in Textile Coloration

The effect of plasma treatment is confined within the surface of the substrate, and a thin near-surface layer whose thickness ranges from 100 Å to multiple micrometers, leaving the bulk of the material unaffected. Plasma technology offers enhanced dyeing of polyester [11], wool [12], and cotton [13] substrates.

9.4.2.1 Plasma Treatment for the Coloration of Polyester

Generally, polyester fibers are dyed in high-temperature, thermosol, or phenol-based carrier methods that are not environmentally friendly processes. A great eco-friendly replacement for the wet chemical-based methods is using low-temperature plasma, especially for dyeing polyethylene terephthalate (PET) fibers. The implementation of air and argon plasma on PET offers enhanced depth of shade. Plasma treatment was found to enhance surface roughness and area, leading to this consequence.

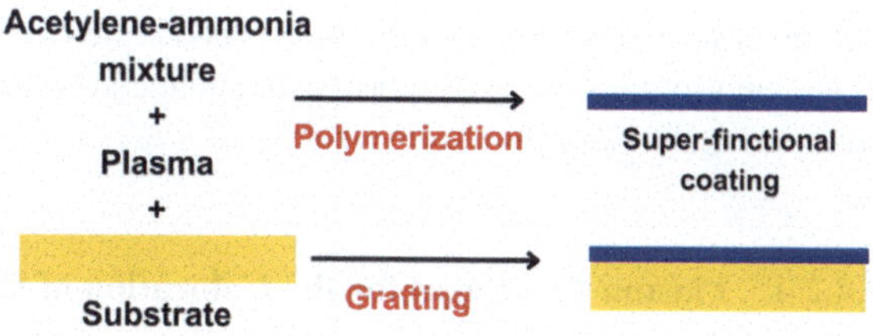

Fig. 9.1 High-performance ultrathin coating on the surface of the PET substrate

Both chemically inert and reactive plasmas have the capability to introduce some hydrophilic groups on the surface of the substrates which attract polar groups of dyes and hence increase the dye-affinity and swelling capacity of PET substrates. Consequently, the K/S values of the dyed PET substrates are increased [11]. High-performance ultrathin coatings can be achieved using plasma polymerization of an acetylene-ammonia mixture on PET fibers. The process comprises deposition and etching, forming a nanoporous, crosslinked network with accessible functional groups, as shown in Fig. 9.1 [74, 75].

Thin multifunctional films can be formed using ammonia/ethylene or acetylene mixture which can introduce active sites for acid dyes and exhibit increased K/S values according to the thickness of the films. The polyester fabrics dyed with plasma exhibit good fastness to rub and wash. The plasma-coated polyesters also show satisfactory abrasion resistance [76].

Ahmed and El-Shishtawy [75] treated PET fibers with $N_2/H_2/He$ plasma and SO_2/O_2 plasma which essentially increased the hydrophilicity and surface area of PET fibers, resulting in an increase in dyeability of the fiber while using water-soluble acid dyes.

9.4.2.2 Plasma Treatment for the Coloration of Wool

Because of having substantial amounts of fatty acids on the surface of the fibers, and disulfide cross-linkages in the A-layer of the exocuticle, the surface of wool fiber is hydrophobic [12, 77]. It is found that using low-temperature plasma instead of wet chemical treatments offers a significant improvement in the coloration of wool substrates. A possible explanation is that the oxidation of cystine caused by plasma may decrease the number of crosslinks on the fiber surface, promoting transcellular and intercellular dye diffusion [78, 79].

9.4.2.3 Plasma Treatment for the Coloration of Cotton

Among the attempts made to strengthen the interaction between reactive dye and cotton fiber, the pretreatment of cotton fibers with plasma deserves attention. Plasma treatment of cotton with amine compounds improves the affinity of reactive dyes toward the fiber [13]. For the purpose of salt-free reactive dyeing of cotton fabric, the substrate can be treated with plasma and chitosan nanoparticles. In addition, this treatment provides the dyed fabric with enhanced color fastness to wash and rub, and color strength (K/S) [80].

9.4.2.4 Plasma Treatment for the Coloration of Cotton/Nylon Blend

When cotton/nylon blend fabrics undergo atmospheric plasma/nano-chitosan treatment, some primary OH groups of the cellulosic material are converted into CONH groups, and hence the cellulosic fibers act like polyamide fibers. Therefore, one bath dyeing of cotton/nylon fabrics with acid dye is possible if the blend fabrics are pretreated with plasma/nano-chitosan [81].

9.4.3 Application of Non-Equilibrium Plasma

The applications of gas discharge plasma are promising eco-friendly alternatives for wet chemical processing [39]. In the surface modification of textile materials, non-equilibrium plasma or non-thermal plasma, containing relatively high-temperature electrons and low-temperature particles is important. This technique enhances the hydrophilicity of cellulosic fibers by adding carboxyl groups generated through the removal of hydrogen from the C6 carbon in the ribose sugar in cellulose and subsequent oxidation, as shown in Fig. 9.2 [4].

In the case of atmospheric pressure plasmas, typically, electrodes are set up in a chamber that contains argon or helium at atmospheric pressure. An AC or DC voltage between 1 and 100 kV, with frequencies ranging from a few hertz to several megahertz, is applied across the electrodes.

The introduction to nanotechnology can be an inventive strategy to utilize the nanoscale substances for creating nanostructure during the finishing and manufacturing process. Some mentionable properties can be soil resistance, antimicrobial, flame retardancy, antistatic behavior, etc. Silver nanoparticles have proven to exhibit excellent antibacterial properties against both gram-positive and gram-negative bacteria [82].

9.5 Sustainability and Plasma Technology

The textile industry is known as an important productive sector worldwide, and everyone, to some range, is connected in its value chain. According to the Global Market Report on Sustainable Textile 2010, the total value of the textile industry worldwide is 1 trillion USD. In total exports, it contributes around 7% and almost 35 million people are employed worldwide. However, this industry is one of the reasons for the environmental pollution worldwide. During the manufacturing process, a large

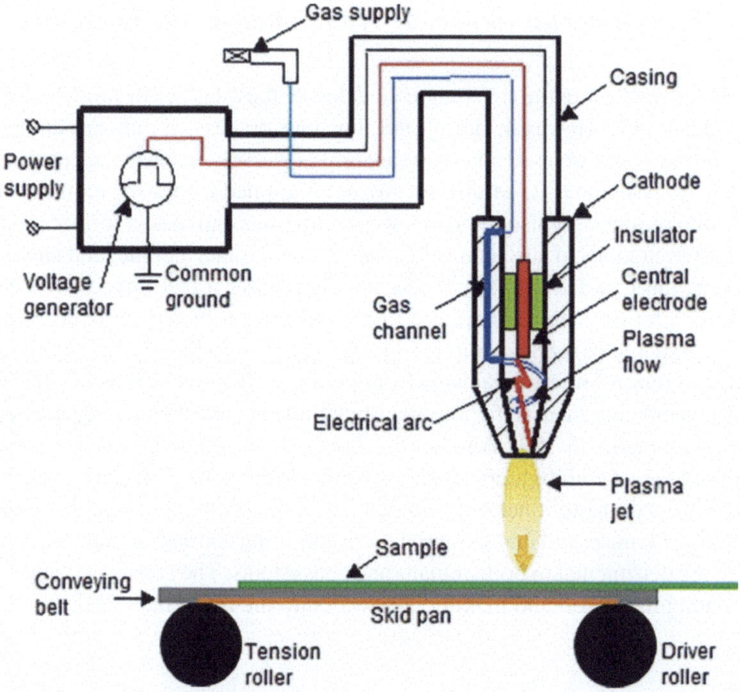

Fig. 9.2 Application of non-equilibrium plasma on textile substrates [4]. Republished with permission from Elsevier. Copyright Elsevier, 2021

amount of chemicals, water, and fuel are used which is alarming for the environment. It can be the reason for climate change and scarcity of resources. For that reason, sustainability in textiles is one of the most important issues [83]. Sustainable textiles must meet or exceed the social, environmental, and economic performance requirements. All the materials must come from any renewable or recycled sources, input and output materials for any process must be safe for human and ecological health, and all the materials must be reusable. Non-toxic chemicals must be used during dyeing and washing [84]. For developing sustainable textiles, the used water must be reusable or using any alternative where water usage is less. Plasma technology has the potential to be used as an alternative. Plasma technology is a dry process and there is no use of any harmful and toxic chemicals. That is why reusing the water after removing the chemicals is easy. It improves the fabric's other properties, such as increasing surface wettability, improving dyeing and printing sustainability, enabling the reuse of effluents, cutting dye and water consumption, and so on [2].

9.5.1 Environmental Impact of Traditional Textile Finishing

The main purpose of textile finishing is to enhance the qualities of fabrics or overall final products [85]. However, during the finishing process, significant amounts of chemicals and water are used and their disposal negatively impacts the environment. Softening, an essential textile finishing process is applied to give the required handle to the textile products. It also improves wear properties and makes further processes easier. Different kinds of surfactants like cationic, nonionic, anionic, and amphoteric are used as fabric softeners [86]. These softeners have a bad impact on the environment as a lot of chemicals are used for producing softeners and some are also non-biodegradable [87, 88]. Antimicrobial finishing is one kind of textile finishing that helps to remove microorganisms from products. Different kinds of salts, detergents, and chemicals are used for antimicrobial finishing and those create a hazardous situation for aquatic life [89]. Another finishing is flame retardant finishing which is mainly used to make firefighters' apparels, work clothing, military garments, institutional upholstery, institutional and commercial carpet, professional racers' garments, and bedding. Flame retardants, essential for fireproofing textiles, include harsh chemicals that are detrimental to both humans and ecosystems. They resist decomposition, accumulating in the soil and making their way into the food chain [90].

9.5.2 Sustainability Benefits of Plasma Technology

In recent years, plasma technology has been getting more attention in the textile sector as an alternative to traditional wet processing and finishing due to its low environmental impact, high efficiency, and simplicity [72]. It is a dry process that reduces the pressure on ETP [91]. In the conventional dyeing processes, almost 50% of dye can be lost in the effluents which causes a huge environmental problem [92]. If the wastewater is not treated, it can create environmental pollution that harms both plants and living organisms. Plasma technology can be used as an alternative for removing synthetically or naturally occurring greases from textile fibers. It also enhances the dye diffusion rate of the fibers by improving the diffusion of dye molecules into the fibers [92]. It increases the wettability of the fabrics and facilitates the proper usage of water [72]. Plasma technology can be effective for surface treatment. It can optimize the surface properties without changing any inherent properties of textile materials. It is a green process as there is no use of harmful chemicals, substances, or solvents, making it very environment friendly [91]. Plasma technology is also effective for the antibacterial finishing of textiles. It eliminates the use of heat and chemical immersion while activating monomers and substrates through grafting reaction and hence it reduces the consumption of chemicals and water and resolves various environmental issues [93].

9.5.3 *Lifecycle Impact and Resource Efficiency*

The lifecycle theory inspects a product's complete life from conception, design, production, sale, customer use, and final disposal [94]. Any textile product's lifecycle typically starts with the 'fiber production' stage and ends with the 'end-of-life' stage. In between, there are textile production, design, clothing production, commercialization, and use stages. Every stage is connected, so it is hard to initiate actions systematically and all the stages have a significant social and environmental impact. So it is necessary to promote such kinds of materials and technologies that contribute to reducing the consumption of resources and toxic substances, integrate recycled materials into the production chain, and establish a sustainable process [95]. Plasma is such a kind of technology that is completely a dry process and there is no use of any toxic and harmful chemicals. It helps to enhance the quality of the products by optimizing the surface properties of the materials. It increases surface wettability and dying consistency, decreases dye and water consumption, and enables effluent reuse, making it more environment-friendly and sustainable [2].

9.5.4 *Plasma for Recycling and Circular Economy in Textiles*

The circular economy is an industrial system that elevates sustainability via restorative or regenerative design and objects. The 'End-of-life' concept is replaced by the circular economy and shifts toward the use of renewable energy, removing the use of toxic chemicals. By using superior design of products, materials, and systems, it aims to eliminate waste [96]. In circular economy approach, waste could be lessened by the continuous flow of materials in a closed loop [97].

Plasma technology can be helpful in establishing a circular economy. According to UNDP's sustainable development goals, there is a great relationship between the circular economy and plasma technology, as shown in Fig. 9.3 [2]. In plasma technology, harmful chemicals and toxic substances are not used, making the used water reusable. To achieve a circular economy, plasma gasification can be a viable technology that can convert solid waste into monetary and social value [2].

9.6 Challenges and Future Directions

The general process of creating plasma is creating an electrical field with an anode and a cathode while placing a gas in between. This system can be kept in a closed chamber or out at atmospheric pressure. The resulting plasma contains a meager fraction of charged particles. Again, fabricating a coating on a material with the help of plasma requires physico-chemical reactions, by overcoming a certain energy barrier created on the substrate's surface [98]. On the contrary, the traditional textile

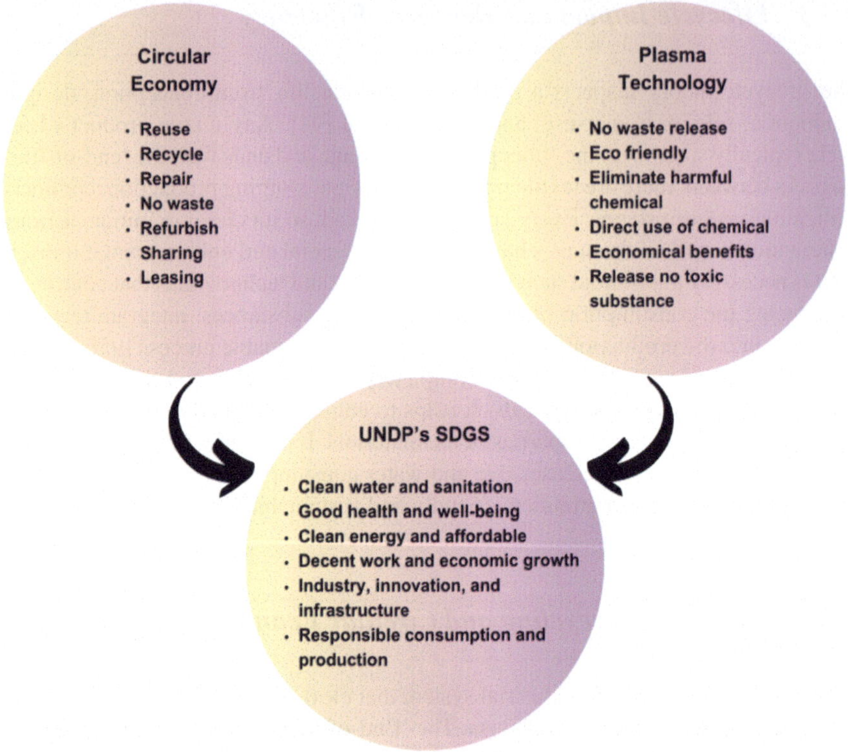

Fig. 9.3 The Interconnection between plasma technology and circular economy in fulfilling UNDP's sustainable development goals

finishing processes possess a high amount of energy and water consumption. Textile wastewater contains both organic and inorganic compounds in high concentration. Many steps need to be taken to remove these substances from water bodies in different steps. But all the existing filtering processes are very expensive yet can not fully recover all the wastages from effluent [99]. This shows a huge challenge for textile finishing in the future in order to save the environment. The only method is to employ systems that can reduce the usage or avoid the energy and water consumption [100]. To actualize this, initiation of plasma by air or available gasses such as H_2, N_2, and O_2 at atmospheric pressure is needed to be designed as an inventive prospect to provide the finishing properties such as antistatic, flame retardancy, anti-microbial, UV, and so on, in cotton, nylon, polyester, etc [101]. Modern finishing techniques have popularized different plasmas, such as atmospheric-pressure dielectric barrier discharge (APDBD), corona discharge at atmospheric pressure (CDAP) etc. [102]. But all of them comprise different challenges with huge future opportunities.

9.6.1 Industrial Scale Challenges

In order to produce a larger economic production, integrating plasma technology can be a great addition to the industry in the long run. Yet this can be a challenging as well as discouraging decision to replace the conventional setup that is already in place, with highly plasma technology this substantial modification of technology's initial cost of setup is very high, as well as functional labor [93]. Another tangible challenge can be dealing with enhanced surface area, and porous large three-dimensional structure. While treating plasma, surface cleanliness can be a major drawback as impurities and condition of the surface of the top layer affect plasma treatments, as plasma only influences the topmost surface of textiles. In the case of a porous three-dimensional structure, deep penetration of plasma species can be difficult compared to a wet process system. In the same way, individual fiber composition enlarges the surface as opposed to a flat surface, which is another obstacle to working with plasma [103]. Some of the important industrial challenges for plasma treatment can be demonstrated as-

- Plasma technologies are unable to replace all wet process segments despite having environmental and economic benefits compared to conventional methods.
- Expansive vacuum pump integration can be an excuse for the traditional industry to implement this plasma treatment, though there is plenty of plasma processing available at low atmospheric pressure.
- Atmospheric pressure plasma treatments need helium as the buffer gas to have a stable operation, which again can act as a disadvantage to implementing this at the industrial level [72].

9.6.2 Limitations in Treating Different Polymeric Materials

In the recent textile world, different countries have already started enhancing new functionalities in textile materials with the help of plasma technologies. Therefore, to integrate plasma treatment in the industrialization era, hurdles are still faced by the factories in different ways. One of the important constraints is the involvement in a large surface area created by individual fibers. Other issues can be named as three-dimensional structure and surface cleanliness. Even though in recent years, research works have been deployed to overcome these issues to bring plasma technology to mass production level [104]. In application, one of the limitations can be the chemical conversion of plasma gases, as different gases create specific chemical functionalities on textile surfaces. For instance, oxygen (O_2) Nitrogen (N_2), and Fluorine (F_2) containing gases produces C = O or O–C–O groups, C–N or O = C = NH groups, and a thin hydrophobic layer respectively. All these changes also depend upon the type of textile material is present on the surface [105].

Plasma processing is used to modify a wide range of fibers including natural fibers (like protein-based and cellulosic fibers), and synthetic fibers, which makes it flexible for different types of textiles and applications. Careful selection of plasma gases is the first condition for the functional design of the end product. Different types of gas (e.g., oxygen, nitrogen, hydrogen, argon, fluorocarbons) can bring various desirable properties like hydrophilicity, oleophilic property, improved adhesion property, and so forth. Each of these customized properties needs adjustment in key plasma operating conditions (treatment time, power, pressure, and gas flow rate). However, as this process is a dry, energy-efficient and clean process, this can be considered [72].

9.6.3 Cost vs. Environmental Benefit Analysis

The extensive consumption of water, subsequent wastewater production, and energy-demanding drying methods in traditional processes reinforce the interest of the industries in more sustainable and efficient solutions. In this regard, plasma technology introduces a groundbreaking approach to substantially lowering water and energy consumption in textile finishing [105]. Radetic et al. [72] showed a comparison of cost between conventional chlorination and plasma treatment. In the case of the conventional method, 7 kWh energy is required for 1 kg of wool, and plasma treatment requires only 0.3–0.6 kWh per kg, reducing 95% energy costs. The annual energy saving can be up to 685 MWh, making plasma more suitable and economically beneficial in the long run. Modifying 12 tons of wool by low-pressure plasma treatment can save 44 t of sodium hypochlorite, 16 t of sodium bisulphite, and 11 t of sulphuric acid. This method also showed that 27,000 m^3 of water could be saved by eliminating the intensive water consumed during washing stages during chlorination. So plasma treatment is undoubtedly cost-saving and reduces wastewater production.

9.6.4 Future Prospects of Plasma-Based Textile Finishing

Adopting plasma technology for antimicrobial finishing in textiles could be a sensible choice, especially considering the drawbacks of conventional methods and the heightened demand for hygienic apparels in the wake of the pandemic. Plasma applications in textiles are becoming essential [105]. Despite being a potential and promising treatment, plasma technology still becomes an insignificant part of industry in many cases because most of them do not consider the concept of higher initial investment. But the larger initial investment can also rapidly be brought back by environmental savings and high-value product profits [72].

9.7 Conclusion

The use of plasma technology in textile finishing has been increasing in recent years to facilitate sustainability and innovation. Plasma treatment modifies the textile surface, improving its compatibility with finishing agents. The use of plasma in the bulk processing of textile materials remains a challenge. Moreover, Further studies can be done to resolve this issue to enhance the use of plasma in textile finishing.

9.7.1 Declaration

The authors declare that there are no financial conflicts of interest or personal relationships that could have influenced the work presented in this chapter. Generative AI and AI-assisted technologies were used exclusively to improve the readability and language of the content. Following this, the authors thoroughly reviewed and edited the text and assume full responsibility for the final publication.

References

1. Shahidi S, Ghoranneviss M, Moazzenchi B (2014) New advances in plasma technology for textile. J Fusion Energy 33:97–102
2. Reema, Khanikar RR, Bailung H, Sankaranarayanan K (2022) Review of the cold atmospheric plasma technology application in food, disinfection, and textiles: A way forward for achieving circular economy. Front Phys, 10, p 942952
3. Mia R, Sk MS, Oli ZBS, Ahmed T, Kabir S, Waqar MA (2021) Functionalizing cotton fabrics through herbally synthesized nanosilver. Clean Eng Technol 4:100227. https://doi.org/10.1016/j.clet.2021.100227
4. Saleem M, Naz M, Shoukat B, Shukrullah S, Hussain Z (2021) Functionality and applications of non-thermal plasma activated textiles: A review. materials today: proceedings 47:S74–S82
5. Jelil RA (2015) A review of low-temperature plasma treatment of textile materials. J Mater Sci 50(18):5913–5943
6. Nam S-H, Boo J-H (2016) Enhancement of photocatalytic activity of synthesized ZnO nanoparticles with oxygen plasma treatment. Catal Today 265:84–89
7. Batool F, Iqbal N, Adeel S, Azeem M, Hussaan M, Mia R (2024) Sugar beet (Beta vulgaris L) leaves as natural colorant for cotton dyeing using an ecofriendly approach toward industrial progress. Sci Prog, 107(3), p 00368504241271737, https://doi.org/10.1177/00368504241271737
8. Jazbec K, Šala M, Mozetič M, Vesel A, Gorjanc M (2015) Functionalization of cellulose fibres with oxygen plasma and ZnO nanoparticles for achieving UV protective properties. J Nanomater 2015(1):346739
9. Tsafack MJ, Levalois-Grützmacher J (2006) Flame retardancy of cotton textiles by plasma-induced graft-polymerization (PIGP). Surf Coat Technol 201(6):2599–2610
10. Kan C-W, Yuen C-WM (2008) Static properties and moisture content properties of polyester fabrics modified by plasma treatment and chemical finishing. Nucl Instrum Methods Phys Res, Sect B 266(1):127–132
11. Raffaele-Addamo A et al (2006) Cold plasma-induced modification of the dyeing properties of poly (ethylene terephthalate) fibers. Appl Surf Sci 252(6):2265–2275

12. Kan CW, Yuen CWM (2006) Retracted: Dyeing behaviour of low temperature plasma treated wool. Plasma Process Polym, 3(8), pp 627–635
13. Ozdogan E, Saber R, Ayhan H, Seventekin N (2002) A new approach for dyeability of cotton fabrics by different plasma polymerisation methods. Color Technol, 118(3)
14. Ramkrishna R, Mukesh R, Subroto M (2015) Basics of plasma and its industrial applications in textiles, SK Nema, & PB Jhala (1st ed), Plasma Technol. Text. Appare, Woodhead Publishing India Pvt. Ltd, New Delhi, pp 1–27
15. Chen FF (1984) Introduction to plasma physics and controlled fusion. Springer
16. Nakasugi Y, Himeno M, Kobayashi H, Ikarashi M, Maeda H, Sato K (2017) Experimental and mathematical analyses of bio-electrochemical conversion of carbon dioxide to methane. Energy Procedia 114:7133–7140
17. Eslami E, Jafari R, Momen G (2021) A review of plasma-based superhydrophobic textiles: theoretical definitions, fabrication, and recent developments. J CoatS Technol Res, pp. 1–24
18. Grill A (1994) Cold plasma in materials fabrication. IEEE Press, New York
19. Kan C-W (2014) A novel green treatment for textiles: plasma treatment as a sustainable technology. CRC Press
20. Takahashi K (2019) Helicon-type radiofrequency plasma thrusters and magnetic plasma nozzles. Rev Mod Plasma Phys 3(1):3
21. Racka-Szmidt K, Stonio B, Żelazko J, Filipiak M, Sochacki M (2021) A review: Inductively coupled plasma reactive ion etching of silicon carbide. Mater 15(1):123
22. Ssekasamba H et al (2024) Review of plasma surface engineering technology toward sustainable textile materials. Mater Circ Econ 6(1):27
23. Manos DM, Flamm DL (1989) Plasma etching: an introduction
24. Adeel S et al (2023) Microwave-assisted sustainable coloration of wool fabric using Rheum emodi-based natural dye. Surf Innov 12(3–4):147–158
25. Kogelschatz U (2003) Dielectric-barrier discharges: their history, discharge physics, and industrial applications. Plasma Chem Plasma Process 23(1):1–46
26. Wagner H-E, Brandenburg R, Kozlov K, Sonnenfeld A, Michel P, Behnke J (2003) The barrier discharge: basic properties and applications to surface treatment. Vacuum 71(3):417–436
27. Mandal R, Singh A, Singh AP (2018) Recent developments in cold plasma decontamination technology in the food industry. Trends Food Sci Technol 80:93–103
28. Jimenez C, Martin S, Gherardi N, Durand J, Cot D, Massines F (2002) Study of the formation of nano and micro particles in a discharge at atmospheric pressure. Mater
29. Salge J (1996) Plasma-assisted deposition at atmospheric pressure. Surf Coat Technol 80(1–2):1–7
30. Schutze A, Jeong JY, Babayan SE, Park J, Selwyn GS, Hicks RF (1998) The atmospheric-pressure plasma jet: a review and comparison to other plasma sources. IEEE Trans Plasma Sci 26(6):1685–1694
31. Bussiahn R et al (2010) The hairline plasma: An intermittent negative dc-corona discharge at atmospheric pressure for plasma medical applications. Appl Phys Lett, 96(14)
32. Sayedi S, Landsberger L, Kahrizi M (1996) Electrical characterization of thin SiO/sub 2/ films created by negative-point oxygen corona discharge processing, In: Proceedings of 1996 canadian conference on electrical and computer engineering, 1 IEEE, pp 64–67
33. Pipliya S, Kumar S, Babar N, Srivastav PP (2023) Recent trends in non-thermal plasma and plasma activated water: Effect on quality attributes, mechanism of interaction and potential application in food & agriculture. Food Chem Adv 2:100249
34. Kan CW, Yuen C (2006) Textile modification with plasma treatment. Res J Text Appar 10(1):49–64
35. Okazaki S, Kogoma M (1993) Development op atmospheric pressure glow discharge plasma and its application on a surface with curvature. J Photopolym Sci Technol 6(3):339–342
36. Kogoma M, Koiwa K, Okazaki S (1994) Deposition of SiNx thin films by atmospheric pressure glow discharge plasmas. J Photopolym Sci Technol 7(2):341–344
37. Kogoma M, Okazaki S (1994) Raising of ozone formation efficiency in a homogeneous glow discharge plasma at atmospheric pressure. J Phys D Appl Phys 27(9):1985

38. Buyle G, Schneider J, Walker M, Akishev Y, Napartovich A, Perucca M (2010) Plasma systems for surface treatment. Plasma Technol Hyperfunct SurfS: Food, Biomed, TextE Appl, pp 33–61
39. Dave H, Ledwani L, Nema S (2019) Nonthermal plasma: a promising green technology to improve environmental performance of textile industries. In: Impact Prospect Green Chem TextE Technol: Elsevier, pp 199–249
40. Naebe M, Denning R, Huson M, Cookson PG, Wang X (2011) Ageing effect of plasma-treated wool. J Text Inst 102(12):1086–1093
41. Naebe M, Cookson PG, Denning R, Wang X (2011) Use of low-level plasma for enhancing the shrink resistance of wool fabric treated with a silicone polymer. J Text Inst 102(11):948–956
42. Naebe M et al (2010) Effects of plasma treatment of wool on the uptake of sulfonated dyes with different hydrophobic properties. Text Res J 80(4):312–324
43. Ratnapandian S, Wang L, Fergusson S, Naebe M (2011) Effect of atmospheric plasma treatment on pad-dyeing of natural dyes on wool
44. Sarra-Bournet C, Turgeon S, Mantovani D, Laroche G (2006) A study of atmospheric pressure plasma discharges for surface functionalization of PTFE used in biomedical applications. J Phys D Appl Phys 39(16):3461
45. Kan C-W, Yuen C-WM (2007) Plasma technology in wool. Text Prog 39(3):121–187
46. McCoustra MR, Mather RR (2018) Plasma modification of textiles: understanding the mechanisms involved. Text Prog 50(4):185–229
47. Gorjanc M, Gorenšek M, Jovančić P, Mozetič M (2013) Multifunctional textiles—modification by plasma, dyeing and nanoparticles. chapter
48. Rakowski W (1997) Plasma treatment of wool today. Part 1–Fibre properties, spinning and shrinkproofing. J Soc Dyers Colour 113(9):250–255
49. Chappell P, Brown J, George G, Willis H (1991) Surface modification of extended chain polyethylene fibres to improve adhesion to epoxy and unsaturated polyester resins. Surf Interface Anal 17(3):143–150
50. Inagaki N (2014) Plasma surface modification and plasma polymerization. CRC press
51. Chan C-M, Ko T-M, Hiraoka H (1996) Polymer surface modification by plasmas and photons. Surf Sci Rep 24(1–2):1–54
52. Chu PK, Chen J, Wang L, Huang N (2002) Plasma-surface modification of biomaterials. Mater Sci Eng R Rep 36(5–6):143–206
53. Sun G (2016) Antimicrobial finishes for improving the durability and longevity of fabric structures. In: Antimicrobial textiles: Elsevier, pp 319–336
54. Gorjanc M, Mozetič M, Vesel A, Zaplotnik R (2018) Natural dyeing and UV protection of plasma treated cotton. Eur Phys J D 72:1–6
55. Mankodi DH, Agarwal DB (2011) Studies on nano–UV protective finish on apparel fabrics for health protection. Res J Text Appar 15(3):11–20
56. Abdel-Aziz M, Eid B, Ibrahim N (2014) Biosynthesized silver nanoparticles for antibacterial treatment of cellulosic fabrics using O2-plasma. AATCC J Res 1(1):6–12
57. Ibrahim NA, Eid BM, Abdel-Aziz MS (2017) Effect of plasma superficial treatments on antibacterial functionalization and coloration of cellulosic fabrics. Appl Surf Sci 392:1126–1133
58. Ruddaraju LK, Pammi SVN, sankar Guntuku G, Padavala VS, Kolapalli VRM (2020) A review on anti-bacterials to combat resistance: From ancient era of plants and metals to present and future perspectives of green nano technological combinations. Asian J Pharm Sci, 15(1), pp 42–59
59. Ražić SE, Čunko R, Bukošek V, Matica B (2011) Antimicrobial modification of cellulose fabrics using low-pressure plasma and silver compounds. Tekst: J TextE & Cloth Technol, 60(9)
60. Ražić SE, Čunko R, Bautista L, Bukošek V (2017) Plasma effect on the chemical structure of cellulose fabric for modification of some functional properties. Procedia Eng 200:333–340
61. Senthilkumar P, Karthik T (2016) Effect of argon plasma treatment variables on wettability and antibacterial properties of polyester fabrics. J Inst Eng (India): Ser E, 97, pp 19–29

62. Rani KV, Sarma B, Sarma A (2019) Plasma pretreatment on tasar silk fabrics coated with Zno nanoparticles against antibacterial activity. Surf Rev Lett 26(05):1850193
63. Intarasang W, Thamjaree W, Boonyawan D, Nhuapeng W (2013) Effect of coating time on LPP treated silk fabric coated with ZnO_2 nanoparticles
64. Şahan G, Demir A (2016) A green application of nano sized chitosan in textile finishing. TextE Appar 26(4):414–420
65. Gorjanc M, Jazbec K, Mozetič M, Kert M (2014) UV protective properties of cotton fabric treated with plasma, UV absorber, and reactive dye. Fibers Polym 15:2095–2104
66. Shahidi S, Moazzenchi B (2019) Comparison between mordant treatment and plasma sputtering on natural dying and UV protection properties of wool fabric. Fibers Polym 20:1658–1665
67. Nguyen Thi H, Vu Thi Hong K, Ngo Ha T, Phan D-N (1575) Application of plasma activation in flame-retardant treatment for cotton fabric. Polymers, 12(7), p 1575
68. Ömeroğulları Z, Kut D (2012) Application of low-frequency oxygen plasma treatment to polyester fabric to reduce the amount of flame retardant agent. Text Res J 82(6):613–621
69. Palaskar S, Desai A, Shukla S (2016) Development of multifunctional cotton fabric using atmospheric pressure plasma and nano-finishing. J TextE Inst 107(3):405–412
70. Kan C-W (2007) Evaluating antistatic performance of plasma-treated polyester. Fibers Polym 8:629–634
71. Ghoranneviss M, Moazzenchi B, Shahidi S, Anvari A, Rashidi A (2006) Decolorization of denim fabrics with cold plasmas in the presence of magnetic fields. Plasma Process Polym 3(3):316–321
72. Radetic M, Jovancic P, Puac N, Petrovic ZL (2007) Environmental impact of plasma application to textiles. In: Journal of physics: Conference series, 71(1): IOP Publishing, p 012017.
73. Buschle-Diller G, Zeronian S, Pan N, Yoon M (1994) Enzymatic hydrolysis of cotton, linen, ramie, and viscose rayon fabrics. Text Res J 64(5):270–279
74. Hegemann D, Hossain MM, Balazs DJ (2007) Nanostructured plasma coatings to obtain multifunctional textile surfaces. Prog Org Coat 58(2–3):237–240
75. Ahmed NS, El-Shishtawy RM (2010) The use of new technologies in coloration of textile fibers. J Mater Sci 45:1143–1153
76. Hossain MM, Herrmann AS, Hegemann D (2007) Incorporation of accessible functionalities in nanoscaled coatings on textiles characterized by coloration. Plasma Processes Polym 4(2):135–144
77. Mia R, Mofasser A, Bhat MA, Howlader MI, Sayed NI, Bakar MA (2024) Utilizing neem wood waste as a natural dye in the development of bio-based functionalized colored jute fabric. Sustain Chem Pharm 38:101489. https://doi.org/10.1016/j.scp.2024.101489
78. Kan CW, Chan K, Yuen C, Miao M (1998) The effect of low-temperature plasma on the chrome dyeing of wool fibre. J Mater Process Technol 82(1–3):122–126
79. Kan CW, Chan K, Yuen C, Miao M (1998) Plasma modification on wool fibre: Effect on the dyeing properties. J Soc Dyers Colour 114(2):61–65
80. Karthikeyan K, Ramachandran T (2017) Effect of plasma and chitosan nanoparticles on cotton reactive dyeing
81. Kaliyamoorthi K, Thangavelu R (2015) Union dyeing of cotton/nylon blended fabric by plasma-nano chitosan treatment. Fash Textes 2:1–10
82. Raslan WM, El-Halwagy AA, Elsayad H (2020) Recent advances in plasma/nanoparticles treatments of textile fibers. J TextEs, Color Polym Sci 17(2):87–105
83. Desore A, Narula SA (2018) An overview on corporate response towards sustainability issues in textile industry. Environ Dev Sustain 20:1439–1459. https://doi.org/10.1007/s10668-017-9949-1
84. Muthu SS (2017) Sustainability in the textile industry. Springer
85. Pratibha M, Verma M, Sheoran MP, Santosh M (2024) Advance textile finishing used in textile industry

86. Juodsnukytė D, Gutauskas M, Krauledas S (2005) Influence of fabric softeners on performance stability of the textile materials. Mater Sci 11(2):179–182
87. Gupta D (2013) Softening treatments for technical textiles. In: Advances in the Dyeing and Finishing of Technical Textiles: Elsevier, pp 154–176
88. Le Marechal AM, Križanec B, Vajnhandl S, Valh JV (2012) Textile finishing industry as an important source of organic pollutants. In: Org Pollut Ten Years After Stock Conv-Environ Anal Updat, IntechOpen Rijeka, pp 29–54
89. Uddin F (2014) Environmental concerns in antimicrobial finishing of textiles. Int J Texte Sci 3(1A):15–20
90. Weil ED, Levchik SV (2008) Flame retardants in commercial use or development for textiles. J Fire Sci 26(3):243–281
91. Shah J, Shah S (2013) Innovative plasma technology in textile processing: a step towards green environment. Res J Eng Sci 2278:9472
92. Zille A (2020) Plasma technology in fashion and textiles. In: Sustainable technologies for fashion and textiles: Elsevier, pp 117–142
93. Naebe M, Haque ANMA, Haji A (2022) 纺织品等离子体辅助抗菌整理技术概述. Eng 12(5):145–163
94. Cao H, Folan P (2012) Product life cycle: the evolution of a paradigm and literature review from 1950–2009. Prod Plan & Control 23(8):641–662
95. Luján-Ornelas C, Güereca LP, Franco-García M-L, Heldeweg M (2020) A life cycle thinking approach to analyse sustainability in the textile industry: A literature review. Sustain 12(23):10193
96. Ekins P, Domenech Aparisi T, Drummond P, Bleischwitz R, Hughes N, Lotti L (2020) The circular economy: What, why, how and where
97. Jia F, Yin S, Chen L, Chen X (2020) The circular economy in the textile and apparel industry: A systematic literature review. J Clean Prod 259:120728
98. Buyle G (2009) Nanoscale finishing of textiles via plasma treatment. Mater Technol 24(1):46–51
99. Correia VM, Stephenson T, Judd SJ (1994) Characterisation of textile wastewaters-a review. Environ Technol 15(10):917–929
100. Weydts D, De Smet D, Vanneste M (2015) Processes for reducing the environmental impact of fabric finishing. In Sustain Appar: Elsevier, pp 35–48
101. Yusuf M (2019) Synthetic dyes: a threat to the environment and water ecosystem. Textes Cloth, pp 11–26
102. El-Sayed E, Hassabo AG (2021) Recent advances in the application of plasma in textile finishing (A Review). J TextEs, Color Polym Sci 18(1):33–43
103. Zille A, Oliveira FR, Souto AP (2015) Plasma treatment in textile industry. Plasma Processes Polym 12(2):98–131
104. Yilma BB, Luebben JF, Nalankilli G (2020) Cold plasma treatment in wet chemical textile processing. Fibres & Textiles in Eastern Europe, 28(6) (144)
105. Naebe M, Haque ANMA, Haji A (2022) Plasma-assisted antimicrobial finishing of textiles: A review. Eng 12:145–163

Chapter 10
Thermal Regulation in Textiles: Phase Change Materials and Sustainability

Borhan Uddin Khan, Jarin Anan Tanha, Jarjis Amin, Tanvir Hasan Fahim, and Md. Rashedul Islam

Abstract Integrating phase change materials (PCMs) into textiles is primarily driven by their exceptional thermo-regulating properties, which arise from their ability to absorb, store, and release thermal energy during phase transitions. This unique capability makes PCMs highly effective in maintaining temperature stability and enhancing the comfort of textile products. This chapter provides a comprehensive review of PCMs and their role in thermal regulation within textiles. It begins with an overview of PCMs, including their classification and the mechanisms underlying their phase change behavior. It further discusses the process of microencapsulation, a key technique used to enhance the stability and performance of PCMs, along with various fabrication methods for integrating PCMs into textile materials. In addition to addressing the technical aspects, the chapter examines characterization techniques used to evaluate the performance of PCM-incorporated textiles and highlights their diverse applications across industries, such as apparel, bedding, and technical textiles. It also outlines the criteria for selecting suitable PCMs based on specific requirements, including thermal properties, durability, and compatibility with textile substrates, while providing a detailed cost analysis in light of current market trends and demands. The chapter concludes by addressing the environmental impacts and challenges of using PCMs, such as sustainability concerns and lifecycle considerations.

Keywords Phase change materials · Thermal regulation · Textile integration · Microencapsulation · Sustainability

B. U. Khan · J. A. Tanha · J. Amin · T. H. Fahim
Department of Apparel Engineering, Bangladesh University of Textiles, Dhaka, Bangladesh

Md. R. Islam (✉)
Department of Wet Process Engineering, Bangladesh University of Textiles, Dhaka, Bangladesh
e-mail: rashedul.islam@wpe.butex.edu.bd

S. Biranje et al. (eds.), *Sustainable Finishing Techniques in Textiles*,
https://doi.org/10.1007/978-981-96-4860-3_10

10.1 Introduction

The human body is a complex thermoregulatory system that continuously generates heat through metabolic processes, cellular respiration, and muscular activity. The body primarily regulates the release of thermal energy by blood vessel dilation temperature. The optimal skin temperature for human comfort is approximately 37 °C, and a deviation of 4.5 °C can lead to thermal discomfort [1]. The body employs various physiological mechanisms, such as sweating, shivering, and vasoconstriction, to maintain thermal equilibrium. However, the body constantly faces challenges in maintaining a comfortable temperature, especially in extreme conditions [2]. Therefore, incorporating phase change materials (PCMs) into textiles has emerged as a promising approach to enhance both textile performance and thermal comfort for wearers.

Phase change materials (PCMs) are thermal energy storage devices that absorb or release latent heat during phase transitions, providing temporary cooling or heating effects [3]. The National Aeronautics and Space Administration (NASA) first used PCM to improve the thermal performance of their astronaut spacesuit. The growing interest in PCM-incorporated textiles can be attributed to the unique ability to absorb, store, and release thermal energy, enhancing thermal regulation and comfort in fabrics. Maintaining stable temperatures, PCM-integrated textiles contribute to energy efficiency by reducing reliance on heating and cooling systems. The versatility of PCMs allows their integration into a wide range of areas, such as spacesuits, sportswear, shoes and accessories, medical applications, bedding and accessories, automotive textiles, nonwoven protective garments, etc. Moreover, using PCMs can extend the lifespan of garments by minimizing wear and tear, which aligns with sustainability goals [4]. However, in-depth research into PCMs can pave the way for advancement in scalable material processing techniques, ultimately fulfilling the demand for high-performance and sustainable textiles.

This chapter aims to provide insights into the phase change materials (PCMs) and their influence on thermal regulation within textiles. In addition, this chapter will present a brief classification of PCMs, microencapsulation of PCMs, and fabrication techniques to incorporate PCMs into textiles. Moreover, it delves into characterization techniques and application areas of PCMs in textiles. The chapter also addresses the selection criteria of suitable PCMs and the cost analysis of PCMs considering the current market demand. This chapter concludes by discussing the environmental impact and challenges associated with PCMs and suggesting future research direction for enthusiastic scholars.

10.1.1 Thermo-Regulation Effect of Phase Change Materials in Textiles

Integrating phase change materials (PCMs) into textiles is primarily driven by their inherent thermo-regulating effect. This effect results from their ability to absorb, store, and release thermal energy during phase transitions. This capability allows PCMs to maintain a stable temperature in textile and clothing applications. Figure 10.1 shows the fundamental mechanism of PCM-incorporated thermal protective textile.

When the temperature of the environment or the human body rises, PCMs absorb excess heat as they undergo transition from solid to liquid. This process occurs at a specific melting point, allowing the PCM to store thermal energy without a significant increase in temperature. On the other hand, when the temperature drops, the PCMs solidify, releasing the stored heat back into the environment; the release of thermal energy helps maintain a comfortable microclimate between the fabric and the skin. This also enhances thermal comfort by preventing the cold sensation experienced by the wearer. Therefore, PCMs act as thermal buffers, smoothing temperature fluctuations. Wang et al. studied the influence of PCMs on smart thermal-protective garments and addressed the same phenomena [5]. The interaction of PCMs with the environment and their phase transition ability within the textile structure is significantly influenced by the thermal characteristics and performance of textile-based PCMs. These mechanisms also impact the design and engineering of such textiles [6, 7].

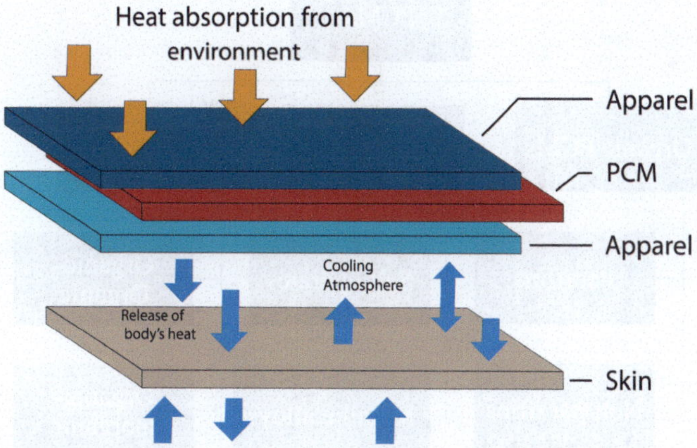

Fig. 10.1 Basic mechanism of PCM-incorporate thermal protective apparel. Adapted from Hossain et al. [5], Copyright © 2023 The Authors

10.2 Classification of Phase Change Materials

Phase change materials (PCMs) can be classified into three groups depending on their natures; organic, inorganic, and eutectic PCMs. They are further categorized based on the chemical composition of the PCM. Figure 10.2 shows the detailed classification of phase change materials.

10.2.1 Organic Phase Change Materials

Organic phase change materials (PCMs) can store and release thermal energy during phase transitions, typically between solid-to-liquid states. These PCMs are composed of organic compounds and further can be classified based on their molecular structures.

10.2.1.1 Paraffins

Paraffin PCMs consist of a mixture of the n-alkenes chain, with a common formula of C_nH_{2n+2}. . Here, n denotes the number of carbon atoms present in the substance. Paraffins with shorter hydrocarbon chains (between 1 to 4 carbon atoms) are gaseous,

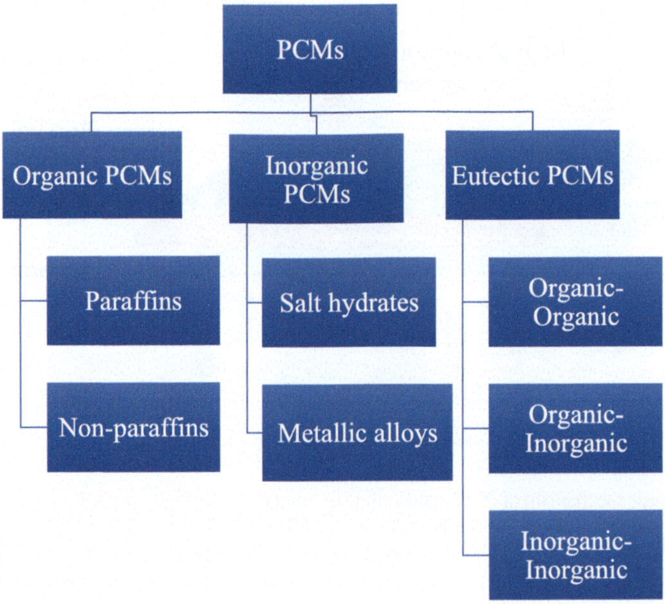

Fig. 10.2 Classifications of phase change materials (PCMs) [8]

medium-length chains (5 to 17 carbon atoms) are liquids, and longer-length chains (18 or more carbon atoms) are solids. A significant amount of heat is released during the crystallization of hydrocarbon chains in paraffin. The melting temperature and the latent heat of fusion increase with increasing hydrocarbon chain length. Considering cost constraints, only technical-grade paraffins are utilized as phase change materials (PCMs) in the latent heat storage (LHS) system. Paraffins are less expensive, noncorrosive, and work in a large temperature range (5 °C to 80 °C) [9]. Paraffin phase change materials offer a wide range of applications, including electrical sectors, renewable solar energy systems, medical sectors, etc. [10].

10.2.1.2 Non-paraffins

Non-paraffin organic phase change materials are the widely used PCMs. This group comprises fatty acids, glycols, poly-alcohols, and sugar alcohols. Though non-paraffin PCMs possess high latent heat capacity, they often suffer from limitations such as low thermal conductivity, low combustion temperatures, flammability, and potential toxicity. Fatty acids have a long hydrocarbon chain (from 10 to 30 carbon atoms) with the typical formula of the $CH_3(CH_{2)_{2n}}COOH$. They are derived from different natural sources. Fatty acid-based PCMs have minimal toxicity, low corrosivity, and an unpleasant odor [9]. They are also proven to be thermally and chemically durable because of the carboxyl group (COOH) at the end of their chain. Moreover, sugar alcohol can be used as PCM. They are the hydrogenated form of carbohydrates and have a hydroxyl group attached to each carbon atom. The chemical formula for them is $HOCH_2[(CH(OH)]_nCH_2OH$. They can store more latent heat because of their dense composition. Their melting temperature also lies between 90 °C to 200 °C. Polyethylene glycols (PEG) or polyethylene oxides are also organic PCMs. In a chemical sense, they are $[OH - (CH_2 - CH_2 - O)M_n - H]$. Here, the subscript Mn indicated the number of average molecular weights. They are hydrophilic and show a negligible degree of supercooling. They have a high flash point, exceeding 182 °C. They also possess remarkable thermal conductivity, approximately 0.205 W/mK [8].

10.2.2 Inorganic Phase Change Materials

Inorganic phase change materials have better thermal energy storage capacity and thermal conductivity than organic PCMs. These materials are frequently integrated with textiles for sports, military, and medical applications [5]. They can be classified based on their nature.

10.2.2.1 Salt Hydrates

$A_xB_y.\,nH_2O$ Is the common formula of salt hydrates. Here, A and B represent cation, and anion, respectively and n indicates the water molecule's number per salt. The terms "water of hydration" and "water of crystallization" are synonymous and refer to the water molecules incorporated into the crystal lattice of salt hydrates. These salts absorb a certain amount of energy, known as the enthalpy of dehydration (ΔHdehyd) which causes loss of their water of crystallization [5, 11]:

$$A_xB_y.nH_2O \rightarrow A_xB_y.mH_2O + (n - m)H_2O \tag{10.1}$$

Or to its anhydrous form:

$$A_xB_y.nH_2O \rightarrow A_xB_y + nH_2O \tag{10.2}$$

Salt hydrates have high latent heat of fusion per unit volume, low corrosiveness, and almost double thermal conductivity compared to paraffin. These materials break into anhydrous salts, releasing water vapor when heated, and storing the absorbed energy. Their positive characteristics have made them popular in various heat storage applications [12]. Salt hydrates include magnesium sulfate ($MgSO_4.7H_2O$), cupric sulfate ($CuSO_4.5H_2O$), gypsum sulfate ($CaSO_4.2H_2O$), etc. [11]. Salt hydrates are mostly used for low-temperature storage applications, due to their melting temperatures under 100 °C. Therefore, nanoparticle-enhanced salt hydrate PCMs have a high potential for smart textiles [13].

10.2.2.2 Metallic Alloys

Metallic PCMs and their alloys offer high thermal conductivity, low melting points, high density, and a high charging and discharging rate in thermal energy storage systems. They also operate at high temperatures (>560°) and have high thermal diffusivity [14, 11]. Binary and ternary alloys of Aluminium (Al), Copper (Cu), Magnesium (Mg), and Zinc (Zn) are suitable as phase-change materials. Aluminum alloys have an adequate phase-change temperature, good thermal stability, and high latent heat density. For example, aluminium–silicon alloy ($AlSi_{12}$) has a melting temperature of 577 °C and a heat of fusion of 560 J/g, making it an ideal candidate for phase change materials. Aluminium–silicon alloys also have a precise eutectic composition and are relatively stable through multiple heating and cooling cycles. Metallic alloy PCMs have potential where high thermal performance and efficiency are critical, such as in thermal energy storage systems, waste heat recovery, building materials, electronics cooling, etc. [15].

10.2.3 Eutectic PCMs

Eutectic phase change materials (PCMs) are a promising category of thermal energy storage materials. These materials consist of a mixture of two or more components that solidify and melt at a specific eutectic point. They offer high thermal conductivities and densities. These properties enable them to store a significant amount of thermal energy in a compact form, making them particularly attractive for applications in building heating and cooling, solar energy systems, and other thermal management solutions [11]. The weight percentage of different PCM materials within the eutectic mixture determines the resultant melting points [9]. While inorganic-inorganic eutectics are the most common type, organic-organic, organic–inorganic mixtures have become increasingly popular in recent research. The commercialization of eutectic PCMs is hindered by high cost and lack of comprehensive thermophysical data. However, ongoing research aims to address these gaps to enhance their designs and make them more efficient [5, 11]. Table 10.1 represents an overview of the three main types of phase change materials which indicates their unique properties at a glance.

Table 10.1 Comparison of different types of phase change materials (PCMs)

Type of PCMs	Advantages	Disadvantages	References
Organic	A broad range of operating temperatures No super-cooling High heat of fusion Reduced vapor pressure Inert to corrosion High latent heat per gram Chemically stable and recyclable Long-term functionality	Low thermal conductivity (about 0.2 W/mK) Substantial volumetric changes Flammability	[16, 17, 11]
Inorganic	Exceptionally high latent heat capacity per liter Superior thermal conductivity (around 0.5 W/mK) Low volume change Non-flammable and poses no fire hazard Availability in low cost	Super-cooling Corrosive Thickening and nucleating agents are essential to mitigate phase separation	[18, 11, 19]
Eutectics	Expensive Efficient thermal energy storage in a minimal volume Relatively high melting temperature	Limited thermo-physical property data is currently available	[20, 18]

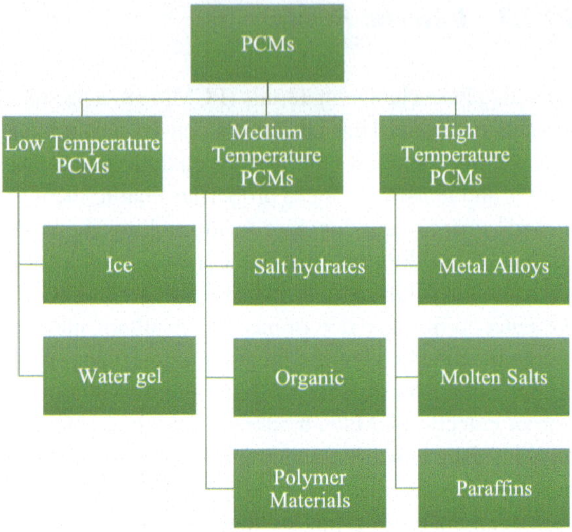

Fig. 10.3 Phase change material classification by melting point [22]

There is another classification of PCMs based on their temperature range (Fig. 10.3). The temperature range of PCMs is an important criterion for understanding their suitability for different applications. A lot of thermal energy storage applications of PCMs are measured by their performance at a particular temperature range [21].

10.3 Thermophysical Characterization of Phase Change Materials

The thermophysical properties of phase change materials (PCMs) are important parameters for effective applications. These properties include melting temperature, latent heat, thermal conductivity, specific heat conductivity, and density. Pasarkar et al. [8] and Sarbu et al. [11] listed the thermophysical properties of possible PCMs, indicating their suitability in various application sectors.

10.3.1 Melting Temperature

The melting temperature determines the operational range of phase change materials (PCMs). This also aligns with the thermal requirements of the specific applications of PCMs [23]. Materials that are melted at 15 °C are best for air conditioning and excess cooling states where temperature fluctuation does not occur frequently. Melting temperature greater than 90 °C can be used to get protection against fire

[24]. PCMs with a temperature range between 15 °C to 90 °C are mostly used in solar heating and heat load leveling applications [23].

10.3.2 Latent Heat of Fusion

Latent heat is the energy absorbed or released by a system during a phase transition at a constant temperature. Latent heat storage can occur through solid–solid, solid–gas, liquid–solid, and liquid–gas interactions [25]. However, for PCMs, only solid–liquid change is used for thermal storage [24]. This characteristic allows PCMs to continue efficient energy storage and release, effectively managing thermal loads [11].

10.3.3 Thermal Conductivity

The thermal conductivity of a PCM influences the rate of heat transfer during the melting and solidification processes. Higher thermal conductivity can enhance the phase change rate, leading to more efficient energy storage and retrieval [26]. Due to the poor thermal conductivity of PCMs, their ability to carry heat falls. The thermal conductivity can be improved by including more thermally conductive materials or encapsulating phase transition elements [7].

10.3.4 Specific Heat Capacity

The amount of energy required to change the temperature of a PCM depends on its specific heat capacity. PCMs with higher specific heat capacities can provide better thermal stability during melting and solidification. In applications where temperature stability is critical, PCMs with higher specific heat capacity can absorb more heat without experiencing a noticeable temperature rise [11].

10.3.5 Density

The density of a PCM impacts its volumetric energy storage capacity. PCMs with higher density can store more energy in a given volume. Higher-density PCMs are particularly helpful in applications where space is limited, such as building integrated thermal energy storage systems or portable thermal management solutions. Thermal conductivity and the heat transfer rate of PCMs are also influenced by density. PCMs with higher density may exhibit different thermal conduction properties which can affect the overall performance of the thermal energy storage system of PCMs [11].

10.4 Microencapsulation of Phase Change Materials

10.4.1 Basic Microencapsulation Principle

Phase change materials (PCMs) can sense environmental stimuli by storing thermal energy in warm environments and releasing it in cooler environments [27]. Due to reversible phase transition (between liquid and solid states) during melting and crystallization, PCMs face challenges in direct textile applications. As a result, PCMs must be encapsulated at the micro or nanoscale to ensure their suitability and longevity [28]. Microencapsulation is a technique that involves the containment of core materials like PCMs within the protective capsule [29]. Microcapsules are formed by encapsulating small solid particles or liquid droplets within a thin, resilient polymer coating. This is an effective technique for preventing PCMs from leaking and reacting with the external environment. The thickness of the capsule walls can be as thin as 1 μm. Depending on the encapsulation method, the particle size of PCM microcapsules can range from less than 1 μm to over 300μm (typically 20–40μm in diameter). The PCM content in microcapsules may be up to 80%–85%. The microencapsulation technique is selected based on factors like the chemical properties of the shell material, the processing costs, the desired capsule size, and solvent usage. The selection of a suitable polymer for the capsule shell material is influenced by the intended application area of that PCM, production processes, cost, and availability [30–32, 29]. Typically, microencapsulation involves two primary steps [33–35]:

- Emulsification: Emulsification impacts particle size and size distribution. This process is influenced by stirring speed, interfacial tension, volumetric ratio, and the chemical properties of the two phases.
- Capsule Formation: The type and quality of surfactant are crucial factors influencing capsule formation, emulsion stability, and microcapsule particle size. The monomer's or prepolymer's crosslinking capabilities control the formation of the capsule particles.

10.4.2 Methods of Microencapsulation

General technologies for microcapsule formation can be categorized into physical and mechanical, chemical, and physicochemical methods [36]. This section discusses three major microencapsulation methods used in textiles: phase coacervation (physicochemical methods), in-situ polymerization (chemical methods), and emulsion and solvent evaporation (chemical methods). The microencapsulation method is selected depending on its compatibility with the textile substrate and subsequent finishing processes.

10.4.2.1 Phase Coacervation Technique

Phase coacervation involves the addition of two immiscible materials (a coating material and a core material) to an aqueous phase to create a dispersion. The coating material then adheres to the surface of the core materials and solidifies through chemical or thermal cross-linking. When the overall free energy of the system decreases, solid capsules are generated (Fig. 10.4) [37].

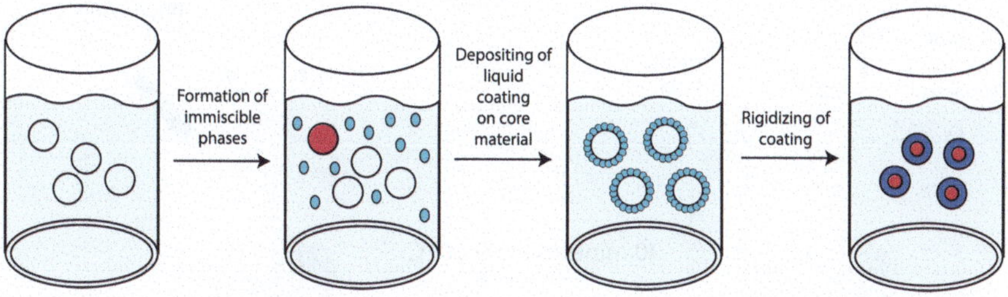

Fig. 10.4 Microencapsulation process by phase coacervation technique. Adapted from Iqbal et al. [38], Copyright © 2019 The Textile Institute

Phase coacervation technique can be categorized into two types: simple coacervation and complex coacervation. Simple coacervation can be achieved by adding a single colloidal solute into an aqueous polymer solution. In the case of complex coacervation, two colloids with opposite charges are added to a polymer solution [39]. Uddin et al. [40] employed the simple coacervation technique to encapsulate paraffin wax. They evaluated the heat storage performance of the synthesized microencapsulated phase change material. The thermal energy storage capacity was found 56–58 J/kg after multiple cycles. Hawlader et al. [41] experimented with Arabic gum/gelatin power, and paraffin wax to perform complex coacervation techniques. They successfully microencapsulated paraffin wax with a thermal energy storage capacity of about 145–240 J/g. Multiple studies have proved the potential of the phase coacervation technique as an encapsulation method [42–44].

10.4.2.2 In Situ Polymerization Process

In situ polymerization refers to the method of combining two immiscible liquids, such as an organic solvent and water, which each contain complementary chemical intermediates. They directly interact with each other in terms of forming solid pre-condensate [45]. In situ polymerization is a frequently employed chemical technique for encapsulating phase change materials (PCMs) (Fig. 10.5) [39]. This approach allows for the formation of micrometer and nanometer-sized capsules and particles. The in-situ process is particularly effective in producing the best possible quality of microcapsules characterized by superior wall diffusion-tightness [46]. In-situ polymerization is a common technique for manufacturing microcapsules using shell materials like Urea Formaldehyde (UF) and Melamine Formaldehyde (MF). The reason behind selecting UF and MF is their excellent shear and thermal stability during application. Additionally, their high reactivity is beneficial, facilitating a rapid microencapsulation process and high-loading content [47]. In terms of using the in-situ polymerization technique in the encapsulation process, the following steps are followed.

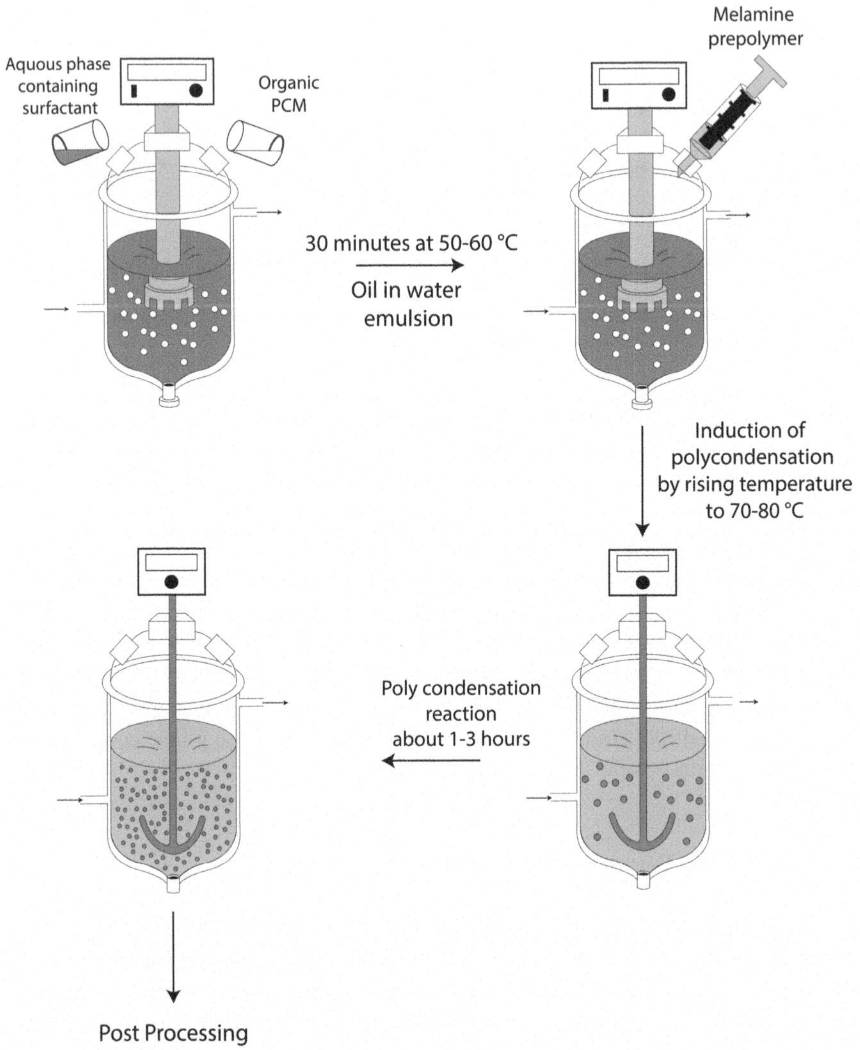

Fig. 10.5 In-situ polymerization process for microencapsulation of PCMs. Adapted from Fabien [39]. Copyright © 2011 The Author(s)

In the first step, an oil-in-water emulsion is formed by dispersing molten PCM within an aqueous continuous phase. The dispersion process is facilitated by adding a suitable surfactant and high-speed stirring system. In the second stage, a pre-polymer is synthesized by dissolving melamine formaldehyde at a temperature range of 50–60 °C using a magnetic stirrer. In the final step, the pre-polymer is gradually added to the emulsion. The stirring continues to encapsulate the PCM droplets [48].

10.4.2.3 Solvent Evaporation Techniques:

Various solvent evaporation techniques have been employed for encapsulation. The choice of these techniques is dependent on the hydrophilicity or hydrophobicity of the encapsulated active core materials [49]. The selection criteria of solvents include high volatility, minimal toxicity, lower boiling point, and capacity to easily dissolve polymers [50].

The solvent evaporation techniques involve several steps (Fig. 10.6) [50]. Initially, the core material is dissolved into an organic solvent containing a polymer. The dispersed phase is then emulsified in an aqueous phase. Later the solvent is removed from the dispersion through evaporation which results in the formation of solid particles from the dispersion droplets. Finally, the solid capsules are further processed to get the powdered form [50].

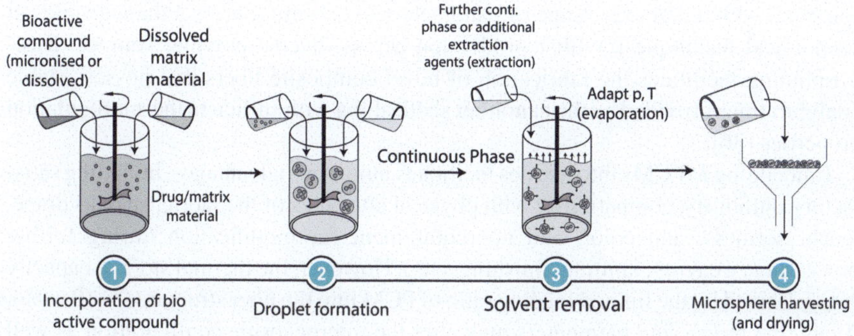

Fig. 10.6 Solvent evaporation technique for microencapsulation of PCMs. Adapted from Freitas et al. [50]. Copyright © 2004 Elsevier B.V

10.5 Basic Techniques to Incorporate Phase Change Materials in Textiles

A significant volume of studies and research works have been carried out by numerous researchers to investigate favorable processes to incorporate microencapsulated PCMs (MPCM) into textiles. Integration of MPCM into synthetic fibers was successfully carried out by Bryant et al. [37] in 1988 after several years of experimentation. Following that, a variety of widely accepted techniques have emerged for incorporating PCM into textiles, such as lamination, coating, injection molding, fiber technology–spinning (melt/dry/wet), bi-component synthetic fiber extrusion, and so forth [28].

10.5.1 Fiber Technology

Microencapsulated PCMs are incorporated into the polymer solutions before fiber extrusion and subsequently extruded polymer is spun following a suitable spinning process. Integrating PCM microcapsules in such a manner enables heat storage over extended periods [28]. Configuring the core–shell structure of PCM-loaded melt spun fiber core facilitates thermo-regulatory properties along with improved scalability, stability, and reproducibility as demonstrated in several studies [51–54]. Electrospinning, with a diameter range of nanometers to sub-microns, is a manageable yet economical technique for fiber production across diverse polymer sources. Electrospinning facilitates the fabrication of novel composite fibers that possess stable configurations provided by the nanofiber shell, along with sufficient thermoregulation properties [36].

Embedding MPCMs into textiles facilitates numerous advantages, including satisfactory durability, compatibility with physical attributes of the fabric, such as drape, touch, softness, and color, and no requirement for modification during further processing (weaving, knitting, finishing, etc.). However, the thermal storage capacity is constrained by the limited loading ratio of PCM into the fiber structure (5–10 wt%). Again, the processing parameters may alter the microcapsule configuration as well as the thermo-mechanical attributes [55–61]. Table 10.2 summarizes the findings of incorporating PCMs into different polymer-based textile fibers in various research works.

Table 10.2 Outcomes of incorporating PCM microcapsules into textile fibers using fiber technology

Materials Utilized	Findings	References
PP (polypropylene) monofilaments with paraffin microcapsules (1%, 2%, and 5%)	Increased microcapsule content raised fiber density, decreased orientation, and increased elongation without altering phase change temperature	[62]
Composite fibers (PE core, PP sheath) with n-octadecane microPCMs via melt spinning	Consistent heat absorption/emission (~32 °C/15 °C); higher enthalpy with more PCM content. Applicable for fabric production	[53]
PAN/VDC fibers with n-octadecane microPCMs (4–40 wt%)	Smooth spinning with $\leq$30 wt% PCM; good enthalpy (30 kJ/kg), Tg = 89 °C$\sim$108 °C, elongation $\sim$7% and tensile strength (0.7–2.0 cN/dtex)	[63]
PVA with TEOS- in situ encapsulated paraffin	High latent heat (23.7 kJ/kg) and good thermal stability; good enough mechanical properties for textile applications	[64]
SEBS (styrene-ethylene/butylene–styrene) with commercial micro PCMs (Tm = 52 °C) by injection molding	Improved thermal regulation near PCM melting point, increasing thermal inertia	[65]

10.5.2 Coating

The coating is prepared by dispersing the PCM microcapsules first and wetting within a dispersion of water solution that contains necessary elements (such as thickener, surfactant, anti-foaming agent, dispersant, polymer mixture, etc.) [2]. In the production of phase-changing fabric, polymers like acrylic [66] and polyurethane-based [67] macromolecules are frequently used. Pad-dry-cure [68], knife-over-roll [69], knife-over-air [70], gravure, transfer, and dip coating are some of the methods available for coating PCM capsules. These techniques are frequently used in the production of shoe accessories, cars, textiles, spacesuits, sportswear, and medical appliances.

In general, various coating parameters influence properties such as tactile comfort, insulating capacity, durability, and thermal stability [29]. When incorporating microcapsules into textile materials, size uniformity, and particle size are crucial considerations [71]. Several authors claim that microcapsules with particle sizes ranging widely from 0.5 to 100 μm can be used in textile applications [72–74]. It was suggested by Sarier and Onder [75] that wax-based particles smaller than 10 μm can be coated on textiles. Table 10.3 displays recent research on incorporating PCM microcapsules into textiles by coating technique.

Table 10.3 Outcomes of incorporating PCM microcapsules into textiles using the coating method

Materials Utilized	Findings	References
Wax-based particles (<10 μm) coated onto textiles	Suggested optimal particle size for textile application, enhancing comfort	[75]
PCM microcapsules with acrylic binder	Four-layer garment system with PCM coatings for improved thermal comfort	[76]
Coacervates II and III with 100% cotton	Explored cotton coating to combine coacervates, enhancing thermal properties	[44]
PTA (poly(tetradecyl acrylate))/BN-x coated fabric	Improved thermal management with boron nitride coating	[77]
Microcapsules with and without clay nanoparticles on cotton fabric	The pad-dry-cure technique was applied; clay nanoparticles improved thermal performance and flame retardancy	[78]

10.5.3 Lamination

Bryant and Colvin [79] initially reported the process of creating foam with microencapsulated alkanes in 1996 integrating the foam utilizing lamination into the textiles. PCM-integrated thin films of polymeric substrates are incorporated into the fabric interior using a lamination process to facilitate the thermo-physiological comfort of textiles [2]. The lamination process involves mixing 20–60% PCM capsules with a water-blown polyurethane foam mix and then applying the mixture to the fabric. The water is then evaporated with hot air [80]. Additionally, this technique increases the passive insulation of fabric because it creates an excellent honeycomb structure following foam creation, which traps a significant amount of air within the foam configuration and makes it less difficult to distinguish the micelles of PCM. Therefore, the cooling effect of PCM can remarkably postpone the microclimatic temperature and moisture rise of the clothing, promoting considerable longevity of the garment usability without posing a serious health concern and ultimately increasing efficiency [2]. When laminated into the fabric, the foam was proposed to be a helpful PCM composite in appliances with thermal insulation, for instance, the inner boots lining, gloves, healthcare items, vehicle interiors, as well as outerwear [74]. Table 10.4 displays the research on incorporating PCM into textiles by lamination technique.

Table 10.4 Outcomes of incorporating PCM microcapsules into textiles using lamination method

Materials Utilized	Findings	References
Microencapsulated n-alkanes in foam	First use of laminated foam for textiles to improve thermal insulation	[79]
Thin polymer film laminated onto fabric	Enhanced thermo-physiological comfort for protective clothing	[28]
PCM capsules with polyurethane foam	Lamination improved insulation due to the honeycomb structure, extending garment usability and comfort	[80]

10.5.4 Impregnation

Apart from lamination and coating, impregnation is also used for PCM incorporation in textiles. Several methods have been employed to impregnate textiles with PCMs. One common approach of impregnation takes place in an aqueous solution containing PCM microcapsules and a binding agent [81]. Researchers also prove the effectiveness of the impregnation technique by conducting plasma treatment. Table 10.5 displays recent research related to the impregnation technique.

Table 10.5 Outcomes of incorporating PCM microcapsules into textiles using the impregnation method

Researcher	Materials Utilized	Findings	References
Younsook Shin	Aqueous solution of microcapsules with polyurethane binder	Impregnation technique for textile surfaces to enhance thermal comfort	[81]
Gökhan Erkan	Microcapsules in water solution, untreated or DBD (di-electric barrier discharge) plasma-treated	Developed methods to treat fibers and enhance thermal regulation	[82]
Wi et al	Shape-stabilized PCMs with palm and coconut oils and exfoliated graphite nanoplatelets	Created PCMs suitable for energy-saving applications by impregnating PCM into graphite nanoplatelets	[83]
Boh and Knez	Microencapsulated essential oils and PCMs	Developed microcapsules with antimicrobial properties applied via immersion impregnation onto textiles	[84]

10.6 Characterization Techniques of Microencapsulated Phase Change Materials

10.6.1 Differential Scanning Calorimeter

Differential scanning calorimeter (DSC) examines thermal effects during heating and cooling cycles when materials undergo thermal transitions. This method may precisely quantify and assess energy absorption and emission in a material within the designated temperature range. The DSC apparatus can measure the thermal energy absorbed during material melting and the energy released during crystallization upon cooling. This method analyzes the material's temperature buffering range by examining the melting and crystallization peaks, as well as the quantity of latent heat absorbed, stored, and released in Joules [37]. The dynamic mode was established as a standard pattern, considering the melting temperature (Tm). The heating rate is established at 0.5 K min^{-1}, commencing from 10 degrees below the material's Tm to 10 degrees above it. The identical technique is executed at a constant cooling rate. For additional comparisons between the two procedures, a suitable temperature range was designated for each sample and applied to both DSC modes. The requirement for this temperature range was established in dynamic mode, defined as the range (Tm $\pm$ 10 °C). In step mode, a variable heating rate is employed, set at 0.1 K min^{-1} inside the phase change temperature range (between Tm $\pm$ 5 °C) and 0.5 K min^{-1} for the remainder of the measurement. When the DSC signal does not revert to zero, it indicates that the thermal equilibrium has not been achieved, the rendered data is unrepresentative and therefore, unsuitable for analysis [85].

When working with PCM-incorporated textiles, different models and methods of DSC are used such as a differential scanning calorimeter model equipped with a refrigerated cooling system and nitrogen as the purge gas [86]. Other models also examine various DSC approaches such as the thermal stability of microcapsules [78], the thermal properties of PTA and PTA/BN-x composite PCMs [77], etc.

10.6.2 Fourier Transform Infrared Spectroscopy

Fourier transform infrared spectroscopy (FTIR) can be employed to identify functional groups in materials, irrespective of their state—liquid, solid, or gaseous. It employs beams of infrared light to do this. Infrared spectroscopy produces a spectrum by quantifying the absorption of infrared radiation by each bond, often depicted as transmittance% vs wavenumber (per cm). The infrared spectrum possesses longer wavelengths and lesser energy compared to ultraviolet–visible light, but microwave radiation exhibits shorter wavelengths and higher energy. Infrared activity is essential for a molecule to discern its functional groups. An infrared-active molecule possesses a dipole moment. Covalent bonds in materials possessing an electric dipole interact with infrared radiation by absorbing energy from the molecule, resulting in the oscillatory motion of the bond. Consequently, infrared radiation must be absorbed by the oscillation that induces a variation in the molecule's net dipole moment [87].

FTIR can be used to analyze and identify the chemical bonds and chemistry of PCMs [37]. FTIR was employed to characterize the chemical and crystalline structures of specific photoluminescence phase-change microcapsule samples [88]. In a separate investigation, researchers studied the impact of dielectric barrier discharge (DBD) plasma treatment on the adhesion of phase-change material (PCM) microcapsules to wool fabric using FTIR spectra [89].

10.6.3 Thermal Gravimetric Analysis

This analytical technique is straightforward and determines the increase or decrease in the weight of a material as a result of temperatures [37]. The thermal stability of microcapsule samples and purified n-eicosane was assessed using thermogravimetric analysis (TGA) [88]. Demirbag et al. explored the temperatures and thermal stability of the microcapsules with the help of a thermal gravimetric analysis system [78]. In another study, Yan et al. used to characterize the thermal stability of PCM fibers using TGA under a nitrogen atmosphere [90].

10.6.4 Infrared Thermography

Induction thermography or pulsed eddy current thermography uses electromagnetic pulses to induce eddy currents in electrically conductive materials. Eddy currents generate heat through resistive losses and then dissipate heat. An infrared camera can detect heat on the surface. Surface or subsurface fissures induce localized alterations in electrical current densities, which are discernible in thermographic pictures [91]. Infrared thermography employs thermographic cameras capable of detecting radiation in the infrared (IR) spectrum, resulting in images of the observed radiation known as thermograms [37].

A hot disk thermal constant analyzer and an infrared thermal imager were employed to measure the thermal properties of PTA and PTA/BN-x composite PCMs [77]. In another study, the infrared camera was utilized to observe thermal behaviors and to derive heat release curves of fabric samples [52]. In a different study, the thermographic images of the samples were photographed by the infrared imager after irradiating by IR lamp.

10.7 Utilizations of Phase Change Materials in Textiles

PCM has gained significant research attention for its prospective applicability since its first attempt at a patent for thermal storage [92]. The quantity of PCMs to be applied to garments needs to be optimized considering the garment usage duration and the level of activity as the thermoregulation of PCM is influenced by textile design factors. Textiles with PCMs are being utilized in diverse application areas, including outerwear, casual garments (such as winter jackets, vests, thermal garments, ski suits, and bottoms), undergarments, socks and stockings, accessories, footwear as well as bedspread, sleeping bags, quilts, comforters, mattresses, and pillow slip. PCMs are also projected to smart textiles, such as protective vests, automobiles, medical, or particular industrial applications [82, 93]. A few application areas are discussed below:

10.7.1 Spacesuits

The early-stage application of PCMs was in spacecraft suits and protective gloves to safeguard astronauts from extreme freezing conditions during space operations [28]. To maintain a comfortable temperature, NASA utilized PCMs to develop thermoregulated garments, while PCM-treated smart fabrics were applied in apparel, interior design, and technological devices. The PCM composite illustrated reasonable thermal stability in those applications [94, 95].

10.7.2 Sportswear

The inclusion of PCMs into athletic apparel facilitates absorbing the excess body heat generated during physical activity and releasing it as required to sustain thermal equilibrium and mitigate heat stress. PCM applications in sportswear encompass undergarments, snowboard gloves, active wear (such as cycling or running clothings, and gym suits), and ice-climbing gear [93]. A research group explored the development of smart knitted fabrics for pre-sport warm-up by incorporating PCM. Their research demonstrated final fabric products with enhanced attributes, including increased UV resistance, water permeability, and bursting resistance. On the other hand, the end product has lower fabric thickness, air permeability, and thermal insulation [96].

10.7.3 Shoes and Accessories

Presently, PCMs are applied in varied footwear, such as ski boots, mountaineering boots, and racing driver boots. Designated temperature is specified based on the end use, such as ~36 °C for motorcycle helmets and ~6 °C for gloves. PCMs in ski boots absorb excess heat produced around the feet and release it back to cooler regions as required, hence maintaining comfort. Plausible items to be PCM integrated include golf shoes, footwear, and ski boots [97].

10.7.4 Medical Applications

PCM microcapsule applications have potential in clinical clothing, patient beddings, gauges, and intensive care unit patients' temperature-regulating products [97]. Shahid et al. have designed microencapsulated phase change material (MPCM) incorporated medical bed sheets utilizing cotton fabric. The developed prototype demonstrated a capacity to absorb additional moisture from the surroundings, effectively controlling the humidity and temperature of the respective fabric. This study has the potential for innovative hospital bed designs that protect patients from bed sores [95].

10.7.5 Bedding and Accessories

Temperature control during sleep while keeping the body warm might be facilitated by inserting microcapsules of PCM in quilts, pillows, and mattresses, and is already in use in sleeping bags [97].

10.7.6 Automotive Textiles

Micro PCMs present promising opportunities in automotive textiles, seat covers in particular. Higher heat storage capacity, lower toxicity, lack of hygroscopic properties as well as cost-effectiveness attributes make paraffin-based PCMs suitable for such applications. PCM-incorporating textiles used in car interiors and seat covers offer excellent thermal control with enhanced comfort for the passengers [28, 98]. PCMs can be installed in the roofs, doors, and side walls of automobiles and buses to improve thermal performance, reduce energy consumption and gas emissions, and mitigate energy gain or loss. This prospective implementation might positively influence global transportation emissions and energy consumption [99].

10.7.7 Nonwoven Protective Garments

The construction of nonwoven protective garments imposes a strict barrier function against dust, liquids, or gas penetration along with limiting the transportation of heat and moisture in the human body [28]. Integrating PCM into a thin polymeric film and laminating it onto the inner surface of nonwoven protective garments can enhance the thermo-physiological comfort of the wearer. The experimental outcomes demonstrate that the cooling impact of the PCM is capable of substantially prolonging the temperature increase, therefore effectively restricting moisture formation in the microenvironment [74]. A PCM-incorporated mathematical model of heat transport by firefighter protective clothing (FFPC) is developed that successfully predicts the phase change effect, accurately reflecting the impact of various PCMs, across different specimen configurations and exposures, excluding the expense of tests [100]. The PCM vest becomes less effective during excessive perspiration and in extremely dry or humid environments. Ghali et al. proposed a design alteration by creating a hybrid PCM-desiccant vest as well as a hybrid PCM-fan for the cooling vest for moderate and elevated ambient temperatures. The use of hybrid PCM-Fan vests is recommended for reduced relative humidity and elevated temperatures (40 °C and 45 °C), while both hybrid vests are suggested for moderate relative humidity. The hybrid PCM-Desiccant vest is suitable in case of greater ambient relative humidity [37].

10.8 Selection Criteria of Suitable Phase Change Materials

The selection of suitable PCMs is determined by temperature range, thermal conductivity, latent heat, durability, cost-effectiveness, environmental impact, specific heat capacity, lifecycle, chemical compatibility, ease of integration, etc. [101, 102].

Organic PCMs have already been extensively explored because of their characteristics of enhanced latent heat capacity, non-corrosiveness, non-toxicity, and a broader range of melting temperatures. N-octadecane, n-hexadecane, and n-eicosane organic PCMs have drawn much attention owing to their phase change temperature of approximately 28 °C, 36 °C, and 18 °C respectively, compatible with human body comfort temperature [39, 103–105]. Regardless, thermoregulatory responses in footwear, advanced technical textiles, and construction materials where the temperature may go beyond 37 °C require phase-changing temperatures beyond such temperature as well. To address such a concern, microencapsulated n-docosane with a phase change transition temperature range of 36 − 45 °C has been applied to impregnate cotton fabrics for thermo-regulation applications, including advanced potential use in high-technical textiles, footwear, and building thermo-regulating covers and paints for thermal energy storage [106].

The method of processing microcapsules and the size of microcapsules varies depending on the specific application and desired properties. However, research shows that microcapsules of 1–30 μm in diameter are capable of withstanding mechanical abrasion, heat, and chemicals [28]. The thermal efficiency of PCM fabric is primarily determined by the quantity of PCM incorporated into the fabric, verified in different studies [107]. The thermoregulatory response is also reliant upon the surface-deposited weight and the mass ratio of binder to microcapsules [108]. For instance, the heat storage capacities of the thermoregulating fabrics ranged from 0.91 to 4.44 J/g, reliant upon the density of the microcapsules used. The treated fabrics maintained 40% of their heat storage capacity following five laundering cycles and microcapsules with elevated core/shell ratios might enhance the thermoregulating effectiveness of fabrics [81].

Laundering durability of PCM-treated textiles is a fundamental requirement of regular-use applications and such resiliency of PCM for sufficient wash cycles under rigorous domestic laundering is a selection criterion for utilizing in textiles [109, 110]. The wash temperature imposes a significant impact on the latent heat capacity of the incorporated PCM in textiles and it is validated that multiple laundering and high-temperature washes reduce thermoregulation efficacy [111]. Moreover, the flammability of PCMs is also a crucial consideration. Paraffin-based PCMs are effective for thermal energy storage but they are prone to combustion. This limits their use in certain applications where excessive heat is involved [112].

Finally, the cost and availability in the market are also important criteria for selecting PCMs. The cost of PCM-treated textiles can be influenced by factors such as processing methods, purchased volume, and purity of PCMs. Table 10.6 illustrates the approximate cost of PCM and PCM-incorporated fabrics.

Table 10.6 Cost analysis of PCM chemicals and PCM-incorporated Fabrics

Type	Description	Price (USD)	References
PCM Chemicals	Paraffin wax	$1.88–2.00 per kg	[113–115]
	TiO_2	$1342.3 per kg	
	Oleic acid	$1.67–1.76 per kg	
	Palmitic acid	$1.61–1.72 per kg	
	Calcium chloride	$0.20 per kg	
PCM Incorporated Fabric	PCM microencapsulated fabric for temperature control	$5.00–50.00 per yard	[116]
	PCM cooling vest (soft and comfortable)	$5.00–50.00 per yard	
	Nonwoven special functional fabric	$3.80–4.50 per yard	
	Sportswear	$5.00–12.00 per yard	
	Cooling polyester	$5.00–7.50 per yard	

10.9 The Environmental Impact of PCM-Incorporated Textiles

The environmental impact of PCM-incorporated textiles has both sustainable and harmful aspects (Fig. 10.7). These impacts must be addressed to identify challenges and limitations and improve the overall quality and market of PCM-incorporated textiles.

10.9.1 Sustainable Aspects

10.9.1.1 Energy Efficiency

PCM textiles can significantly enhance energy efficiency by regulating temperature. Studies show that PCM materials can effectively absorb and retain excess body heat in warm environments and promote thermal comfort to the wearer. This reduced the reliance on air conditioning or other cooling or other cooling systems which leads to lower energy consumption and electricity usage [117, 118]. Therefore, they contribute to reducing carbon emissions, conserving energy, and acting in the context of the global energy crisis.

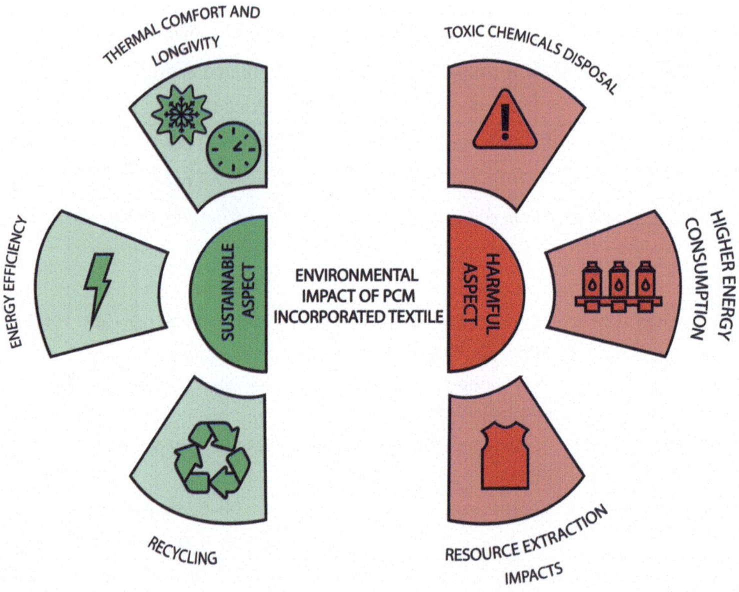

Fig. 10.7 Environmental impacts of PCM-incorporated textiles

10.9.1.2 Thermal Comfort and Longevity

The ability of PCM textiles to provide thermal comfort contributes to the longevity of the garments. When consumers experience greater satisfaction with such garments, they can use them for an extended period. Moreover, PCM textiles can regulate body temperature and enhance overall comfort, reducing unnecessary movements and garment alterations. Therefore, PCM textiles may extend the lifespan of clothing and other textile products by reducing wear and tear resistance, a critical factor in sustainability [119, 120].

10.9.1.3 Recycling

Since PCM textiles can be designed to have a longer functional life, they can reduce textile waste. Moreover, incorporating PCMs can lead to the development of multifunctional textiles that serve multiple purposes, further minimizing the need for excess products and reducing overall waste. There is a growing consumer demand for sustainable and eco-friendly products. PCM textiles can offer both functional benefits and sustainability features which can meet the market demand for sustainable products.

10.9.2 Harmful Aspects

10.9.2.1 Higher Energy Consumption

The production of PCM textiles involves additional processing steps, such as microencapsulation, PCM solution preparation, and fabrication methods for incorporating PCM. These processes can lead to higher energy consumption and increased carbon emissions compared to conventional textiles [5].

10.9.2.2 Toxic Chemicals Disposal

During the production of PCM textiles, certain chemical additives are used to enhance their properties, such as flame retardants or stabilizers. Many of these additives can be harmful to the environment. For example, certain flame retardants like chlorinated phosphates and polybrominated diphenyl ethers have toxic effects [121, 122]. Moreover, some PCMs and their additives may leach into the environment when exposed to moisture or heat. This leaching can contaminate soil and water sources, posing risks to ecosystems and human health [123].

10.9.2.3 Resource Extraction Impacts

The raw materials used to produce PCMs, especially organic PCMs derived from petroleum can have significant environmental impacts. The extraction and processing of fossil fuels can lead to habitat destruction, pollution, and greenhouse gas emissions. Additionally, the sourcing of inorganic materials also involves environmentally harmful mining practices [124].

10.10 Challenges of Phase Change Materials and Future Directions

10.10.1 Challenges

PCMs were initially developed to safeguard astronauts from significant temperature fluctuations in space. However, their (PCMs) application in day-to-day textiles brings several challenges.

10.10.1.1 Limited Thermal Conductivity

A key obstacle encountered with PCM-based materials is their poor thermal conductivity. For instance, paraffin-based PCM materials have a thermal conductivity of 0.22 W/m K, whereas graphite powder has a value of 2–90 W/m K. Due to its low conductivity, it is challenging to swiftly store and release heat at the time of melting and crystallization. However, it might be required to improve the PCM microcapsule's shell walls and cores' thermal conductivity to obtain quickly responding qualities to temperature fluctuations [98].

10.10.1.2 Limited Duration of Cooling and Warming Effect

The human body generates heat as a byproduct of metabolic processes and this heat production varies depending on the level of physical activity. As metabolic rates increase, the body generates excess heat. To maintain thermal comfort and prevent overheating, the body must effectively dissipate this excess heat to the environment [125]. However, the duration of the cooling or warming effect provided by PCMs is limited. Gao et al. [126] showed that the cooling or warming effect duration decreased during high-intensity physical activity. They addressed the finite amount of PCM incorporated into the textiles and the rate of heat transfer is responsible for this limitation.

10.10.1.3 Imbalance Temperature Gradient

The temperature difference between the PCM and the human body influences the rate at which the PCM absorbs or releases heat. A larger temperature difference increases the heat transfer rate between the PCM and the body. This leads to faster heat absorption or release by the PCM, leading to quicker cooling or warming effects [127].

In hot conditions, the required temperature gradient is strongly encouraged to be at least 6 °C to produce a noticeable cooling effect. It is preferred to have at least a 12 ^{0}C (6 ^{0}C above the phase change temperature) temperature gap to gain the desired output. However, it is quite challenging to maintain this temperature difference throughout varying environmental conditions, considering both cooling and warming effects. For instance, to achieve an acceptable cooling rate, the skin temperature should be at least 34 °C or higher if the PCM has a melting or solidifying temperature of 28 °C, and it is quite easy to get the desired rate when the PCM-textile is worn in close contact with the body. However, it is not recommended to wear the same clothes that have PCM integrated into them when the temperature drops to 22 °C because it is cold enough for the body [128].

10.11 Future Directions

PCMs offer great opportunities to enhance the performance and comfort of textiles in various applications. Extensive research is required to mitigate challenges associated with PCM textiles, promote sustainable and eco-friendly approaches, and introduce smart functionalities into PCM textiles.

10.11.1 Advanced Encapsulation Techniques

The advancement of nanotechnology may lead to improved PCM encapsulation, longevity of PCM textiles, durability, performance, integration systems into textiles, etc. Researchers have initiated the development of nanocapsules in place of micro-capsules to improve the features and application area of PCM textiles [129]. Studies show that the nanoencapsulation technique can improve thermal comfort and be used as a PCM/plaster wallboard in buildings to store thermal energy [130].

10.11.2 Intelligent Textile Systems

The integration of smart functionalities into PCM textiles is becoming popular among researchers. By integrating sensors, actuators, and electronic components, these textiles can actively monitor and respond to changes in environmental conditions and user needs. As a result, this leads to the development of self-regulating garments that can adapt to different climates and physical activities [131].

10.11.3 Sustainable Phase Change Materials

The textile industry is exploring innovative materials to meet consumer demands and ensure sustainability within the supply chain. Bio-based phase change materials align with the growing demand for sustainable textiles. Studies have shown that bio-PCMs derived from natural sources, such as animal fats and plant oils, reduce reliance on synthetic materials and contribute to lower environmental impact [132, 133]. However, researchers are still working to improve their integration in textiles.

10.11.4 Cost-Effective Phase Change Material Integration

The cost of phase change materials and their integration into textiles is still a matter of concern. The widespread adoption of PCM textiles depends upon this factor. The upcoming challenges for researchers might be the development of low-cost PCM textiles with scalable manufacturing processes [37].

10.11.5 IoT-Enabled PCM Textiles

Integrating PCM textiles with the Internet of Things (IoT) has the potential to revolutionize the way we interact with clothing. A wearer can monitor their thermal comfort, track their health metrics, and receive personalized recommendations by connecting smart devices to the PCM-incorporated garments [134]. Therefore, there is a great prospect for future researchers to work on this type of application.

10.12 Conclusion

The incorporation of phase change materials (PCMs) into textiles presents a pivotal role in the aspects of sustainable and functional clothing. The thermoregulation property of PCM facilitates diverse applications, ranging from regular wear to advanced technical textiles. Different types of PCMs—organic, inorganic, and eutectic—each introduces specific attributes compatible with specific requirements of the end product. However, PCM textiles have both sustainable and harmful effects on the environment. Challenges include limited thermal conductivity, limited cooling and warming effect duration, and imbalance temperature gradient associated with PCM textiles. Since the consumer demand for advanced, eco-friendly textiles is growing, integrating PCMs presents significant opportunities for innovation and development within the textile industry, making them a valuable asset for future textile technologies. Therefore, more intensive study is required to functionalize more smart properties into PCM textiles, and more research is mandatory to improve the current limitations and ensure their long-term performance.

Declaration

The authors declare that they have no competing financial interests or personal relationships that could have influenced the work presented in this chapter. Generative AI and AI-assisted technologies were used solely to enhance the readability and language of the text. The authors have thoroughly reviewed and edited the content as necessary and take full responsibility for the final version of the publication.

References

1. Maughan R (2003) Impact of mild dehydration on wellness and on exercise performance. Eur J Clin Nutr 57(2):S19–S23
2. Prajapati DG, Kandasubramanian B (2020) A review on polymeric-based phase change material for thermo-regulating fabric application. Polym Rev 60(3):389–419
3. Ye S, Zhang Q, Hu D, Feng J (2015) Core–shell-like structured graphene aerogel encapsulating paraffin: shape-stable phase change material for thermal energy storage. J Mater Chem A 3(7):4018–4025
4. Shahid MA, Hossain MT, Hossain I, Limon MGM, Rabbani M, Rahim A (2024) Research and development on phase change material-integrated cloth: a review. J Ind Text 54:15280837241262518
5. Hossain MT, Shahid MA, Ali MY, Saha S, Jamal MSI, Habib A (2023) Fabrications, classifications, and environmental impact of PCM-incorporated textiles: current state and future outlook. ACS Omega 8(48):45164–45176
6. Ghaddar N, Kishore RA, Menberg K, Ndukwu MC, Hou H, Islam R, Sahin EB, Sett S, Mandal J (2024) Sustainable cooling solutions. One Earth 7(8):1315–1319
7. Pritom MM, Islam MA, Moshwan MM, Hossain A, Hasan MZ, Siddiquee MAB, Ayatullah Hosne Asif A (2024) Phase change materials in textiles: synthesis, properties, types and applications–a critical review. Textile Res J, 00405175241246822
8. Pasarkar NP, Yadav M, Mahanwar PA (2023) A review on the micro-encapsulation of phase change materials: classification, study of synthesis technique and their applications. J Polymer Res (Vol. 30): Springer Science and Business Media B.V.
9. Sharma A, Tyagi VV, Chen CR, Buddhi D (2009) Review on thermal energy storage with phase change materials and applications. Renew Sustain Energy Rev 13(2):318–345
10. Al-Yasiri Q, Szabó M (2021) Paraffin as a phase change material to improve building performance: an overview of applications and thermal conductivity enhancement techniques. Renew Energy Environ Sustain 6:38
11. Sarbu I, Dorca A (2019) Review on heat transfer analysis in thermal energy storage using latent heat storage systems and phase change materials. Int J Energy Res 43(1):29–64
12. Al-Abbasi O, Abdelkefi A, Ghommem M (2017) Modeling and assessment of a thermochemical energy storage using salt hydrates. Int J Energy Res 41(14):2149–2161
13. Wong-Pinto L-S, Milian Y, Ushak S (2020) Progress on use of nanoparticles in salt hydrates as phase change materials. Renew Sustain Energy Rev 122:109727

14. Gil A, Medrano M, Martorell I, Lázaro A, Dolado P, Zalba B, Cabeza LF (2010) State of the art on high temperature thermal energy storage for power generation. Part 1—Concepts, materials and modellization. Renew Sustain Energy Rev 14(1):31–55
15. Fernandez AI, Barreneche C, Belusko M, Segarra M, Bruno F, Cabeza LF (2017) Considerations for the use of metal alloys as phase change materials for high temperature applications. Sol Energy Mater Sol Cells 171:275–281
16. Bista S, Hosseini SE, Owens E, Phillips G (2018) Performance improvement and energy consumption reduction in refrigeration systems using phase change material (PCM). Appl Therm Eng 142:723–735
17. Gholamibozanjani G, Farid M (2021) A critical review on the control strategies applied to PCM-enhanced buildings. Energies 14(7):1929
18. Kazemi Z, Mortazavi SM, Shahmoradi Ghaheh F (2023) Developing a novel thermoregulating cotton fabric using inorganic eutectic phase change material. Cellulose 30(2):1287–1303
19. Ushak S, Gutierrez A, Galleguillos H, Fernandez AG, Cabeza LF, Grágeda M (2015) Thermophysical characterization of a by-product from the non-metallic industry as inorganic PCM. Sol Energy Mater Sol Cells 132:385–391
20. Dinesh R, Hussain SI, Roseline AA, Kalaiselvam S (2021) Experimental investigation on heat transfer behavior of the novel ternary eutectic PCM embedded with MWCNT for thermal energy storage systems. J Therm Anal Calorim 145:2935–2949
21. Su W, Darkwa J, Kokogiannakis G (2015) Review of solid–liquid phase change materials and their encapsulation technologies. Renew Sustain Energy Rev 48:373–391
22. Ge H, Li H, Mei S, Liu J (2013) Low melting point liquid metal as a new class of phase change material: an emerging frontier in energy area. Renew Sustain Energy Rev 21:331–346
23. Farid MM, Khudhair AM, Razack SAK, Al-Hallaj S (2004) A review on phase change energy storage: materials and applications. Energy Convers Manage 45(9–10):1597–1615
24. Rostami S, Afrand M, Shahsavar A, Sheikholeslami M, Kalbasi R, Aghakhani S, Shadloo MS, Oztop HF (2020) A review of melting and freezing processes of PCM/nano-PCM and their application in energy storage. Energy 211:118698
25. Mishra A, Shukla A, Sharma A (2015) Latent heat storage through phase change materials. Resonance 20:532–541
26. Karaipekli A, Biçer A, Sarı A, Tyagi VV (2017) Thermal characteristics of expanded perlite/paraffin composite phase change material with enhanced thermal conductivity using carbon nanotubes. Energy Convers Manage 134:373–381
27. Pauken M, Emis N, Watkins B (2007) Thermal energy storage technology developments. AIP Conference proceedings
28. Mondal S (2008) Phase change materials for smart textiles–An overview. Appl Therm Eng 28(11–12):1536–1550
29. Sarier N, Onder E (2012) Organic phase change materials and their textile applications: an overview. Thermochim Acta 540:7–60
30. Alay S, Alkan C, Göde F (2011) Synthesis and characterization of poly (methyl methacrylate)/n-hexadecane microcapsules using different cross-linkers and their application to some fabrics. Thermochim Acta 518(1–2):1–8
31. Nelson G (2002) Application of microencapsulation in textiles. Int J Pharm 242(1–2):55–62
32. Sarier N, Onder E (2007) The manufacture of microencapsulated phase change materials suitable for the design of thermally enhanced fabrics. Thermochim Acta 452(2):149–160
33. Loxley A, Vincent B (1998) Preparation of poly (methylmethacrylate) microcapsules with liquid cores. J Colloid Interface Sci 208(1):49–62
34. Sánchez L, Sánchez P, de Lucas A, Carmona M, Rodríguez JF (2007) Microencapsulation of PCMs with a polystyrene shell. Colloid Polym Sci 285:1377–1385
35. Sundberg EJ, Sundberg DC (1993) Morphology development for three-component emulsion polymers: Theory and experiments. J Appl Polym Sci 47(7):1277–1294
36. Kizildag N (2023) Smart textiles with PCMs for thermoregulation. In: Multifunctional phase change materials, pp 445–505. Elsevier

37. Itani M, Bachnak R, Ghaddar N, Ghali K (2019) Evaluating performance of hybrid PCM-fan and hybrid PCM-desiccant vests in moderate and hot climates. J Build Eng 22:383–396
38. Iqbal K, Khan A, Sun D, Ashraf M, Rehman A, Safdar F, Basit A, Maqsood HS (2019) Phase change materials, their synthesis and application in textiles—a review. J Textile Inst 110(4):625–638
39. Fabien S (2011) The manufacture of microencapsulated thermal energy storage compounds suitable for smart textile. Develop Heat Transfer 10:171–198
40. Uddin M, Zhu H, Hawlader M (2002) Effects of cyclic operation on the characteristics of a microencapsulated PCM storage material. Int J Solar Energy 22(3–4):105–114
41. Hawlader M, Uddin M, Khin MM (2003) Microencapsulated PCM thermal-energy storage system. Appl Energy 74(1–2):195–202
42. Deveci SS, Basal G (2009) Preparation of PCM microcapsules by complex coacervation of silk fibroin and chitosan. Colloid Polym Sci 287:1455–1467
43. Malekipirbazari M, Sadrameli S, Dorkoosh F, Sharifi H (2014) Synthetic and physical characterization of phase change materials microencapsulated by complex coacervation for thermal energy storage applications. Int J Energy Res 38(11):1492–1500
44. Onder E, Sarier N, Cimen E (2008) Encapsulation of phase change materials by complex coacervation to improve thermal performances of woven fabrics. Thermochim Acta 467(1–2):63–72
45. Pasarkar NP, Yadav M, Mahanwar PA (2023) A review on the micro-encapsulation of phase change materials: classification, study of synthesis technique and their applications. J Polym Res 30(1):13
46. Sarier N, Onder E (2012b) Organic phase change materials and their textile applications: an overview. In: Thermochimica Acta (Vol 540, pp 7–60)
47. Fabien S, The manufacture of microencapsulated thermal energy storage compounds suitable for smart textile. www.intechopen.com
48. Iqbal K, Khan A, Sun D, Ashraf M, Rehman A, Safdar F, Basit A, Maqsood HS (2019a) Phase change materials, their synthesis and application in textiles—a review. J Textile Inst 110:625–638. Taylor and Francis Ltd.
49. Berkland C, King M, Cox A, Kim K, Pack DW (2002) P recise control of PLG microsphere size provides enhanced control of drug release rate a a a b (Journal of Controlled Release, Issue. www.elsevier.com/locate/jconrel
50. Freitas S, Merkle HP, Gander B (2005) Microencapsulation by solvent extraction/evaporation: reviewing the state of the art of microsphere preparation process technology. J Control Release 102(2):313–332
51. Chen S, Tao L, Cai W, Liu H, Wang X, Zhang X (2021) Fabrication and characterization of electrospun poly (acrylonitrile-co-vinylidene Chloride) Copolymer/Poly (n-tetradecyl acrylate-co-n-hexadecyl Acrylate) Sheath/Core nanofiber-wrapped thermo-regulated filaments. ACS Appl Energy Mater 4(5):5359–5366
52. Yan Y, Li W, Zhu R, Lin C, Hufenus R (2021) Flexible phase change material fiber: a simple route to thermal energy control textiles. Materials 2021, 14, 401. In: s Note: MDPI stays neu-tral with regard to jurisdictional clai-ms in ….
53. Zhang X, Wang X, Tao X, Yick KL (2005) Energy storage polymer/MicroPCMs blended chips and thermo-regulated fibers. J Mater Sci 40:3729–3734
54. Zhang X, Wang X, Zhang H, Niu J, Yin R (2003) Effect of phase change material content on properties of heat-storage and thermo-regulated fibres nonwoven
55. Boh B, Knez E, Staresinic M (2005) Microencapsulation of higher hydrocarbon phase change materials by in situ polymerization. J Microencapsul 22(7):715–735
56. Bryant Y (1992) Fibers with enhanced, reversible thermal energy storage properties. Techtextil-symposium
57. Bryant YG, Colvin DP (1988) Fiber with reversible enhanced thermal storage properties and fabrics made therefrom. In: Google patents
58. Colvin DP, Bryant YG (1997) Thermally enhanced foam insulation. In: Google patents

59. Cox R (2001) FIBERS-Re-positioning acrylic fibers for the new millennium. Chem Fibers Int 51(2):118–120
60. Lennox-Kerr P (1999) 1,000 frames per second: revealing the secrets of meltblown nonwovens. Chem Fibers Int 49(2):166–167
61. Nuckols M (1999) Analytical modeling of a diver dry suit enhanced with micro-encapsulated phase change materials. Ocean Eng 26(6):547–564
62. Leskovšek M, Jedrinović G, Stankovič Elesini U (2004) Properties of polypropylene fibres with incorporated microcapsules. Acta Chim Slov 51(4):699–715
63. Zhang X-X, Wang X-C, Tao X-M, Yick K-L (2006) Structures and properties of wet spun thermo-regulated polyacrylonitrile-vinylidene chloride fibers. Text Res J 76(5):351–359
64. Jiang M, Song X, Ye G, Xu J (2008) Preparation of PVA/paraffin thermal regulating fiber by in situ microencapsulation. Compos Sci Technol 68(10–11):2231–2237
65. Juárez D, Ferrand S, Fenollar O, Fombuena V, Balart R (2011) Improvement of thermal inertia of styrene–ethylene/butylene–styrene (SEBS) polymers by addition of microencapsulated phase change materials (PCMs). Eur Polymer J 47(2):153–161
66. Lee E, Han S, Lee K-H, Lee J, Cho G (2018) Thermal properties of combat uniforms treated with microencapsulated octadecane and change in clothing microclimate via thermal manikin. J Textile Inst 109(5):585–595
67. Yan C, Yu Z, Yang B (2013) Improvement of thermoregulating performance for outlast/silk fabric by the incorporation of polyurethane microcapsule containing paraffin. Fibers and Polymers 14:1290–1294
68. Jantang S, Chaiyasat P (2018) High performance poly (methyl methacrylate-acrylic acid-divinylbenzene) microcapsule encapsulated heat storage material for thermoregulating textiles. Fibers and Polymers 19:2039–2048
69. Pause B (1995) Development of heat and cold insulating membrane structures with phase change material. J Coated Fabrics 25(1):59–68
70. Tao X (2015) Handbook of smart textiles. Springer Singapore
71. Shin Y, Yoo DI, Son K (2005b) Development of thermoregulating textile materials with microencapsulated phase change materials (PCM). IV. Performance properties and hand of fabrics treated with PCM microcapsules. J Appl Polymer Sci 97(3):910–915
72. Colvin DP, Bryant YG (1998) Protective clothing containing encapsulated phase change materials. ASME international mechanical engineering congress and exposition
73. Ghosh SK (2006) Functional coatings and microencapsulation: a general perspective. Functional Coatings: by polymer microencapsulation, 1–28
74. Pause B (2003) Nonwoven protective garments with thermo-regulating properties. J Ind Text 33(2):93–99
75. Sarier N, Onder E (2007) Thermal characteristics of polyurethane foams incorporated with phase change materials. Thermochim Acta 454(2):90–98
76. Kim J, Cho G (2002) Thermal storage/release, durability, and temperature sensing properties of thermostatic fabrics treated with octadecane-containing microcapsules. Text Res J 72(12):1093–1098
77. Li S, Wang H, Mao H, Li L, Shi H (2021) Enhanced thermal management performance of comb-like polymer/boron nitride composite phase change materials for the thermoregulated fabric application. J Energy Storage 40:102826
78. Demirbağ S, Aksoy SA (2016) Encapsulation of phase change materials by complex coacervation to improve thermal performances and flame retardant properties of the cotton fabrics. Fibers Polymers 17:408–417
79. Bryant YG, Colvin DP (1994) Fabric with reversible enhanced thermal properties. In: Google patents
80. Paul R (2014) Functional finishes for textiles: improving comfort, performance and protection. Elsevier
81. Shin Y, Yoo DI, Son K (2005a) Development of thermoregulating textile materials with microencapsulated phase change materials (PCM). II. Preparation and application of PCM microcapsules. J Appl Polymer Sci 96(6):2005–2010

82. Erkan G (2004) Enhancing the thermal properties of textiles with phase change materials. Res J Text Appar 8(2):57–64
83. Wi S, Seo J, Jeong S-G, Chang SJ, Kang Y, Kim S (2015) Thermal properties of shape-stabilized phase change materials using fatty acid ester and exfoliated graphite nanoplatelets for saving energy in buildings. Sol Energy Mater Sol Cells 143:168–173
84. Boh B, Knez E (2006) Microencapsulation of essential oils and phase change materials for applications in textile products
85. Barreneche C, Solé A, Miró L, Martorell I, Fernández AI, Cabeza LF (2013) Study on differential scanning calorimetry analysis with two operation modes and organic and inorganic phase change material (PCM). Thermochim Acta 553:23–26
86. Sánchez L, Lacasa E, Carmona M, Rodriguez JF, Sánchez P (2008) Applying an experimental design to improve the characteristics of microcapsules containing phase change materials for fabric uses. Ind Eng Chem Res 47(23):9783–9790
87. Khan SA, Khan SB, Khan LU, Farooq A, Akhtar K, Asiri AM (2018) Fourier transform infrared spectroscopy: fundamentals and application in functional groups and nanomaterials characterization. Handbook of materials characterization, 317–344
88. Liu H, Shen H, Zhang H, Wang X (2022) Development of photoluminescence phase-change microcapsules for comfort thermal regulation and fluorescent recognition applications in advanced textiles. J Energy Storage 49:104158
89. Oliveira FR, Fernandes M, Carneiro N, Pedro Souto A (2013) Functionalization of wool fabric with phase-change materials microcapsules after plasma surface modification. J Appl Polym Sci 128(5):2638–2647
90. Yan Y, Li W, Zhu R, Lin C, Hufenus R (2021) Flexible phase change material fiber: a simple route to thermal energy control textiles. Materials 14(2):401
91. Netzelmann U, Walle G, Lugin S, Ehlen A, Bessert S, Valeske B (2016) Induction thermography: principle, applications and first steps towards standardisation. Quant InfraRed Thermogr J 13(2):170–181
92. Liu H, Wang X, Wu D (2019) Innovative design of microencapsulated phase change materials for thermal energy storage and versatile applications: a review. Sustain Energy Fuels 3(5):1091–1149
93. Salaün F (2019) Phase change materials for textile application. Textile industry and environment, 1–27
94. Hossain MA (2023) Simultaneous thermoregulated and sensorial effect of smart textiles with artificial composite phase change materials (CPCM) incorporated with Carbon Nano Conductive Materials for special workers and extreme weather conditions. School of Fashions and Textiles, RMIT University, Victoria 3056, Australia (engr. anowar@ yahoo. com)
95. Shahid MA, Saha C, Miah MS, Hossain MT (2023) Incorporation of MPCM on cotton fabric for potential application in hospital bed sheet. Heliyon 9(6)
96. Hassabo AG, Eid MM, Mahmoud ER, Asser NA (2023) Innovation of smart knitted fabrics for functional performance of sportswear upon treatment using phase change material. Egypt J Chem 66(3):133–156
97. Zhang X (2001) Heat-storage and thermo-regulated textiles and clothing. Smart Fibres, Fabrics and Clothing 2001:34–57
98. Mondal S (2011). Thermo-regulating textiles with phase-change materials. In Functional textiles for improved performance, protection and health, pp 163–183. Elsevier
99. Ismail KA, Lino FA, Machado PLO, Teggar M, Arıcı M, Alves TA, Teles MP (2022) New potential applications of phase change materials: a review. J Energy Storage 53:105202
100. McCarthy LK, Di Marzo M (2012) The application of phase change material in fire fighter protective clothing. Fire Technol 48:841–864
101. Mattila H (2006) Intelligent textiles and clothing. Woodhead Publishing
102. Md Suhaib MK, Sanjay B, Nikhil S, Trivendra S, Sanjay C (2024) A comprehensive review on phase change materials: properties
103. Khan S, Kim J, Acharya S, Kim W (2021) Review on the operation of wearable sensors through body heat harvesting based on thermoelectric devices. Appl Phys Let 118(20)

104. Mohaddes F, Islam S, Shanks R, Fergusson M, Wang L, Padhye R (2014) Modification and evaluation of thermal properties of melamine-formaldehyde/n-eicosane microcapsules for thermo-regulation applications. Appl Therm Eng 71(1):11–15
105. Song C, Duan G, Wang D, Liu Y, Du H, Chen G (2021) Study on the influence of air velocity on human thermal comfort under non-uniform thermal environment. Build Environ 196:107808
106. F De Castro P, Minko S, Vinokurov V, Cherednichenko K, Shchukin DG (2021). Long-term autonomic thermoregulating fabrics based on microencapsulated phase change materials. ACS Appl Energy Mater 4(11):12789–12797
107. Siddiqui MOR, Sun D (2015) Computational analysis of effective thermal conductivity of microencapsulated phase change material coated composite fabrics. J Compos Mater 49(19):2337–2348
108. Salaün F, Devaux E, Bourbigot S, Rumeau P (2010) Thermoregulating response of cotton fabric containing microencapsulated phase change materials. Thermochim Acta 506(1–2):82–93
109. Abuzaid AI, Reichard G (2016) An assessment of utilizing phase change materials (PCM) towards energy performance in building enclosures. 3rd Residential building design & construction conference
110. Ahrari M, Khajavi R, Dolatabadi MK, Toliyat T, Rashidi A (2017) A review on application of phase change materials in textiles finishing. Int J Mater Metall Eng 11(5):400–405
111. Scheepers R (2022) The effect of laundering on the physical and thermal properties of phase change textile materials. In: Phase change materials-technology and applications. IntechOpen
112. Rossi RM, Bolli WP (2005) Phase change materials for the improvement of heat protection. Adv Eng Mater 7(5):368–373
113. Bland A, Khzouz M, Statheros T, Gkanas EI (2017) PCMs for residential building applications: a short review focused on disadvantages and proposals for future development. Buildings 7(3):78
114. Kosny J, Shukla N, Fallahi A (2013) Cost analysis of simple phase change material-enhanced building envelopes in southern US climates
115. Sisco Research Laboratories Pvt. Ltd. (SRL)—India (2024). https://www.srlchem.com/
116. Alibaba.com. Manufacturers, Suppliers, Exporters & Importers from the world's largest online B2B market- place. https://www.alibaba.com/
117. Faruk MO, Ahmed A, Jalil MA, Islam MT, Adak B, Hossain MM, Mukhopadhyay S (2021) Functional textiles and composite based wearable thermal devices for Joule heating: progress and perspectives. Appl Mater Today 23:101025
118. Peng Y, Cui Y (2020) Advanced textiles for personal thermal management and energy. Joule 4(4):724–742
119. Colvin DP, Colvin VS, Bryant YG, Hayes LG, Spieker MA (2000) Development of a cooling garment with encapsulated PCM. ASME Int Mech Eng Congress Exposition
120. Hamdan H, Ghaddar N, Ouahrani D, Ghali K, Itani M (2016) PCM cooling vest for improving thermal comfort in hot environment. Int J Therm Sci 102:154–167
121. La Guardia MJ, Hale RC, Harvey E, Mainor TM, Ciparis S (2012) In situ accumulation of HBCD, PBDEs, and several alternative flame-retardants in the bivalve (Corbicula fluminea) and gastropod (Elimia proxima). Environ Sci Technol 46(11):5798–580
122. Pan L, Bian W, Zhang J (2013) The effect of explicit solvent on photodegradation of decabromodiphenyl ether in toluene: insights from theoretical study. J Phys Chem A 117(25):5291–5298
123. Su Y, Fan Y, Ma Y, Wang Y, Liu G (2023) Flame-retardant phase change material (PCM) for thermal protective application in firefighting protective clothing. Int J Therm Sci 185:108075
124. Magendran SS, Khan FSA, Mubarak N, Vaka M, Walvekar R, Khalid M, Abdullah E, Nizamuddin S, Karri RR (2019) Synthesis of organic phase change materials (PCM) for energy storage applications: a review. Nano-structures & Nano-objects 20:100399
125. Abreu-Vieira G, Xiao C, Gavrilova O, Reitman ML (2015) Integration of body temperature into the analysis of energy expenditure in the mouse. Molecular Metabolism 4(6):461–470

126. Gao C (2014b) Phase-change materials (PCMs) for warming or cooling in protective clothing. In: Protective clothing, pp 227–249. Elsevier
127. Gao C, Kuklane K, Holmér I (2010) Cooling vests with phase change material packs: the effects of temperature gradient, mass and covering area. Ergonomics 53(5):716–723. https://doi.org/10.1080/00140130903581649
128. Gao C (2014a) Fashion and function: challenges faced by textiles incorporated with phase change materials
129. Nikpourian H, Bahramian AR, Abdollahi M (2020) On the thermal performance of a novel PCM nanocapsule: the effect of core/shell. Renew Energy 151:322–331
130. Maleki B, Khadang A, Maddah H, Alizadeh M, Kazemian A, Ali HM (2020) Development and thermal performance of nanoencapsulated PCM/plaster wallboard for thermal energy storage in buildings. J Build Eng 32:101727
131. Cherif C, Tran N, Kirsten M, Bruenig H, Vogel R (2018) Environmentally friendly and highly productive bi-component melt spinning of thermoregulated smart polymer fibres with high latent heat capacity. Express Polymer Letters 12(3)
132. El Majd A, Younsi Z, Youssef N, Belouaggadia N, El Bouari A (2023) Experimental study of thermal characteristics of bio-based textiles integrating microencapsulated phase change materials. Energy Build 297:113465
133. Simonsen G, Ravotti R, O'Neill P, Stamatiou A (2023) Biobased phase change materials in energy storage and thermal management technologies. Renew Sustain Energy Rev 184:113546
134. Tabor J, Chatterjee K, Ghosh TK (2020) Smart textile-based personal thermal comfort systems: current status and potential solutions. Adv Mater Technol 5(5):1901155
135. Applications, and Future Prospects [Comprehensive Review]. Int J Rese Publ Rev 5(1):998–1002. https://ijrpr.com/uploads/V5ISSUE1/IJRPR21654.pdf

Chapter 11
Evolution of Flexible Electronics in Textiles

Neha Rani Kumar

Abstract Devices that are bendable, stretchable, and conformable and made using flexible and conductive materials are referred to as flexible devices. This property allows them to maintain their functionality when subjected to mechanical stress. Various fields, like healthcare, energy storage, thermal management, etc., have witnessed remarkable advancements resulting from blending flexible electronics with textiles. The evolution of wearable electronics signifies a significant shift from rigid devices to adaptable, conformable solutions that seamlessly fit the human body's contours and movements. This transition can be attributed to innovations in the development of smart fibres and textiles that are lightweight, breathable, and possess stretchable nature. These materials are the foundation for weaving flexible and stretchable electronics, enabling innovations like soft sensors, breathable substrates, and conductive materials. However, despite these strides, challenges persist in effectively amalgamating scalable textile manufacturing with smart device integration. The need for scalable, diverse textile manufacturing technologies seamlessly integrating with smart devices is a significant area to explore. This chapter discusses the evolution of flexible electronics in textiles and provides insights into the future development of novel flexible textile structures and high-performance textile devices.

Keywords Flexible electronics · Stretchable materials · Wearable electronics · Smart textiles · Integration challenges

11.1 Introduction

The combination of electronics and textiles has led to a new era of innovation, where traditional fabrics are transformed into platforms that can sense, compute, and communicate [1]. This integration, known as smart or e-textiles, involves adding flexible electronics to textile materials. This has opened up new possibilities and

N. R. Kumar (✉)
Department of Chemistry, Dhemaji College, Dhemaji, Assam, India
e-mail: nehakumar0926@gmail.com

S. Biranje et al. (eds.), *Sustainable Finishing Techniques in Textiles*,
https://doi.org/10.1007/978-981-96-4860-3_11

applications in various fields, such as sports, healthcare, and fashion [2, 3]. Developing flexible electronics within textiles reflects an important shift from conventional rigid circuitry to adaptable, stretchable, and washable components. Early advances in the discipline, including the development of conductive polymers and carbon-based materials, laid the groundwork for flexible electronics capable of conforming to the contours of the human body or integrating discreetly into everyday clothing. Unprecedented conductivity is now possible with materials like graphene, conductive yarns, and nanowires, all of which retain the comfort and flexibility required for textile applications [4]. Advanced printing technologies (such as inkjet and screen printing) and creative knitting and weaving processes that directly incorporate electrical materials into textile structures exemplify how fabrication methods have evolved. These methods ensure textiles' durability and washability while preserving their aesthetic qualities [1]. The merging of electronics with textiles has unlocked numerous applications. Smart textiles with biometric sensors can monitor vital signs, providing personalized health insights and early disease detection [5]. Athletes in sports and fitness benefit from real-time performance data gathered by smart gear, which improves training effectiveness and reduces injury risk [6]. Fashion designers are exploring interactive garments that respond to environmental stimuli or user interactions, blurring the lines between technology and style [7]. Sustainable practices, such as recycling electronic components and reducing energy consumption during production, are essential for mitigating the ecological footprint of smart textiles. E-textiles, which are sufficiently soft to wear and smart enough to interact with their surroundings, hold great potential to transform healthcare from a disease-management model to one focused on disease prevention and enhancing well-being. They can be discreetly and comfortably attached to the body or clothing to provide intelligent assistance. Sustainability remains a cornerstone of the evolution of flexible electronics in textiles [8]. The field of e-textiles has gained rapid momentum since the 2000s, aided by the advancement of the Internet of Things, intelligent hardware, big data, and potential applications in the military, healthcare, energy, etc. [9]. E-textiles are ideal for wearable bioelectronics due to their many advantages, including flexibility, lightweight, breathability, and ease of integration with conventional clothing [10]. These electronics can be worn externally as skin-interfaced devices or internally as implantable devices. Regular and continuous real-time monitoring of various human health indicators—such as heart rate, blood pressure, glucose level, and body temperature—can noninvasively alert users and healthcare providers to abnormal or urgent conditions. This enables cost-effective, personalized healthcare at any time and place [11]. The field of wearable electronics has witnessed significant progress in developing new materials, processes, and sensing mechanisms. However, despite progress, most published results in e-textiles are still far from being commercially adopted for practical applications. The remaining challenges include materials, manufacturing techniques, integration, testing standards, and wearable technology methods [12]. This chapter discusses the historical development, fabrication techniques and components, integration methods, applications, sustainability considerations, and future prospects of flexible electronics in textiles. It provides a thorough overview of how these technologies are transforming industries and paving the way towards a more connected and sustainable future for smart textiles, using examples of several products that are finding practical applications.

11.2 Historical Insights

The synergistic progress in the area of traditional textiles and advanced electronic technologies has paved the way for continuous innovation in the area of smart textiles. Here, we look into important milestones in the evolution of smart textiles, emphasizing noteworthy discoveries and the individuals and institutions that have pushed this area forward [13–15].

11.2.1 Early Innovations and Foundations

The fundamental materials required to create conductive threads, fabrics, and e-textiles have existed for more than a millennium; they were present as early as the 1600s. For generations, artists have twisted delicate metal foils, frequently made of gold or silver, around fabric threads. Early examples of conductive threads in fashion can be seen in the gold-wrapped embroidery on several of Queen Elizabeth I's dresses [16].

11.2.2 The Advent of Electrical Integration

With the advent of electricity in the late nineteenth century, designers and engineers began fusing electrical components into jewelry and clothes. This era saw the creation of illuminated and motorized accessories, such as necklaces, hats, brooches, and costumes. For instance, the Electric Girl Lighting Company provided entertainment services with young ladies decked out in evening gowns studded with lights [17, 18]. In 1986, the Museum of Contemporary Craft in New York City held an innovative Body Covering exhibition that examined the link between technology and textiles. The main attraction of this exhibition was space suits and clothing that could inflate, deflate, light up, and regulate temperature [19]. Designer Diana Dew displayed alarm-sounding belts and electroluminescent party dresses [20]. In the 1990s, inventor Harry Wainwright made another innovation by creating the first fully animated sweatshirt using fibre optics, LEDs, and a microprocessor to control animation frames, resulting in a full-colour cartoon on apparel. In 1997, Wainwright produced the first computer numerical controlled (CNC) machine that automatically inserted fibre optics into flexible textiles, a development that helped create animated coats for Disney Parks. This was just one of Wainwright's many discoveries [21].

11.2.3 Technological Advancements and Commercialization

In the 1990s, researchers from MIT, including Sandy Pentland, Thad Starner, and Steve Mann, started creating wearable computers [22]. This work evolved into more integrated solutions with Maggie Orth and Rehmi Post exploring digital electronics and conductive fabrics, leading to embroidered electronic circuits. With the introduction of Moondial in 1998, Sabine Seymour combined "silicon and style" to create fashion-tech hybrids. Harry Wainwright's animated coats garnered attention that same year, which prompted partnerships with MIT researchers and more wearable technology breakthroughs [18]. The 2000s saw significant advances in the commercialization of smart textiles. The launch of resources like E-textile Lounge and companies like Plug and Wear offered materials for knitting and sewing conductive fabrics. Notable developments were the Motherboard shirt from Georgia Tech and the Lilypad microcontroller for textiles created by Leah Buechley. The production of interactive clothing was encouraged by the development of the Lilypad, a microcontroller created especially for textiles by Leah Buechley in 2007. Forster Rohner launched the Climate Dress in 2009, exhibiting cutting-edge embroidery methods for the integration of technology into textiles [18, 21].

11.2.4 Recent Innovations and Industry Collaboration

A period of enhanced industry collaboration and advanced research in smart textiles was observed in the 2010s and 2020s. Important innovations included the Haute Tech Lab at Drexel University, the Midi Controller Jacket by Machina, and the stretchable conductive ink by DuPont. Companies like Bebop Sensors and Google's Project Jacquard explored commercial applications for e-textiles. In 2015, notable collaborations occurred, such as Ralph Lauren and OMSignal's Polotech Shirt and Google's Project Jacquard, showing the potential of smart fabrics in everyday design. In 2016, the Department of Defense and M.I.T. announced a $302 million agreement, highlighting the strategic necessity of creating new fabrics for defense applications. In a report offering predictions for wearable technology, IDTechEx highlighted the expanding use of conductive threads, inks, and textiles [23]. Xenoma introduced the E-Skin, a smart shirt embedded with 14 stretchable sensors that can detect motion and vital signs, with applications encompassing fitness tracking, rehabilitation, and gaming. Hexoskin released an updated version of its biometric smart shirts capable of providing detailed health metrics, including heart rate variability, respiratory rate, and sleep patterns. The information gathered is beneficial for clinical research as well as for monitoring personal health. Myant introduced the Skin line of smart apparel, consisting of undergarments with built-in sensors to track activity levels, respiration patterns, and heart rate. The apparel connects data and notifications about health with a mobile application. DuPont unveiled Intexar Heat, a thin, stretchable heating technology that can be embedded into fabrics. This innovation is particularly useful for outdoor clothing, providing adjustable warmth to the wearer without bulky layers [24]. The second generation of the Jacquard smart jacket, released by Google and Levi's, featured improved functionality, more responsive touch-sensitive fabric, extended battery life, and methods to control connected devices. The integration of electronics with textiles has also been beneficial for the space sector. For example, researchers from MIT developed BioSuit, a skin-tight spacesuit that offers an improvement over modern gas-pressurized spacesuits. It can provide astronauts with enhanced mobility and real-time health monitoring. The BioSuit's flexible design significantly improved traditional rigid spacesuits [25]. Lumenus unveiled a line of smart clothes with integrated LED lights to enhance pedestrian and cycling safety [26]. In conjunction with a smartphone app, the clothing can provide turn signals and brake lights, enhancing safety and visibility in poor light. Researchers have made significant advancements in creating flexible, printable electronics that may be incorporated into fabrics. These developments make it possible to integrate more intricate and robust electronic circuits into textiles, improving their applications. Nextiles launched a smart fabric platform that uses custom threads to measure motion and biomechanics [27]. With its precise insights into muscle activity and mobility, this technology is very helpful for physical therapy and sports performance analysis. During the COVID-19 pandemic, Myant developed textiles capable of monitoring vital body signs such as temperature and respiratory rate, thus assisting in the early diagnosis of symptoms and remote monitoring of patients [28]. Energy-harvesting materials capable of harnessing the wearer's movements to generate power have made significant advancements. These fabrics integrate piezoelectric or triboelectric materials to convert kinetic energy into electrical energy, powering embedded sensors and devices [29]. Researchers from MIT and Harvard have created smart bandages that track the healing of wounds using built-in sensors [30]. These bandages can provide real-time data on infection, inflammation, and tissue regeneration, improving patient outcomes and reducing recovery times. Another significant leap in the field of e-textiles took place with the integration of artificial intelligence with smart textiles. AI algorithms embedded in fabrics are capable of analysing data from sensors to provide personalized health insights, enhancing health monitoring. Advanced sensors integrated with textiles are also capable of monitoring air quality. These fabrics can detect pollutants such as particulate matter and volatile organic compounds, providing real-time feedback to the wearer and promoting a healthier living environment. These breakthroughs demonstrate the rapid growth and expanding range of uses for smart textiles, underscoring their potential to completely transform several sectors, including sports, fashion, healthcare, and environmental monitoring [31]. The focus on sustainability led to the development of eco-friendly smart textiles made from biodegradable and recycled materials. These textiles integrate advanced sensors and circuits while minimizing environmental impact, aligning with global sustainability goals. Future-focused research aims to improve smart textiles' functionality, dependability, and sustainability. It is anticipated that developments in printable electronics, wearable computers, and nanomaterials will enhance the potential of electronic textiles, opening up new uses and functionalities. The integration of artificial intelligence and machine learning algorithms with textile-based sensors holds significance in the healthcare industry [32]. The development of smart textiles has been fuelled by advances in manufacturing, interdisciplinary cooperation, and materials research. Each milestone, from conductive threads to sophisticated wearable electronics, has contributed to the overall evolution of electronic textiles, with potential applications in healthcare, sports, fashion, and beyond.

11.3 Materials Employed

Integrating electronics into textiles demands materials that deliver electrical conductivity while ensuring flexibility, durability, and comfort. Amongst the other properties that are to be considered, some important characteristics include thermal and mechanical properties, thermal and chemical resistance, density and weight, interconnections with traditional wires, heat removal efficiency, reliability, durability, etc. [33]. Significant advancements in materials science have led to the development of diverse materials specifically tailored for flexible electronics in textiles [34]. Below is a detailed overview of key functional materials used in the smart-textile sector (Fig. 11.1).

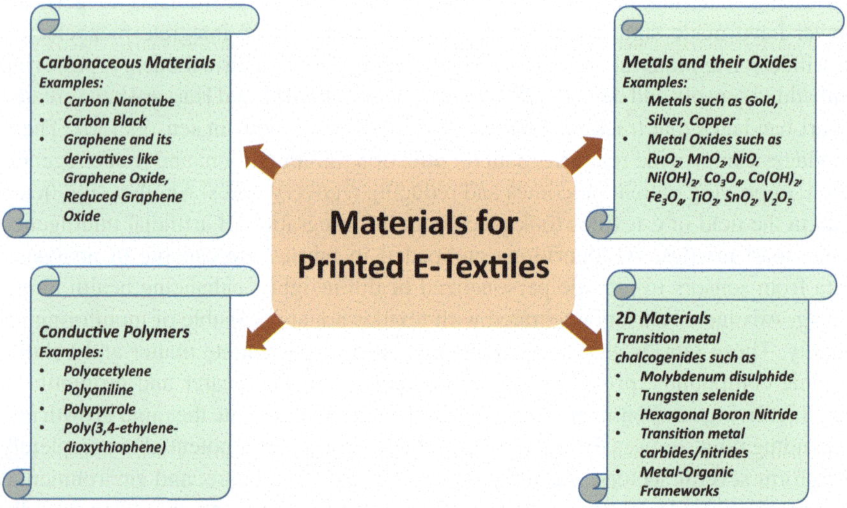

Fig. 11.1 Materials for printed e-textiles

11.3.1 Carbon-based Materials

Carbon-based materials are fundamental to the development of smart and flexible electronics due to their exceptional electrical, mechanical, and thermal properties. They are also easy to process, cheap, and compatible with other materials. As these materials offer a unique combination of high conductivity and robustness, they have emerged as perfect candidates for integration into textiles where durability, flexibility, and stability are essential. The carbonaceous materials that are widely used include carbon nanotubes (CNTs), carbon black, activated carbon, graphene, graphene oxide, reduced graphene oxide, etc. CNTs which possess cylindrical nanostructures and are created from rolled graphene sheets can exist in two forms: single-walled carbon nanotubes (SWCNTs) and multi-walled carbon nanotubes (MWCNTs). These structures can offer both flexibility and strength when incorporated into textiles either as coatings or woven within fibres. These properties render CNTs potential candidate for use in artificial muscles, actuators, and lightweight electromagnetic shields, presenting intriguing opportunities for nanoscale electronic applications in e-textiles [35]. Carbon black, a paracrystalline carbon also exhibits excellent conductive and electro-catalytic properties and high porosity making it efficient and economical for sensor fabrication [36]. Activated carbon, known for its substantial specific surface area, serves as an electrode material in energy storage devices like supercapacitors. The compatibility and high surface area of these materials facilitate their application in various wearable electronic devices [37]. Graphene consists of a monolayer of carbon atoms arranged in a two-dimensional honeycomb pattern. It is widely used in electronic applications owing to its exceptional mechanical, thermal, and electrical properties. Graphene can be applied to textiles as a coating or incorporated into fibres, providing extraordinary flexibility and strength. It can chemically bind with textiles which enhances its potential for use in wearable smart e-textiles such as sensors and energy storage equipment [38]. Another derivative of graphene, i.e., graphene oxide offers robust electronic, mechanical, and optical properties, along with a large surface area, high stability, chemical functionalization capability, and a layered structure. Graphene oxide can act both as a semiconductor or an insulator, which broadens its application in the textile industry [39].

11.3.2 Conductive Polymers

Conductive polymers (CPs) are organic polymers capable of conducting electricity through a conjugated bond system along their polymer chain. They are extensively researched for wearable electronics due to their reversible faradaic redox nature, high charge density, and cost-effectiveness compared to metal oxides. CPs can be easily processed and applied to textiles, making them versatile materials for flexible electronics [1, 40]. Polyaniline (PANI), an intrinsically conductive polymer, is noted for

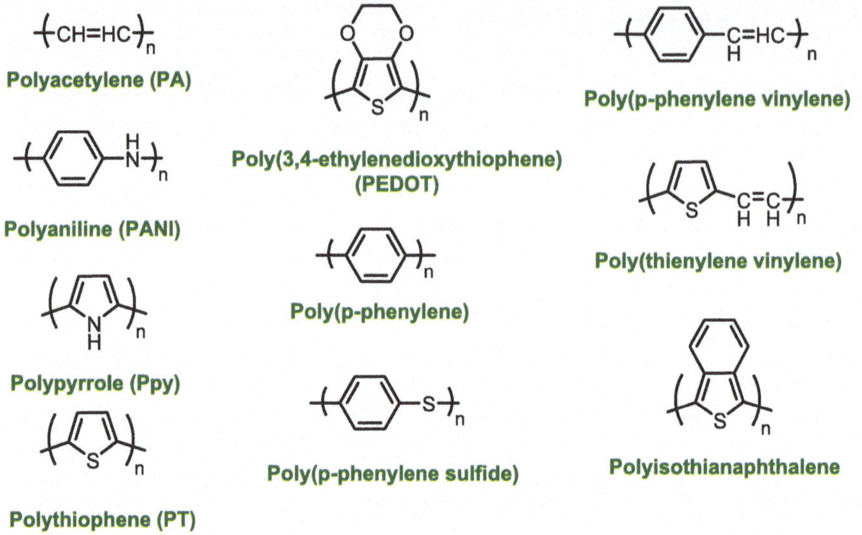

Fig. 11.2 Commonly employed conducting polymer in smart textiles

its high conductivity, specific capacitance, and flexibility. It is used in flexible electronics, energy storage, and conversion devices. PANI can undergo multiple redox reactions, making it suitable for applications requiring high charge storage capacity. However, its instability in certain environments often necessitates combining it with other materials to improve its performance. Polypyrrole (PPy) is another conjugated polymer known for its large theoretical capacitance, good redox properties, and high environmental stability. It is biocompatible and non-toxic, making it apt for wearable electronics. PPy is frequently blended with other polymers or used to form composites to enhance its mechanical properties and processability. These composites can exhibit electrical characteristics akin to metals or semiconductors. Poly(3, 4-ethylenedioxythiophene) (PEDOT) is widely utilized due to its superior conductivity and processability. In its oxidized state, PEDOT is highly conductive and serves as an electrode material in wearable electronics. Although its low stability can be a drawback, blending PEDOT with polystyrene sulfonate results in a highly conductive and practical electrode material for wearable electronics, exhibiting both ionic and electronic conductivity (Fig. 11.2).

11.3.3 Metals and Metal Oxides

The high conductivity of metals and metal oxides makes them potential conductive components in printed electronics. The excellent electrical properties displayed by metals like gold and silver, and the variable oxidation states displayed by metal oxides make them suitable for applications such as energy storage and sensors [41]. The high electrical conductivity and compatibility with textile substrates make silver nanoparticles excellent conducting materials in the domain of e-textiles. They can be applied using multiple printing techniques, including inkjet printing, or integrated into textile fibres during production [42]. However, its commercial application is limited due to its high cost. Copper presents a cost-effective alternative with slightly lower electrical conductivity. However, copper nanoparticles are susceptible to oxidation which hinders their potential. But their lower cost and high conductivity make them ideal candidates for further research [43]. Gold nanoparticles are also widely utilized in high-performance applications owing to their high electrical conductivity, ductility, and reliable performance. Their stability and resistance to oxidation also make gold nanoparticles suitable for use in wearable electronics [44]. Metal oxides, such as ruthenium dioxide (RuO_2), manganese dioxide (MnO_2), nickel oxide (NiO), and nickel hydroxide [$Ni(OH)_2$], are recognized for their high theoretical specific capacitance, good thermal stability, and high electronic conductivity [45]. These materials are employed in supercapacitors and other energy storage devices, albeit with challenges like high production cost and poor conductivity that need to be addressed for practical applications.

11.3.4 2D Materials

The discovery of graphene has led to the discovery of a variety of graphene-like two-dimensional materials (2DM) with exceptional properties [46]. These materials, characterized by their single or few atomic layer thickness, possess unique electronic, mechanical, and thermal properties, making them apt for next-generation electronics and wearable applications [47]. Transition Metal Dichalcogenides, such as MoS_2, exhibit properties comparable to graphene and are suitable for electronic devices due to their good electrochemical properties [48]. They can be synthesized using varied methods and are increasingly considered for flexible electronics integration. Hexagonal boron nitride (h-BN), an isomorph of graphene, offers excellent mechanical robustness, thermal stability, and chemical inertness [49]. While it is an insulator, its properties can be adjusted through doping, substitution, or hybridization, making it useful in applications like electrodes and energy storage. MXenes, a family of two-dimensional metal carbides and nitrides, possess unique properties, including high conductivity, large surface area, and good electrochemical stability. They are utilized in numerous applications, such as sensors, energy storage, and wearable electronics, and are potential candidates for next-generation electronic applications. These materials, with their unique properties, are crucial for developing printed e-textiles, enabling the creation of flexible, durable, and highly efficient electronic textiles [50].

11.4 Textile Substrates for Printed E-textiles

Flexible substrates are crucial in integrating electronic components into textiles, maintaining alignment with the fabric's mechanical properties and comfort levels. The robust thermal stability (withstand temperatures up to 400 °C) and mechanical flexibility of polyimide (PI) make it an ideal candidate for creating flexible circuits in challenging environments. Polyethylene terephthalate (PET) is also widely used as a substrate in wearable sensors and flexible displays attributed to its chemical resilience, transparency, ease of fabrication, etc. Textile materials such as nylon, cotton, and polyester can also function as substrates when blended with conductive fibres, such as silver-coated threads or coated with conductive polymers like polyaniline or PEDOT. This enables them with electronic properties [51]. Adhesives and functional coatings are necessary for the long-term durability and effective integration of flexible electronics into textiles. Barrier films like parylene coatings offer outstanding moisture resistance and biocompatibility, which are crucial for wearable medical devices. Silicone-based adhesives offer robust bonding without compromising comfort or functionality by preserving the stretchability and flexibility of electrical components [52]. In this context, materials science is advancing with an emphasis on developing stretchable electronics with 100% elastomeric substrates, biocompatible materials for medical implants, such as silk fibroin, and self-healing materials with internal or external healing processes to repair small damages. These advancements will lead to a significant increase in the longevity and dependability of electronic textiles.

11.4.1 Fabrication Techniques

The innovative properties displayed by smart textiles can be attributed to varied type of advanced fabrication techniques which enable the seamless integration of electronic components such as sensors, actuators, and conductive pathways into textiles. These advanced fabrication techniques include printing, electrospinning sewing, coating, weaving, 3D printing, etc. Each method is characterized by a unique set of advantages in terms of durability, flexibility, and performance, contributing to the development of wearable electronics that are both functional and comfortable [53]. An overview of some key fabrication techniques in the smart textile sector and their impact on the future of wearable technology are discussed below.

11.4.1.1 Printing Techniques

Printing is an efficient method for integrating electronic elements into textiles, facilitating the creation of intricate designs and working circuits. The screen Printing Technique is an effective method for the application of conductive inks and the creation of flexible, large-area circuits [54]. This technique involves pushing ink through a mesh screen onto the fabric, and the development of stretchable and conductive inks involves using materials like silver nanoparticles, graphene, etc. They not only improve the durability but also enhance the conductivity of printed circuits. Inkjet printers deposit high-precision ink droplets onto textiles, allowing for complex patterns and customizable designs. Conductive inks, including carbon-based and metallic nanoparticles, print flexible electronic circuits. Printed electronics are now reliable and efficient because of the developments in ink formulation technology [55].

11.4.1.2 Electrospinning

Nanofibre mats created by electrospinning can be used as substrates or functional layers in smart textiles. An electric field is applied to a polymer solution, resulting in fine fibres being ejected through a nozzle. It is possible to embed conductive or sensing materials into nanofibres, thus providing an extensive surface area for interacting with other textile components besides enhancing their electronic properties and flexibility. This technique permits the integration of nanomaterials such as carbon nanotubes or metallic nanoparticles, which improve conductivity and sensor performance [56].

11.4.1.3 Sewing and Embroidery

Flexible circuitry can be created by employing sewing and embroidery techniques, effectively integrating conductive threads into the fabric structure. Textile conductive paths are created using threads made from materials like silver or stainless steel for sewing or embroidery. Electronic components such as actuators and sensors can be directly integrated. This method offers a robust and adaptable way to make wearable electronics since the threads resist stretching and washing. The performance and dependability of embroidered circuits have been increased recently by using high-conductivity yarns and improved stitching methods [57].

11.4.1.4 Coating and Laminating

Coating and laminating techniques are now widely employed for applying thin layers of conductive or functional materials onto textiles to enhance their electronic properties [58]. The Chemical Vapor Deposition (CVD) technique deposits thin films of conducting materials, like graphene or metallic oxides, onto fabric surfaces [59]. The advantages of this method include uniform coverage and high adhesion that assists in maintaining the integrity of electronic components. Another important technique similar to CVD is sputtering, which involves depositing thin film from a target material onto a substrate. This technique is used to apply metallic coatings that enhance conductivity besides protecting underlying electronics. The main application of coatings and laminations is to create flexible sensors, thus augmenting the durability of smart textiles against environmental factors such as moisture and abrasion.

11.4.1.5 Weaving and Knitting

Weaving and knitting are important techniques in fabric production, each with its unique sets of advantages. Weaving involves interlacing warp and weft yarns to produce structured and durable fabrics. These fabrics have many applications, including fashion, home furnishings, and industrial textiles. Knitting involves interlooping yarns, and the fabric produced is breathable, inherently stretchable, and comfortable. Such fabrics are suitable for activewear and hosiery clothing. Both weaving and knitting techniques can produce diverse and intricate patterns, thus making the techniques versatile and efficient. Compared to others, the superiority of these techniques encompasses enhanced durability, comfort, and flexibility, making them indispensable for addressing varied consumer and industrial needs [60].

11.4.1.6 3D Printing

3D printing has significantly revolutionized several sectors, including textiles, by facilitating rapid prototyping, customization, and the creation of intricate designs. In the textile sector, 3D printing is used to produce complex patterns, seamless clothing, and unique textile structures, which cannot be accomplished by conventional techniques. The production of bespoke clothing and textiles customized to individual body measurements can be made possible by 3D printing technology. This also minimizes waste and enhances sustainability. Amongst the several advantages of 3D printing technology in the textile sector, some important advantages include flexibility of design, lesser waste generation, faster production times, easy creation of intricate patterns and geometries, enhanced aesthetic appeal, enhanced sustainability, and more innovative products [61]. Revealed at Amsterdam Fashion Week in 2010, Iris van Herpen's groundbreaking Crystallization top is recognized as one of the first 3D-printed couture garments ever exhibited on a runway. This milestone has opened the door to more sustainable and inventive approaches in the textile industry, highlighting the profound impact of this technology. Besides the different fabrication techniques discussed above, hybrid fabrication techniques can be used to enhance the performance and functionality of smart textiles. In the hybrid fabrication method, two or more fabrication techniques discussed above are integrated to allow for the optimization of electronic features and durability. Hybrid approaches offer a versatile solution for advanced smart textile applications, addressing challenges such as flexibility, washability, and integration of multiple components. These fabrication techniques have propelled the development of smart textiles by offering numerous ways to integrate electronic components and textile materials, leading to improved performance, flexibility, and durability.

11.4.2 Components of E-textiles

Smart textiles seamlessly merge several components to blend electronic capabilities with textile comfort. Essential to their function are interconnects and communication elements like wires and antennas that facilitate data transfer and power among components, computers, and users. These interconnects can be produced through extrusion or by utilizing conductive threads in embroidery. Antennas crafted from conductive threads or fabrics support wireless communication and energy transfer [62]. Interconnects must endure wear and tear, puncture, washing, and folding to avoid failure. Traditional methods like hot bar soldering and insulation displacement connections are prone to corrosion or short circuits over time. In contrast, newer methods, such as conductive inks and threads, provide superior flexibility and durability during washing. Other interconnection techniques include sewing, sputtering, soldering, and snaps, with the selection depending on machine compatibility and thread flexibility [63]. Another key aspect of smart textiles is electronic sensors and actuators, which can monitor movement, physiological states, and environmental conditions. Movement sensors, including inertial motion, optical, and strain sensors, transform mechanical deformation into electrical signals. Strain sensors can be resistive or capacitive, made from conductive materials through screen printing, sewing, knitting, or fabric layering [64–66]. However, continuous straining can introduce non-linearity and signal drift, diminishing device longevity. Textile-based sensors can assess various health parameters, including pH levels, sodium, conductivity, and hydration during physical activity, and can detect immune responses and neurodegenerative conditions [67, 68]. These sensors are embedded into textiles using embroidering, patterning, or laminating electrodes. Power and energy generation components are crucial for sustaining the electronic parts of smart textiles over their lifespan [69]. Batteries, ideally replaceable and rechargeable, often face challenges due to their bulk and rigidity. Innovations in thin, flexible, and discreet batteries created by embroidering or printing with conductive materials present promising solutions [70]. Energy can be harvested from light, body heat, motion, and friction through technologies like photovoltaic cells, thermoelectric generators, and piezoelectric materials. Flexible thermoelectric generators embedded in textiles can convert body heat into electrical energy. Solar cells, including dye-sensitized, perovskite, and polymer variants, can be integrated to capture solar energy. Human motion can generate energy through triboelectric or piezoelectric methods that transform mechanical friction or pressure into electricity [71, 72]. The central processing unit (CPU) serves as the control centre of the smart textile system, managing operations, processing data, and storing information [73]. Transistors, characterized by their electron gate, source, and drain, are created using traditional elements, soft lithography, or evaporation techniques. While it is feasible to develop textile-based computers, most applications currently depend on conventional CPUs for their reliability and efficiency. Smart textiles integrate these elements to provide superior functionality, comfort, and adaptability across various uses, including health monitoring and wearable electronics.

Continuous innovation in materials science and engineering propels the advancement of more sophisticated and dependable smart textile technologies.

11.4.3 Applications

Smart textiles represent an innovative convergence of technology and textiles. The technological components embedded in textiles include sensors, actuators, microcontrollers, and conductive threads. These smart textiles have various applications encompassing several domains, such as healthcare, sports, fashion, military, consumer electronics, and industrial sectors [74]. Smart textiles embedded with sensors have the potential to monitor vital indicators such as heart rate, blood pressure, body temperature, and breathing rate. This allows for continuous health data monitoring in real-time, even in environments outside of clinical settings. For instance, clothes equipped with temperature sensors may measure changes in body temperature, while smart shirts with integrated electrocardiogram (ECG) sensors can monitor heart activity. These textiles also aid in physical therapy by electrical stimulation of muscles. Smart textiles in sports are of great significance to the players in tracking physiological information like heart rate, respiration rate, and oxygen level. They can also prevent injuries by analysing movement patterns and providing corrective feedback on posture and impact detection. The fashion industry embraces smart textiles to create interactive clothing that responds to environmental stimuli or user input [75, 76]. The growth of the e-textile market in the military sector is attributed to the concerns over national security fuelled by products that offer crucial capabilities besides safeguarding soldiers. These materials often offer advantages such as communication abilities and protection against external threats. E-textiles are also vital in aiding worker safety by continuously monitoring health factors and environmental conditions in the industrial sector. Monitoring worker movements and exertion levels also optimize workflows and increase efficiency, reducing overexertion and related mishaps. Smart textiles are used in the home and lifestyle industry to improve comfort and convenience by integrating smart home technologies. With features like integrated heating, cooling, or massaging capabilities, these textiles can regulate lighting, sound, and temperature according to user preferences.

Furthermore, smart textiles connect human intentions with machine activities and act as a medium for human–computer interaction. Innovations such as 3D-printed elastomer networks or tactile sensors created by layering conductive fabrics improve soft objects for interactive applications [77]. Tactile learning is enhanced by electronic embroidered books, while "sonification" is the process of turning non-audio information into sound [78]. Smart patches can wirelessly control machines using triboelectric interactions or Micro-Electro-Mechanical Systems (MEMS)-based clenching motion detectors [79]. Incorporating strain sensors into glove-based devices can enhance their control during tasks such as driving, while highly conductive yarn-based touch-sensitive sleeves allow for accurate device operation.

Smart textiles are also an important medium for promoting human interactions, bridging the gap between interactivity and interconnectivity. An example of this is a gown bodice. Such systems can change color in response to the wearer's brainwaves, externalize mood, and encourage interaction. Electronic textiles are of immense significance and consequently have gained interest from government agencies, academic researchers, and the industrial sector, thus reflecting the potential of these revolutionary materials to enhance functionality, improve safety, and greater user interaction. The table below mentions several examples of smart textiles that have gained recognition, illustrating the diverse applications and advancements in this field (Table 11.1).

11.4.4 Sustainability Considerations

As discussed before, e-textiles offer numerous innovative applications. However, the sustainability of these products is a crucial factor to be considered. To ensure that e-textiles positively contribute to environmental sustainability, several elements need to be addressed, ranging from the materials used in manufacture to the lifecycle and disposal at the end of the product's life. A rapid surge in the market of smart textiles capable of reacting to external stimuli has been observed. Such textiles are of great significance in healthcare, sportswear, automotive applications, etc., and are generally obtained through complex combinations of traditional methods and advanced technologies like coating and inkjet printing. Development necessitates collaboration between consumers and manufacturers with an emphasis on sustainability through recyclable materials [80, 81]. Advancements in production technologies, communication systems, and research are essential for creating durable, sustainable smart textiles that enhance quality of life and meet future needs [82].

Waste generation is mostly caused by consumer habits, with the biggest consumption occurring in the initial stages of product use. Recycling textiles is challenging due to fibre blends and the extensive use of chemicals further complicates production. A significant amount of e-waste is generated due to the short lifespan of electronic devices, and combined with textile waste; the problem gets more severe as proper recycling methods are not adopted. Adopting and enhancing sustainability in the case of smart textiles will be an environment-friendly practice due to reduced waste, energy consumption, and material usage [83–85]. The sustainability considerations for textiles in flexible electronics are discussed below.

- **Sustainable Materials**
  - o The use of biodegradable materials such as polylactic acid (PLA), polycaprolactone (PCL), polyglycolic acid (PGA), polybutylene succinate (PBS), polyhydroxybutyrate (PHB), polytrimethylene carbonate (PTMC), polyethylene oxide (PEO) etc. and natural fibres such as cotton, wool, silk etc. should be increased and more researched upon.
  - o Recycled materials can be integrated into textiles and electronic components.

Table 11.1 Examples of some practical applications[24]

Name of material	Company producing the same	Applications
Nadi X yoga pants	Wearable X	Yoga posture correction
OMsignal biometric smartwear	OMsignal	Health monitoring, fitness tracking
Hexoskin smart shirts	Hexoskin	Health monitoring, athletic performance tracking
Athos training apparel	Athos	Muscle activity tracking, fitness performance
Sensoria fitness socks	Sensoria	Running metrics, gait analysis
Neopenda smart baby hat	Neopenda	Newborn vital signs monitoring
Levi's commuter trucker jacket	Levi's, google	Smartphone connectivity, navigation
Ambiotex smart shirt	Ambiotex	Health monitoring, fitness tracking
Ralph lauren polotech shirt	Ralph Lauren, OMsignal	Fitness tracking, biometrics
HexoSkin arctic smart shirt	HexoSkin	Cold weather performance monitoring
Heddoko smart compression wear	Heddoko	Biomechanical analysis, injury prevention
Samsung NFC suit	Samsung, Rogatis	NFC connectivity, access control
Athlete recovery sleepwear by under armour	Under Armour	Sleep quality improvement, recovery
Solar charged jacket by Vollebak	Vollebak	Visibility, safety
Polar team pro shirt	Polar	Team sports performance tracking
Siren diabetic socks	Siren	Foot temperature monitoring
Emglare smart shirt	Emglare	Health monitoring, fitness tracking
BeBop sensors smart fabric	BeBop sensors	Pressure sensing, health monitoring
Sensoria fitness T-shirt	Sensoria	Fitness tracking, heart rate monitoring
Medtronic Zephyr BioPatch	Medtronic, Zephyr technology	Continuous health monitoring
Hexoskin smart vest	Hexoskin	Health monitoring, athletic performance
Zyro health smart shirt	Zyro health	Health monitoring, fitness tracking
Supreme smart jacket	Supreme, google Jacquard	Smartphone connectivity, navigation

(continued)

Table 11.1 (continued)

Name of material	Company producing the same	Applications
LifeShirt by VivoMetrics	VivoMetrics	Continuous health monitoring
AiQ smart clothing	AiQ smart clothing	Health monitoring, sports performance
Xenoma e-skin shirt	Xenoma	Motion capture, fitness tracking
SmartCap	SmartCap Technologies	Fatigue monitoring, safety in industrial applications

  - o The development of non-toxic conductive inks and eco-friendly adhesives should be promoted.

- **Energy Efficiency**
  - o Incorporate energy harvesting technologies such as solar energy, thermoelectric energy, etc.
  - o Design low-power electronics and efficient power management systems.
  - o Manufacturing processes that minimize energy, emissions, and resource consumption should be adopted.
- **Durability and Longevity**
  - o The materials and components should be selected so that they are resistant to wear, washing, and environmental conditions.
  - o The durability of electronic components can be increased by application of protective coatings.
  - o More emphasis should be given to the creation of modular, easily repairable e-textiles.
- **End-of-Life Management**
  - o The components used and textiles should be designed such that easy separation and recycling can be employed.
  - o Implementation of take-back programs can be efficient for proper recycling or disposal.
  - o The development of biodegradable electronic components can also be of significance.
- **Social and Ethical Considerations**
  - o Source materials should be purchased from suppliers who adhere to fair labor practices and environmental standards.
  - o It is imperative to ensure safe working conditions and support local manufacturing.
  - o Engaging with communities to design beneficial e-textiles can also be helpful.
- **Lifecycle Assessment**

- o Lifecycle assessments should be conducted to evaluate environmental impacts from raw material extraction to disposal.
- o Assessment data can also improve sustainability and reduce resource use and environmental impact.

Innovations in smart textiles have led to the development of several advanced materials. For instance, graphene-coated textiles are lightweight and environmentally friendly and offer excellent electrical properties. Sustainable practices in the case of e-textiles have been on the rise since the last decade. MycoWorks developed Mycelium-based leather as a biodegradable alternative to traditional leather [86]. Piñatex, made from pineapple leaf fibres, is now integrated with conductive materials. Researchers from the University of Maryland and the Royal Institute of Technology in Sweden have developed nanocellulose fibres and conductive cellulose, respectively. These materials are biodegradable and have excellent conductive properties [87]. Recycling techniques are also widely being explored, and it has been found that recycled carbon fibre retains the benefits of virgin carbon fibre while minimizing the environmental impact. TENCEL™ Lyocell fibres produced with REFIBRA™ technology use cotton textile waste as raw material and wood. The fibres contain at least 30% recycled material sourced from pre- and post-consumer waste [88]. This is a significant development in using recycled materials, which otherwise could have entered landfills or been incinerated. These developments are a prime example of how sustainability and technological growth may coexist in smart textiles. It is important that we carefully consider the variables that sustain rather than just heedlessly pursuing development. This way, we can sustain and develop for the better.

11.5 Challenges and Future Directions

E-textiles have garnered significant attention owing to a plethora of applications, including wearable technology, sports performance, and health monitoring. However, several obstacles must be overcome to ensure their widespread usage and maximum potential realization [89]. This section examines these issues and suggests possible paths for developing flexible electronics in textiles in the future [90]. An overview of the challenges of the widespread use of e-textiles is described in Fig. 11.3.

To counter the challenges discussed above, the following road map may be followed for adopting sustainable practices on smart textiles.

- **Advanced Materials and Technologies**—The electrical, mechanical, and thermal properties of e-textiles can be enhanced by using nanomaterials like graphene. The environmental impact of electronic components can be reduced by conducting more research on biodegradable materials, which, in turn, will improve the sustainability of the materials. Developing fabrics with intrinsic electronic properties will eliminate the need for separate electronic components.
- **Enhanced Energy Solutions**—Using flexible and stretchable batteries that integrate seamlessly into textiles can ensure comfort for the users. The energy

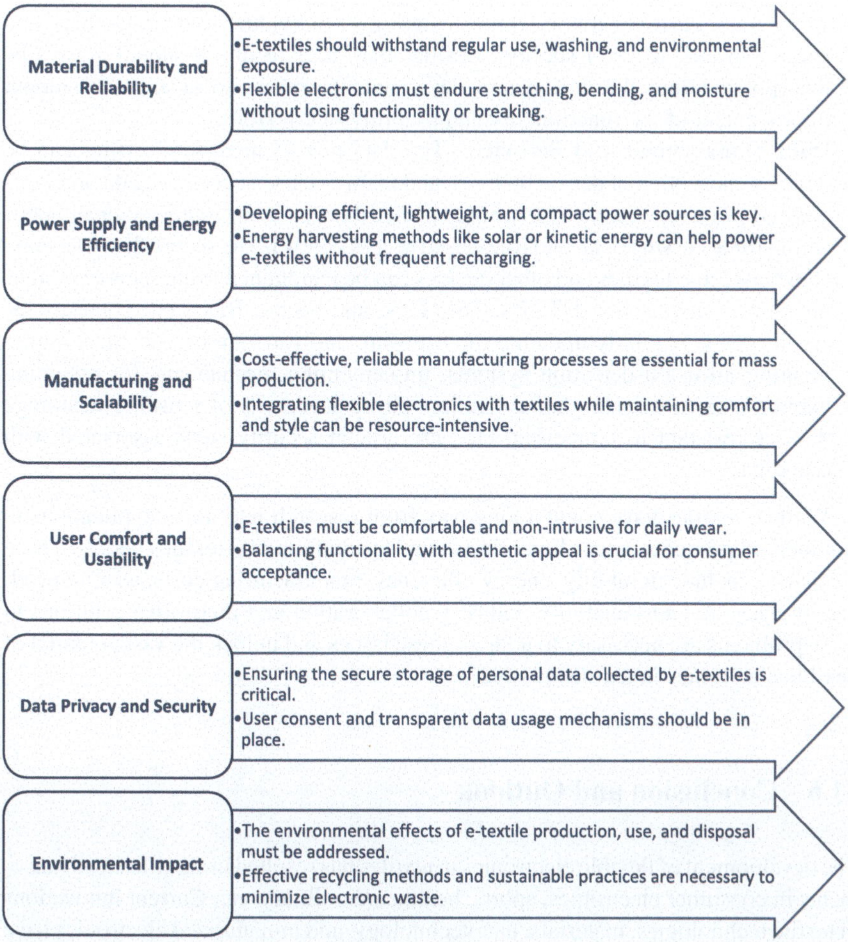

Fig. 11.3 Overview of the challenges of the widespread use of e-textiles

harvesting technologies also need to be advanced to efficiently convert ambient energy such as solar, kinetic, and thermal energy into usable power for e-textiles. The need for physical connectors can be eliminated by making use of wireless charging methods, and this, in turn, will also improve the usability of the product.

- **Integration and Manufacturing Innovations**—The 3D printing practice is significant for precise and scalable integration of electronics into textiles. To ensure mass usage, high-volume and cost-effective methods such as roll-to-roll manufacturing processes can be made use of. Besides that, integrating AI and machine learning is imperative to optimize manufacturing processes, enhance quality control, and reduce defects.
- **User-Centric Design**—The ergonomic design principles should be prioritized to ensure the comfort and user-friendliness of e-textiles. In the healthcare sector,

sports, etc., customizable e-textiles catering to individual user preferences and needs can lead to great success. Besides that, user feedback should always be incorporated into the design and development process to ensure the smooth improvement of the functionality and usability of e-textiles.

- **Data Management and Security**—The data management in e-textiles can be done by encryption of data at both rest and in transit, use of encrypted databases or secured cloud services for storage of collected data, ensuring multi-factor authentication for accessing data, and role-based access control. The secure data transmission between e-textiles and other devices can be maintained using secure communication protocols like HTTPS, SSL/TLS, and secure Bluetooth connections. Additionally, regularly updating the firmware and software of e-textiles, implementing intrusion detection systems, implementing mechanisms for obtaining explicit user consent for data collection, physical security of sensors and storage devices resistant to tampering, etc., are crucial security steps associated with e-textiles.

While e-textiles have come a long way from research into niche applications to commercial products, issues hindering the wide adoption of e-textiles persist. These challenges include durability, energy efficiency, manufacturing costs, user comfort, data privacy, and sustainability. Industry collaboration and innovative solutions to these problems are necessary to address these issues and unlock the vast potential of flexible electronics in textiles [91, 92].

11.6 Conclusion and Outlook

The development of flexible electronics in textiles has resulted in significant advancements in consumer electronics, sports, healthcare, military, etc. Current innovations in textile technologies, materials, nanotechnology, and miniaturized electronics have rendered wearable systems accessible. Nevertheless, issues such as durability, energy economy, manufacturing costs, user comfort, data privacy, sustainability, etc., still persist. A broader consumer market of e-textiles necessitates tackling these problems with creative solutions and interdisciplinary cooperation. The different techniques used to perceive, actuate, and interconnect have been emphasized by surveys of each component, yet commercialization, standardization, and prototyping still pose major challenges. Future prospects for flexible electronics in textiles are quite promising. The innovation in this sector will be driven by ethical manufacturing and legal compliance. It will be imperative to develop standardized electronic textiles, clothing that responds to stimuli without batteries, and environmentally friendly production techniques. The key to gaining user acceptability will be comfort, durability, and mechanical resistance. Furthermore, sustainability will be improved by developing hybrid materials that are easily divided into their component parts for recycling without sacrificing their desired qualities. By overcoming these obstacles, we may improve our clothes' intelligence, utility, and interactivity and open the door to a more sustainable, integrated wearable technology future.

Declaration The authors declare that they have no known competing financial interests or personal relationships that could have appeared to influence the work reported in this chapter. Generative AI and AI-assisted technologies have been used in the writing process to improve the readability and language of the work only. After using this tool, the authors carefully reviewed and revised the content to ensure accuracy and quality. The authors take full responsibility for the final version of this publication.

References

1. Islam MR, Afroj S, Yin J, Kostya S, Novoselov JC, Karim N (2024) Advances in printed electronic textiles. Adv Sci 11(6):2304140. https://doi.org/10.1002/advs.202304140
2. Yang K, McErlain-Nayor SA, Isaia B, Callaway A, Beeby S (2024) E-Textiles for sports and fitness sensing: current state, challenges, and future opportunities. Sens 24(4):1058. https://doi.org/10.3390/s24041058
3. Tat T, Chen G, Zhao X, Zhou Y, Xu J, Chen J (2022) Smart textiles for healthcare and sustainability. ACS Nano 16(9):13301–13313. https://doi.org/10.1021/acsnano.2c06287
4. Tseghai GB, Malengier B, Fante KA, Nigusse AB, Langenhove LV (2020) Integration of conductive materials with textile structures, an overview. Sens 20(23):6910. https://doi.org/10.3390/s20236910
5. Berciu AG, Dulf E (2023) Smart textiles and artificial intelligence for analysis of sleep quality and early disease diagnosis. Appl Med Inform 45:S25
6. Mahmud AA, Wickramarathne TI, Kuys B (2020) Effects of smart garments on the well-being of athletes: a scoping review protocol. BMJ Open 10:e042127. https://doi.org/10.1136/bmjopen-2020-042127
7. Zheng Q, Xu C, Jiang Z, Zhu M, Chen C, Fu F (2021) Smart actuators based on external stimulus response. Front Chem 9:650358. https://doi.org/10.3389/fchem.2021.650358
8. Dulal M, Afroj S, Ahn J, Cho Y, Carr C, Kim ID, Karim N (2022) Toward sustainable wearable electronic textiles. ACS Nano 16(12):19755–19788. https://doi.org/10.1021/acsnano.2c07723
9. (a) Steffens F, Gralha SE, Ferreira ILS, Oliveira FR (2019) Military textiles-an overview of new developments. Key Eng Mat 812:120–126. https://doi.org/10.4028/www.scientific.net/KEM.812.120 (b) Du K, Lin R, Yin L, Ho JS, Wang J, Lim CT (2022) Electronic textiles for energy, sensing and communication. iScience 25(5):104174. https://doi.org/10.1016/j.isci.2022.104174
10. Gong S, Lu Y, Yin J, Levin A, Cheng W (2024) Materials-driven soft wearable bioelectronics for connected healthcare. Chem Rev 124(2):455–553. https://doi.org/10.1021/acs.chemrev.3c00502
11. Libanori A, Chen G, Zhao X, Zhou Y, Chen J (2022) Smart textiles for personalized healthcare. Nat Electron 5:142–156. https://doi.org/10.1038/s41928-022-00723-z
12. Cherenack K, Pieterson LV (2012) Smart textiles: challenges and opportunities. J Appl Phys 112:091301. https://doi.org/10.1063/1.4742728
13. A Timeline of textile history, textile heritage museum. https://textileheritagemuseum.org/textiles-ancient-times-to-modern-day/. Last accessed 05 Aug 2024
14. A look back at the history of textiles (2016) Wilkins Linen team. https://www.wilkinslinen.com/textiles/look-back-history-textiles/. Last accessed 05 Aug 2024
15. Evolution of textile industry (2018) Textile school. https://www.textileschool.com/126/evolution-of-textile-industry/. Last accessed 05 Aug 2024

16. Tale 2-a history of smart fabric (2016) LOOMIA. https://shorturl.at/Ng5Rw. Last accessed 30 Aug 2024
17. Fiaz F (2017) History of e-textile, etechtex. https://etechtex.wordpress.com/2017/05/28/history-of-e-textile/. Last accessed 30 Aug 2024
18. Smith P (1968) Body covering. In: Museum of contemporary crafts. The American Craft Council, New York. https://digital.craftcouncil.org/digital/collection/p15785coll6/id/1237/. Last accessed 30 Aug 2024
19. Flood K (2011) The original creators: Diana dew. https://www.vice.com/en/article/the-original-creators-diana-dew/. Last accessed 30 Aug 2024
20. Post R, Orth M, Russo P, Gershenfeld N (2000) E-broidery: design and fabrication of textile-based computing. IBM Syst J 39:840–860. https://doi.org/10.1147/sj.393.0840
21. Syduzzaman Md, Ullah SP, Farhana K, Ahmed S (2015) Smart textiles and nano-technology: a general overview, J Tex Sci Eng 5(1):1000181. https://doi.org/10.4172/2165-8064.1000181
22. Gregory RV, Samuel RJ, Hanks T (2001) National textile centre annual report, USA
23. Riley TH, Dias T, Cork C (2018) A historical review of the development of electronic textiles. Fibers 6(2):34. https://doi.org/10.3390/fib6020034
24. E-textiles and smart clothing 2020–2030: technologies, markets and players, IDTechEx. https://www.idtechex.com/en/research-report/e-textiles-and-smart-clothing-2020-2030-technologies-markets-and-players/735/. Last accessed 30 Aug 2024
25. Chu J (2014) Shrink-wrapping spacesuits, MIT News. https://news.mit.edu/2014/second-skin-spacesuits-0918/. Last accessed 30 Aug 2024
26. Quick D (2015) Lumenus smart jacket signals a change of direction for cyclists, New Atlas. https://newatlas.com/lumenus-connected-cycling-jacket-led-safety-signals-interbike/39449/. Last accessed 30 Aug 2024
27. Nextiles launches smart fabric that tracks human performance data (2021) Home textiles today staff. https://www.hometextilestoday.com/technology/nextiles-launches-smart-to-track-human-performance-data/. Last accessed 30 Aug 2024
28. Myant Unveils SKIIN smart clothing platform at CES (2018) Fibre2fashion. http://www.fibre2fashion.com/news/apparel-news/myant-unveils-skiin-smart-clothing-platform-at-ces-240023-newsdetails.htm. Last accessed 8 Jul 2024
29. Yu A, Pu X, Wen R, Liu M, Zhou T, Zhang K, Zhang Y, Zhai J, Hu W, Wang ZL (2017) Core-shell-yarn-based triboelectric nanogenerator textiles as power cloths. ACS Nano 11(12):12764–12771. https://doi.org/10.1021/acsnano.7b07534
30. Mostafalu P, Tamayol A, Rahimi R, Ochoa M, Khalilpour A, Kiaee G, Yazdi IK, Bagherifard S, Dokmeci MR, Ziaie B, Sonkusale SR, Khademhosseini A (2018) Smart bandage for monitoring and treatment of chronic wounds. Small 14(33):1703509. https://doi.org/10.1002/smll.201703509
31. Peng H, Li H, Tao G, Xia L, Xu W, Zhai T (2023) Smart textile optoelectronics for human-interfaced logic systems. Adv Funct Mater 34(2):2308136. https://doi.org/10.1002/adfm.202308136
32. Zaman SU, Tao X, Cochrane C, Koncar V (2022) Smart e-textile systems: a review for healthcare applications. Electron 11(1):99. https://doi.org/10.3390/electronics11010099
33. Kaushik V, Lee J, Hong J, Lee S, Lee S, Seo J, Mahata C, Lee T (2015) Textile-based electronic components for energy applications: principles, problems, and perspective. Nanomater 5(3):1493–1531. https://doi.org/10.3390/nano5031493
34. Du K, Lin R, Yin L, Ho JS, Wang J, Lim CT (2022) Electronic textiles for energy, sensing, and communication. iScience 25(5):104174. https://doi.org/10.1016/j.isci.2022.104174
35. (a) Lu W, Zu M, Byun JH, Kim BS, Chou TW (2012) State of the art of carbon nanotube fibers: opportunities and challenges. Adv Mater 24(14):1805–1833. https://doi.org/10.1002/adma.201104672 (b) Zhang X, Lu W, Zhou G, Li Q (2019) Understanding the mechanical and conductive properties of carbon nanotube fibers for smart electronics. Adv Mater 32(5):1902028. https://doi.org/10.1002/adma.201902028
36. (a) Talarico D, Arduini, F, Constantino A, Carlo MD, Compagnone D, Moscone D, Palleschi G (2015) Carbon black as successful screen-printed electrode modifier for phenolic

compound detection. Electrochem Commun 60:78–82. https://doi.org/10.1016/j.elecom.2015.08.010 (b) Marsh H, Rodríguez-Reinoso F (eds) (2006) Activated carbon. Elsevier Science Ltd. Oxford. https://www.sciencedirect.com/book/9780080444635/activated-carbon (c) Pandolfo AG, Hollenkamp AF (2006) Carbon properties and their role in supercapacitors. J Power Sources 157(1):11–27. https://doi.org/10.1016/j.jpowsour.2006.02.065

37. (a) Choi JH (2010) Fabrication of a carbon electrode using activated carbon powder and application to the capacitive deionization process. Sep Purif Technol 70(3):362–366. https://doi.org/10.1016/j.seppur.2009.10.023 (b) Gu W, Yushin G (2013) Review of nanostructured carbon materials for electrochemical capacitor applicatios: advantages and limitations of activated carbon, carbide-derived carbon, zeolite-templated carbon, carbon aerogels, carbon nanotubes, onion-like carbon, and graphene. WIREs Energy Environ 3(5):424–473. https://doi.org/10.1002/wene.102
38. Fang B, Chang D, Xu Z, Gao C (2019) A review on graphene fibers: expectations, advances, and prospects. Adv Mater 32(5):1902664. https://doi.org/10.1002/adma.201902664
39. (a) Sarker F, Potluri P, Afroj S, Koncherry V, Novoselov KS, Karim N (2019) Ultrahigh performance of nanoengineered graphene-based natural jute fiber composites. ACS Appl Mater Interfaces 11(23):21166–21176. https://doi.org/10.1021/acsami.9b04696 (b) Inagaki M, Kang F (eds) (2014) Materials science and engineering of carbon: fundamentals, 2nd edn. Butterworth-Heinemann, Oxford
40. Grancarić AM, Jerković I, Koncar V, Cochrane C, Kelly FM, Soulat D, Legrand X (2018) Conductive polymers for smart textile applications. J Ind Text 48(3):612–642. https://doi.org/10.1177/1528083717699368
41. Ibrahim N, Akindoyo JO, Mariatti M (2022) Recent development in silver-based ink for flexible electronics. J Sci: Adv Mater Devices 7(1):100395. https://doi.org/10.1016/j.jsamd.2021.09.002
42. (a) Boumegnane A, Nadi A, Cherkaoui O, Tahiri, M (2022) Inkjet printing of silver conductive ink on textiles for wearable electronic applications. Mater Today Proc 58(4):1235–1241. https://doi.org/10.1016/j.matpr.2022.01.469 (b) Kim CK, Lee GJ, Lee MK, Rhee CK (2014) A novel method to prepare Cu@Ag core-shell nanoparticles for printed flexible electronics. Powder Technol 263:1–6. https://doi.org/10.1016/j.powtec.2014.04.064
43. Xia X, Xie C, Cai S, Yang Z, Yang X (2006) Corrosion characterstics of copper microparticles and copper nanoparticles in distilled water. Corros Sci 48(12):3924–3932. https://doi.org/10.1016/j.corsci.2006.04.007
44. (a) Li W, Sun Q, Li L, Jiu J, Liu XY, Kanehara M,, Minari T, Suganuma K (2020) The rise of conductive copper inks: challenges and perspectives. Appl Mater Today 18:100451. https://doi.org/10.1016/j.apmt.2019.100451 (b) Lu H, Lei M, Zhao C, Yao Y, Gou J, Hui D, Fu YQ (2015) Controlling Au electrode patterns for simultaneously monitoring electrical actuation and shape recovery in shape memory polymer. Compos Part B: Eng 80:37–42. https://doi.org/10.1016/j.compositesb.2015.05.039
45. (a) Majumdar D, Maiyalagan T, Jiang Z (2019) Recent progress in ruthenium oxide-based composites for supercapacitor applications. ChemElectroChem 6(17):4343–4372. https://doi.org/10.1002/celc.201900668 (b) Xia H, Meng YS, Yuan G, Cui C, Lu L (2012) A symmetric RuO_2/RuO_2 supercapacitor operating at 1.6 V by using a neutral aqueous electrolyte. Electrochem Solid-State Lett 15:A60. https://doi.org/10.1149/2.023204esl (c) Subramanian V, Hall SC, Smith PH, Rambabu B (2004) Mesoporous anhydrous RuO_2 as a supercapacitor electrode material. Solid State Ion 175(1–4):511–515. https://doi.org/10.1016/j.ssi.2004.01.070 (d) Brisse AL, Stevens P, Toussaint G, Crosnier O, Brousse T (2018) $Ni(OH)_2$ and NiO based composites: battery type electrode materials for hybrid supercapacitors devices. Mater 11(7):1178 https://doi.org/10.3390/ma11071178 (e) Cai J, Zhang D, Ding WP, Zhu ZZ, Wang GZ. He JR, Wang HB, Fei P, Si TL (2020) Promising rice-husk derived carbon/$Ni(OH)_2$ composite materials as a high-performing supercapacitor electrode. ACS Omega 5(46):29896–29902. https://doi.org/10.1021/acsomega.0c04117

46. Afroj S, Britnell L, Hasan T, Andreeva DV, Novoselov KS, Karim N (2021) Graphene-based technologies for tackling COVID-19 and future pandemics. Adv Funct Mater 31(52):2107407. https://doi.org/10.1002/adfm.202107407
47. (a) Yun Q, Li L, Hu Z, Lu Q, Chen B, Zhang H (2020) Layered transition metal dichalcogenide-based nanomaterials for electrochemical energy storage. Adv Mater 32(1): 1903826. https://doi.org/10.1002/adma.201903826 (b) Tanwar S, Arya A, Gaur A, Sharma AL (2021) Transition metal dichalcogenide (TMDs) electrodes for supercapacitors: a comprenensive review. J Phys Condens Matter 33(30): 303002. https://doi.org/10.1088/1361-648X/abfb3c
48. Dulal M, Islam MR, Maiti S, Islam MH, Ali I, Abdelkader AM, Novoselov KS, Afroj S, Karim N (2023) Smart and multifunctional fiber-reinforced composites of 2D heterostructure based textiles. Adv Funct Mater 33(40):2305901. https://doi.org/10.1002/adfm.202305901
49. (a) Bao J, Jeppson K, Edwards M, Fu Y, Ye L, Lu X, Liu J (2016) Synthesis and applications of two-dimensional hexagonal boron nitride in electronics manufacturing. Electron Mater Lett 12:1–16. https://doi.org/10.1007/s13391-015-5308-2 (b) Pu J, Zhang K, Wang Z, Li C, Zhu K, Yao Y, Hong G (2021) Synthesis and modification of boron nitride nanomaterials for electrochemical energy storage: from theory to application. Adv Funct Mater 31(48):2106315. https://doi.org/10.1002/adfm.202106315
50. (a) Anasori B, Lukatskaya MR, Gogotsi Y (2017) 2D metal carbides and nitrides (MXenes) for energy storage. Nat Rev Mater 2:16098. https://doi.org/10.1038/natrevmats.2016.98 (b) Li X, Huang Z, Zhi C (2019) Environmental stability of MXenes as energy storage materials. Front Mater 6. https://doi.org/10.3389/fmats.2019.00312
51. (a) Zhang Z, Guo X, Wen F, Shi Q, He T, Dong B, Lee C (2021) In reference module in biomedical sciences, Elsevier, Amsterdam (b) Zhao X, Zhou Y, Xu J, Chen G, Fang Y, Tat T, Xiao X, Song Y, Li S, Chen J (2021) Soft fibers with magnetoelasticity for wearable electronics. Nat Commun 12:6755. https://doi.org/10.1038/s41467-021-27066-1
52. Tian Y, Huang X, Cheng Y, Niu Y, Ma J, Zhao Y, KouX KQ (2022) Applications of adhesives in textiles: a review. Eur Polym J 167:111089. https://doi.org/10.1016/j.eurpolymj.2022.111089
53. (a) Kamiya R, Cheeseman BA, Popper P, Chou TW (2000) Some recent advances in the fabrication and design of three dimensional textile preforms: a review. Compos Sci Technol 60(1):33–47. https://doi.org/10.1016/S0266-3538(99)00093-7 (b) Khair N, Islam R, Shahariar H (2019) Carbon-based electronic textiles: materials, fabrication processes and application. J Mater Sci 54:10079–10101. https://doi.org/10.1007/s10853-019-03464-1
54. Labiano II, Arslan D, Yenigun EO, Asadi A, Cebeci H, Alomainy A (2021) Screen printing carbon nanotubes textiles antennas for smart wearables. Sensors 21(14):4934. https://doi.org/10.3390/s21144934
55. Shah MA, Lee DG, Lee BY, Hur S (2021) Classifications and applications of inkjet printing technology: a review. IEEE Access 9:140079–140102. https://doi.org/10.1109/ACCESS.2021.3119219
56. Bhardwaj N, Kundu SC (2010) Electrospinning: a fascinating fiber fabrication technique. Biotechnol Adv 28(3):325–347. https://doi.org/10.1016/j.biotechadv.2010.01.004
57. Labeeb AM, Sobhy SK, Eltopshy SM, Elgory ZM, Swielam EM (2016) Embroidery on textiles as a smart solution for wearable applications. Int J Des 9(2):53–58. https://doi.org/10.21608/idj.2019.83296
58. Smith WC (2019) Commentary/overview of textile coating and laminating. In: Smith WC (ed) Smart textiles coatings and laminates, 3rd edn. The Textile Institute Book Series, pp 3–10 https://doi.org/10.1016/B978-0-08-102428-7.00001-8
59. Wilson JIB (2009) Textile surface functionalization by chemical vapour deposition (CVD). In: Wei Q (ed) Surface modification of textiles. Woodhead Publishing Series in Textiles, pp 126–138. https://doi.org/10.1533/9781845696689.126
60. Nawab Y, Hamdani STA, Shaker K (eds) (2017) Structural textile design: interlacing and interlooping, 1st edn. CRC Press. https://doi.org/10.1201/9781315390406
61. (a) Chatterjee K, Ghosh TK (2019) 3D printing of textiles: Potenntial roadmap to printing with fibers. Adv Mater 32(4):114. https://doi.org/10.1002/adma.201902086 (b) Dip TM, Emu AS, Nafiz MNH, Kundu P, Rakhi HR, Sayam A (2021) 3D printing technology for textiles and fashion. Text Prog 167–260. https://doi.org/10.1080/00405167.2021.1978223

62. (a) Yang K, Isaia B, Brown LJ, Beeby S (2019) E-textiles for healthy ageing. Sensors 19(20):4463. https://doi.org/10.3390/s19204463 (b) Gonçalves C, da Silva AF, Gomes J, Simoes R (2018) Wearable e-textile technologies: a review on sensors, actuators and control elements. Inventions 3(1):14. https://doi.org/10.3390/inventions3010014
63. (a) Micus S, Kirsten I, Haupt M, Gresser GT (2020) Analysis of hot bar soldering, insulation displacement connections (IDC), and anisotropic conductive adhesives (ACA), for the automated production of smart textiles. Sensors 20(1):5 https://doi.org/10.3390/s20010005 (b) Ayodele E, Zaidi S, Zhang Z, Scott J, Kong Q, McLernon D (2019) Weft knit smart data glove. In: proceedings of the IEEE 16th international conference on wearable and implantable body sensor networks (BSN), Chicago, IL, USA. https://eprints.whiterose.ac.uk/145383/ (c) Stark L (2017) Data vows: reimagining ritual through e-textile practice. In: Proceedings of the ISWC '17: proceedings of the 2017 ACM international symposium on wearable computers. https://doi.org/10.1145/3123021.3123072
64. Chen Y, Xu B, Gong J, Wen J, Hua T, Kan CW, Deng J (2019) Design of high-performance wearable energy and sensor electronics from fiber materials. Appl Mater Interfaces 11(2):2120–2129. https://doi.org/10.1021/acsami.8b16167
65. Bahadır SK, Atalay Ö, Kalaoglu F, Mitilineos SA, Vassiliadis S (2018) High frequency attenuation characterization of knitted e-textile structures. IOP conf ser: Mater Sci Eng 460:012054. https://doi.org/10.1088/1757-899X/460/1/012054
66. Rezaei A, Cuthbert TJ, Gholami M, Menon C (2019) Application-based production and testing of a core–sheath fiber strain sensor for wearable electronics: feasibility study of using the sensors in measuring tri-axial trunk motion angles. Sensors 19(19):4288. https://doi.org/10.3390/s19194288
67. Coyle S, Lau KT, Moyna N, O'Gorman D, Diamond D, di Francesco F, Costanzo D, Salvo P, Trivella MG, De Rossi DE (2010) BIOTEX—biosensing textiles for personalised healthcare management. IEEE Trans Inf Technol Biomed 14:364–370. https://doi.org/10.1109/TITB.2009.2038484
68. Saied IM, Chandran S, Arslan T (2019) Integrated flexible hybrid silicone-textile dual-resonant sensors and switching circuit for wearable neurodegeneration monitoring systems. IEEE Trans Biomed Circuits Syst 13:1304–1312. https://doi.org/10.1109/TBCAS.2009.2038228
69. (a) Fernández-Caramés TM, Fraga-Lamas P (2018) Towards the internet of smart clothing: a review on IoT wearables and garments for creating intelligent connected E-textiles. Electron 7(12): 405. https://doi.org/10.3390/electronics7120405 (b) Ray S, Park J, Bhunia S (2016) Wearables, implants, and internet of things: The technology needs in the evolving landscape. IEEE Trans on Multi-Scale Comput Syst 2:123–128. https://doi.org/10.1109/TMSCS.2016.2553026
70. Kiourti, A (2018) Textile-based flexible electronics for wearable applications: from antennas to batteries. In: Proceedings of the 2018 2nd URSI atlantic radio science meeting (AT-RASC), Gran Canaria, Spain
71. Gonçalves C, da Silva AF, Gomes J, Simoes R (2018) Wearable e-textile technologies: a review on sensors, actuators and control elements. Inventions 3(1):14. https://doi.org/10.3390/inventions3010014
72. Dong K, Peng X, Wang ZL (2019) Fiber/fabric-based piezoelectric and triboelectric nanogenerators for flexible/stretchable and wearable electronics and artificial intelligence. Adv Mater 32(5):e1902549. https://doi.org/10.1002/adma.201902549
73. Lemey S, Agneessens S, Rogier H (2018) Wearable smart objects: microwaves propelling smart textiles: a review of holistic designs for wireless textile nodes. IEEE Microw Mag 19(6):83–100. https://doi.org/10.1109/MMM.2018.2844030
74. Dang T, Zhao M (2021) The application of smart fibers and smart textiles. J Phys Conf Ser 1790:012084. https://doi.org/10.1088/1742-6596/1790/1/012084
75. Koncar V (2016) Introduction to smart textiles and their applications. In: Smart textiles and their applications. Woodhead Publishing Series in Textiles, pp 1–8. https://www.sciencedirect.com/science/article/abs/pii/B9780081005743000011

76. Coyle S, Diamond D (2016) Medical applications of smart textiles. In: Langenhove LV (ed) Advances in smart medical textiles. Woodhead Publishing Series in Textiles, pp 215–237. https://doi.org/10.1016/B978-1-78242-379-9.00010-4
77. (a) Chen Y, Deng Z, Ouyang R, Zheng R, Jiang Z, Bai H, Xue H (2021) 3D printed stretchable smart fibers and textiles for self-powered e-skin. Nano Energy 84:105866. https://doi.org/10.1016/j.nanoen.2021.105866 (b) Eutionnat-Diffo PA, Cayla A, Chen Y, Guan J, Nierstrasz V, Campagne C (2020) Development of flexible and conductive immiscible thermoplastic/elastomer monofilament for smart textiles applications using 3D printing. Polymers 12(10):2300. https://doi.org/10.3390/polym12102300
78. (a) Morgan L (2019) Loughborough University London. Textile intersections exhibition. https://cfpr.uwe.ac.uk/textile-intersections-exhibition/. Last accessed on 5 Aug 2024 (b) Stewart R (2019) Cords and chords: exploring the role of e-textiles in computational audio. Front ICT 6:12. https://doi.org/10.3389/fict.2019.00002
79. Le X, Shi Q, Sun Z, Xie J, Lee C (2022) Noncontact human-machine interface using complementary information fusion based on MEMS and triboelectric sensors. Adv Sci 9(21):221056. https://doi.org/10.1002/advs.202201056
80. Tien ND, Lyngstadaas SP, Mano JF, Blaker JJ, Haugen HJ (2021) Recent developments in chitosan-based micro/nanofibers for sustainable food packaging, smart textiles, cosmeceuticals, and biomedical applications. Molecules 26(9):2683. https://doi.org/10.3390/molecules26092683
81. Provin AP, Dutra ARA, Machado MM, Cubas ALV (2021) New materials for clothing: rethinking possibilities through a sustainability approach-a review. J Clean Prod 282(1):124444. https://doi.org/10.1016/j.jclepro.2020.124444
82. Zhang Y, Xia X, Ma K, Xia G, Wu M, Cheung YH, Yu H, Zou B, Zhnag X, Farha OK, Xin JH (2023) Functional textiles with smart properties: their fabrications and sustainable applications. Adv Funct Mater 33(33):2301607. https://doi.org/10.1002/adfm.202301607
83. Veske P, Ilen E (2020) Review of the end-of-life solutions in electronics-based smart textiles. J Text Inst 112(9):1500–1513. https://doi.org/10.1080/00405000.2020.1825176
84. Hossain MT, Shahid MA, Limon MGM, Hossain I, Mahmud N (2024) Techniques, applications, and challenges in textiles for a sustainanle future. J Open Innov: Technol Mark Complex 10(1):100230. https://doi.org/10.1016/j.joitmc.2024.100230
85. Shi HH, Pan Y, Xu L, Feng X, Wang W, Potluti P, Liangbing H, Hasan T, Huang YYS (2023) Sustainable electronic textiles towards scalable commercialization. Nat Mater 22:1294–1303. https://doi.org/10.1038/s41563-023-01615-z
86. Elsacker E, Vandelook S, Peeters E (2023) Recent technological innovations in mycelium materials as leather substitutes: a patent review. Front Bioeng Biotechnol 11:1–12. https://doi.org/10.3389/fbioe.2023.1204861
87. Fang Z, Zhnag H, Qiu S, Kuang Y, Zhou J, Lan Y, Sun C, Li G, Gong S, Ma Z (2021) Versatile wood cellulose for biodegradable electronics. Adv Mater Technol 6(2):2000928. https://doi.org/10.1002/admt.202000928
88. REFIBRA^{TM} technology (2024) Contributing to the circular economy in textiles. https://www.tencel.com/b2b/technologies/refibra-technology. Last accessed 5 Aug 2024
89. Junior HLO, Neves RM, Monticelli FM, Agnol LD (2022) Smart fabric textiles: recent advances and challenges. Textiles 2(4):582–605. https://doi.org/10.3390/textiles2040034
90. Pandiyan A, Veeramuthu L, Yan ZL, Lin YC, Tsai CH, Chang ST, Chiang WH, Xu S, Zhou T, Kuo CC (2023) A comprehensive review on perovskite and its functional composites in smart textiles: progress, challenges, opportunities, and future directions. Science 140:101206. https://doi.org/10.1016/j.pmatsci.2023.101206
91. Chan M, Esteve D, Fourniols JY, Escriba C, Campo E (2012) Smart wearable systems: current status and future challenges. Artif Intell Med 56(3):137–156. https://doi.org/10.1016/j.artmed.2012.09.003
92. Mokhtari F, Cheng Z, Raad R, Xi J, Foroughi J (2020) Piezofibers to smart textiles: a review on recent advances and future outlook for wearable technology. J Mater Chem 8:9496–9522. https://doi.org/10.1039/D0TA00227E

Chapter 12
Emerging Innovations in Agro-Textile: Raw Materials and Finishing Techniques

Amal Ray, Ashutosh Kumar Srivastava, and Sukhbir Singh

Abstract Technical textiles have become essential beyond clothing, particularly in agriculture, to enhance productivity and sustainability. As global food demand rises, preserving crops and improving quality without excessive pesticide use is crucial. Agro-textiles, including greenhouse coverings, mulching, windshields, and protective fabrics, help safeguard crops, reduce environmental impact, and boost agricultural efficiency. These innovations support sustainable farming by improving yield, quality, and long-term ecological balance.

Keywords Technical textile · Agro-textiles · Cladding materials · Mulch · And Protection fabrics

12.1 Introduction

Climate change, resource shortages, population growth, and global challenges affect agriculture and crop production. Urbanization and industrialization are reducing the amount of farmland, while the use of pesticides and harmful chemicals on crops is affecting human health. To address these issues, it is important to focus on new research that supports food security, reduces agriculture's impact on the environment, and develops sustainable farming methods. Agro-textiles offer a way to grow more crops without using harmful chemicals, introducing a modern and efficient approach to farming [1, 2]. Agro-textile is a key sector within technical textiles, focusing on using agricultural textile materials to increase crop yields while ensuring environmental safety. Agro-textile products are important for improving the growth of crops and vegetables, even with challenges from climate change. They help reduce harmful pesticides and herbicides, making farming healthier and more sustainable.

A. Ray (✉) · A. K. Srivastava
Lovely Professional University, Phagwara, Punjab, India
e-mail: textile.amal@gmail.com

S. Singh
Faculty of Fashion and Design, SGT University, Gurugram, Haryana, India

S. Biranje et al. (eds.), *Sustainable Finishing Techniques in Textiles*,
https://doi.org/10.1007/978-981-96-4860-3_12

By creating a safer environment for crops, agro-textiles help increase yields and protect the environment, meeting the demand for more sustainable farming. These materials are known for being durable, easy to process, and eco-friendly. Agro-textiles prevent soil from drying out, which boosts crop yields and improves product quality [3].

Different shade fabrics can affect how much and how well crops grow [4]. Around the world, agro-textiles have been proven to increase crop production and protect crops from outside factors. Compared to traditional plastic films, agro-textiles offer many benefits, such as breathability, flexibility, easy installation, and better support for crop growth. Their ability to withstand weather and microbes is crucial for protecting plants from temperature changes between day and night. Key features of agro-textiles include strength, flexibility, porosity, protection from sunlight, and biodegradability. Agro-textiles are widely used in agriculture and gardening to protect livestock, provide shade, control weeds and pests, and extend the growing season. These materials, whether knitted, woven, or non-woven, are important for improving agricultural productivity and supporting sustainable farming.

Lightweight, spun-bonded, non-woven fabrics are used for various purposes, including weed control, thermal insulation, and shade. Wind and hail shelters are made of heavy-duty knitted, woven, and nonwoven materials. In horticulture, capillary nonwoven mats are very useful in providing moisture to growing plants [5]. The kind of fiber employed in producing high-performance agro-textiles is crucial to ensure that it will effectively fulfill its protective functions and survive environmental conditions. For many years, synthetic fibers like polypropylene (PP), polyethylene (PE), polyethylene terephthalate (PET), and polyamide (PA) have been the main choice for creating agro-textiles because they meet these requirements. These textile products are useful because they have high tensile strength, protection from microorganisms, biodegradability, abrasion resistance, protection from higher solar radiation, and an increased capacity to retain water. However, these fibers are not biodegradable and negatively influence the environment after their useful lives. The primary goal is to provide an overview and context for agro-textile products, which have a wide range of uses and can increase crop output. Agro-textiles made of natural fibers can be applied when biodegradability, wet strength, and moisture retention are necessary. In addition to its traditional uses, cellulose is employed in agriculture in various ways thanks to recent developments [6]. Its distinct behavior over its service life and the natural degradation of the cellulose material with soil increases the soil's fertility. Thus, this sustainability of soil health using cellulosic natural fiber agro-textile is another important aspect apart from its conventional plastic material application in agriculture. For agro-textile goods, polyolefin fibers are the most chosen man-made fiber owing to their best mechanical property, high strength, low weight, and extended lifespan [7]. Synthetic fibers are favored over natural fibers for their superior service life, affordability, biodegradability, and ease of transport and storage. They are increasingly utilized in agriculture amid growing population pressures on food crops.

This chapter aims to critically discuss agro-textile products' use, materials, and finishing techniques. It is a very crucial and rapidly evolving subfield within technical textiles. Agro-textile includes anything from fishing to uses in agriculture and husbandry. The global community regards the relevance of agro-textile as a major domain [8]. Agro-textile products are categorized according to their intended use, such as greenhouse capillary mats, harvesting fabrics, shade cloth fabrics, crop protection covers, and landscaping fabrics. These materials have been used in many ways with excellent outcomes that positively impact the development and yield of many crops and vegetables. Agro-textile products are made of different textile fabrics to keep the soil at the proper humidity, regulate temperature, and shield agricultural goods from hail and wind damage. Also, agro-textile goods such as harvesting fabrics, bird protection fabrics, sunscreen, and windshields [3].

The Ministry of Textiles has decided to offer an intervention or component under the technology mission on Technical Textiles for the promotion usage of agro-textile in all of India to encourage and increase the use of agro-textile in agriculture, horticulture, floriculture, and related activities [9].

12.2 What is the Agro-Textile?

The term "agro-textile" or "agro-tex" is the same thing that is used to refer to woven, knitted, and non-woven textiles. They are known for various agricultural applications, including horticultural floriculture, aquaculture, animal protection, greenhouse shading, weed and insect control for plant growth, extending the growing season for cultivation, and promoting agro-products [10]. It also helps raise the soil's temperature and maintain adequate humidity. Protect the crop harvests, for instance, from hail damage. This industry is leading the way in the expansion of technological textiles. In the agricultural industry, agro-textiles are utilized for their practical advantages. Stress on agricultural crop output has risen due to population growth. Therefore, given the existing water and space limits, it is imperative to boost the quantity and quality of agro-products [11]. However, it is impossible to completely counter this growing demand with conventional methods such as pesticides and herbicides. The agriculture and horticulture sectors know the need to pursue various technologies to increase overall yields and provide higher-quality crop grain. Strength, elongation, stiffness, resistance to solar radiation, and resilience to hazardous conditions are all crucial features for agro-textiles. These traits aid in developing and harvesting crops and other foods [12].

Moreover, agro-textiles are part of technical textile materials primarily employed to protect plants and animals from environmental factors to enhance the plant's yield. In agriculture, animal husbandry, and horticulture, these textiles are mostly utilized to manage plant nutrient intake, help with postharvest and process operations, and mitigate the negative effects of climate and environmental conditions on crop production [13]. According to preliminary research, the yield of fruits can be increased by 30% in favorable conditions and by up to 50% in unfavorable ones by employing woven return gauze. In agriculture, agro-textiles are also employed to stop harmful weeds from growing.

12.3 Importance of Agro-Textile

Agro textiles are a crucial and growing sector of technical textiles, with applications in fishing, horticulture, and husbandry. They enhance crop yield by maintaining soil humidity, regulating temperature, and protecting crops from environmental factors like hail. Various products are gaining attention, including bird nets, harvesting nets, and mulch mats. Natural fiber-based agro textiles are useful where moisture retention and biodegradability are needed, while polyolefin fibers are preferred for their strength, lightweight nature, and durability.

The benefits of using agro-textiles in agriculture, as outlined in Fig. 12.1, have been established through various studies [2, 14, 15].

12.4 The Essential Properties of Agro-Textile

Fibers used in agro-textiles vary by application, with natural fibers like jute, coir, sisal, and hemp being common, but synthetic fibers such as nylon, polyethylene, polypropylene, and polyester are preferred for their affordability, durability, and ease of transport and storage [13, 16]. Some characteristics essential for ensuring the effectiveness and longevity of agro-textiles in agricultural applications are listed in Fig. 12.2.

12.5 Raw Materials of Agro-Textiles

12.5.1 Fibers Used for Agro-Textiles

Weather and microorganism resistance are two of the most crucial qualities of textiles for agricultural uses. Man-made fiber are the ideal option in this case for creating agro-textile goods. Regarding synthetic fibers, polypropylene and polyethylene are the most suitable for widespread usage [16]. Besides that, jute is a natural fiber that can be used since it is useful, biodegradable, and functions as a natural fertilizer after degradation. The brief classification is given in Fig. 12.3.

Mostly synthetic (or man-made) fibers are chosen for agricultural products over natural fibers due to their excellent mechanical properties, high strength, and durability. This makes them appropriate for agricultural applications. On the other hand, agro-textiles made of natural fibers have a specialized function, but they also break

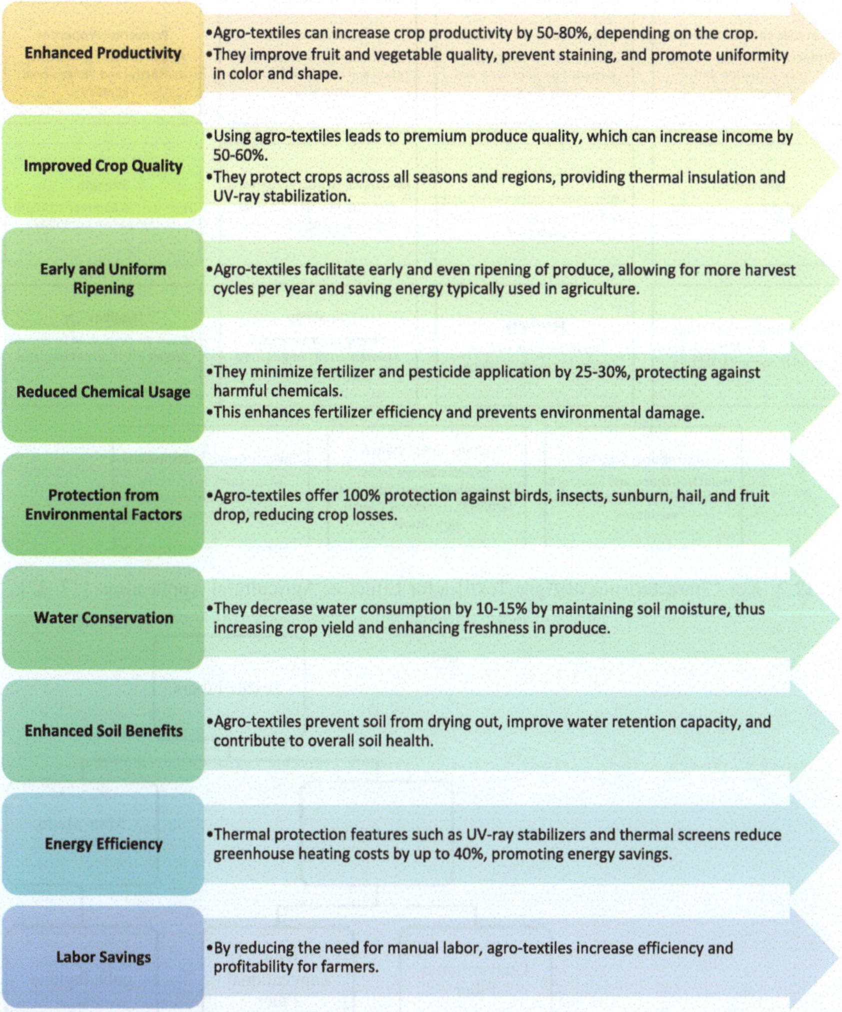

Fig. 12.1 Benefits of the usage of agro-textiles

down and become organic fertilizer in the soil after a few years of usage [2]. Synthetic fibers used in agro-tech sectors to keep product durability in mind are polyester, nylon, polypropylene, polyethylene, and polyolefin. Natural fibers mainly used in this sector for biodegradability and are environmentally friendly are jute, wool, coir, sisal, flax, and hemp. Because of their favorable price-performance ratio, synthetic fibers are favored over natural fibers for agro-textile applications. In addition, natural fibers in agro-textiles have particular uses, where qualities like biodegradability, high moisture retention, and wet strength can be successfully utilized.

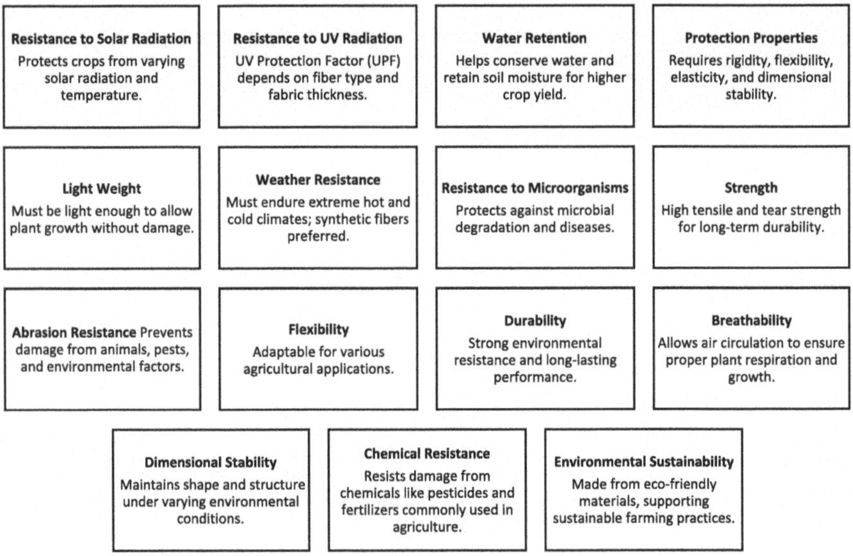

Fig. 12.2 Key Characteristics of Agro-Textiles for Effective Agricultural Applications [17–22]

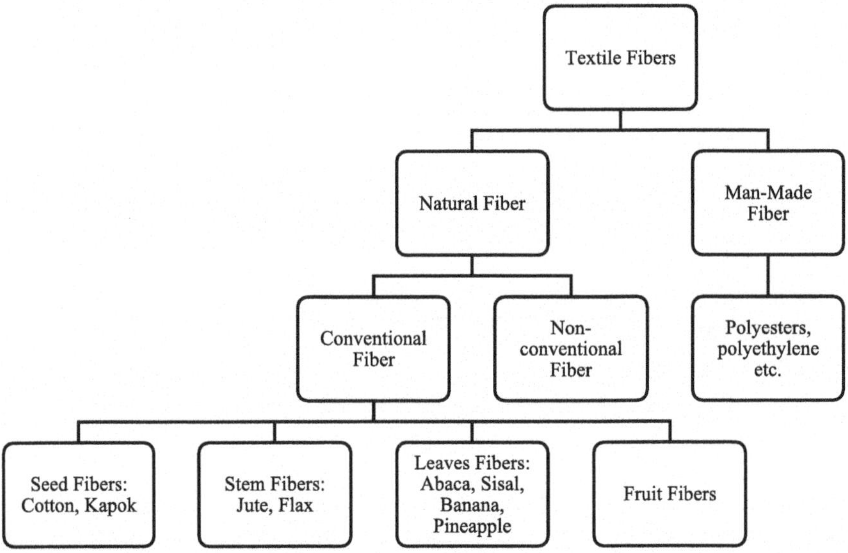

Fig. 12.3 Classification of agro-textiles

12.5.2 Manufacturing Processes of Agro-Textile

Several techniques of fabric production can be used to produce agro-textiles. Each method offers particular advantages for the specific product. Thus, different fabric manufacturing techniques are discussed, as shown in Fig. 12.4.

12.5.2.1 Weaving Process

Today, high-speed weaving machines, i.e., Sulzer projectile weaving machines, make woven fabrics. This type of weaving machine can produce fabrics with a large breadth and various ranges of weights. These weaving machines are available to produce agro-textiles with weaving widths ranging from 540 to 846 cm. It is possible to create fabrics with a mesh width ranging from 1.8 mm to 40 mm. Other popular fabric manufacturing methods, such as rapier and air-jet high-speed weaving machines, are not advised for producing agro-textiles because they do not meet the requirement of weaving breadth [23].

12.5.2.2 Knitting Process

Warp knitting is the most popular technique for agro-textiles. Raschel warp-knitted machines make protective fabric that is warp-knitted and used in many industries. However, agro-knitted fabrics are made with different lapping configurations. Here, separate yarn systems are transformed into textiles through a process known as building or lapping [24].

12.5.2.3 Non-Woven Process

Non-woven textiles can be produced using a variety of methods. This type of agro-textile is mostly made using spun bonding and needle punching techniques. The tensile strength of the spun-bonded non-woven fabric is stronger and more consistent in all directions. It may also tear with high strength. Plant's growing bags made

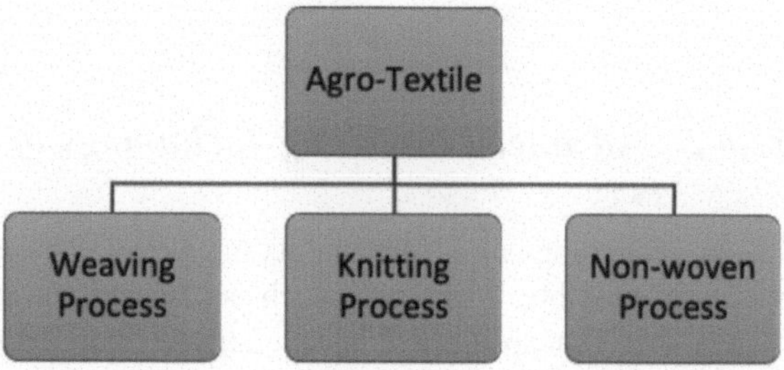

Fig. 12.4 Manufacturing process of agro-textile

of needle-punched non-woven fabrics benefit from conventional clay pots that are burned. An additional benefit of using natural fibers is that the container breaks down after its service when buried. Other important methods to produce non-woven fabrics are wet, non-woven, hydroentangled, thermal, and stitch-bonded, which are extensively used in agriculture. Nonwoven fabrics for the agriculture sector are accessible in customized alternative solutions [5].

12.6 Finishing Technique of Agro-Textile

In the textile industry, many finishing processes have been available to improve or modify agro-textile products' bulk and surface properties. However, the finishing processes are applied to textile fabrics to improve their functional property after production. In agro-textile manufacturing, finish concerns the processes that convert the woven, knitted, and non-woven fabric into a usable material and, more specifically, any process executed after agro-textile production to improve performance. The majority of agro-textiles are ready-to-use, prefabricated goods. They comprise various synthetic and natural fibers with unique qualities and extra uses [25]. The additional features might be included in the fiber production process or added as a unique finish to the finished product or textile fabric. To provide agro-textile with a particular finish, various application techniques such as coating, spraying, cushioning, etc. The completed product might need to be sterilized before use. Thus, these finishes need to be applied in an area free from infections. The finish that will be used might be a key component of the finished product, or it could just be an auxiliary one.

12.6.1 Different Finishing Methods of Agro-Textile

Various common textile finishing methods Fig. 12.5, including the exhaust process, impregnation Method, transfer method, spraying method, coating method, foam application method, etc. are used to apply finishing agents to fabrics, each differing in application technique, liquor ratio, processing time, and material compatibility to achieve desired functional properties in agro-textiles [26–30].

12.6.2 Advanced Finishing Technology of Agro-Textile

Recent technological advancements have made environmentally friendly finishing processes possible. Techniques such as plasma treatment, nanotechnology, biotechnology, microencapsulation, laser treatment, biopolymer application, and ultrasonic finishing are utilized to achieve these results [31].

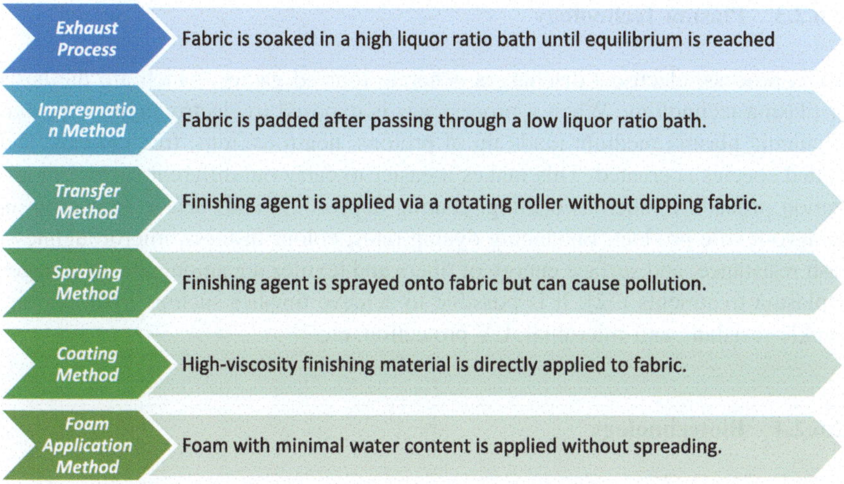

Fig. 12.5 Various common finishing methods applied to agro-textiles

12.6.2.1 Nanotechnology

Less than 100 nm in size and tolerance are the subject of nanotechnology. It's one of the fast-develop techniques of the twenty-first century. Nanoparticles can be employed in the agricultural textile industry as a long-lasting surface coating that is both economical and sustainable. It creates agro-textiles with various useful properties, including flame retardancy, stain resistance, abrasion resistance, and water-repellent properties. By using intricate metallic compounds like copper, silver, or zinc, nano-metal oxide coatings can achieve these. The most commonly used processes in this finishing technique are layer-by-layer processing, padding, and sol–gel methods [31].

12.6.2.2 Microencapsulation Technology

A micro-packaging process called micro-encapsulation involves thinly depositing a polymeric coating on tiny solid or liquid particles. This method is extensively used in the textile industry. The microcapsule wall is called the shell, while the substance within is called the core material. Functional properties, including flame retardancy, antimicrobial coatings, and antistatic finishes, may be obtained using this approach.

12.6.2.3 Plasma Technology

In this process, electrical discharges generate ionized gasses, which are the basis for plasma technology. When a process gas is exposed to electromagnetic power, a dynamic plasma medium made up of protons, neutrons, ions, free radicals, and excited species is created. This makes it easier to carry out different surface modification procedures such as etching, coating deposition, and surface activation on the agro-textile product. Enhancing dyeing rates, colour fastness, microroughness, wash resistance, and surface energy of fibers and textiles are among the advantages of plasma treatments [32]. It is possible to achieve finishes such as hydrophobic, wrinkle-resistant, anti-microbial, UV protection, etc.

12.6.2.4 Biotechnology

The function of enzymes is the most important in biotechnology. The benefits of enzymes versus chemicals are numerous. Most significantly, they are task-specific and do not require a catalyst to speed the reaction or change the chemical equilibrium between reactants and products. The textile industry has been using enzymes since the very early 1980s. They are employed in many phases of agro-textile processing, including de-sizing, scouring, bleaching, and biopolishing. The enzymes can be used for many finishing techniques. These include enhanced hand sensations, surface modification, softening, and anti-pilling properties. For the heavy-weight fabric stone wash look, cellulase enzymes and pumice stones are used. This process is also known as bio-stoning [33].

12.6.2.5 Laser Treatment

In this method, textile fabric finishing processing can be done precisely with laser treatment, a physical treatment that doesn't require water. The efficiency intensity of laser radiation is applied to the textile to be treated. Through this process, converted hydrophilic fibers that were previously hydrophobic become more dye-able without causing harm to the material properties. Carbon dioxide lasers, composed of gases like carbon dioxide, nitrogen, hydrogen, and helium, are the most appropriate lasers for the materials. It can alter the surface and add features, including wettability, optical qualities, and stickiness [34].

12.6.2.6 Biopolymers

Marine food supplies and agricultural feedstock are the origins of different biopolymers. It is well known as polysaccharides, chitosan, and sericin proteins are a few examples of finishing agents. Utilizing biopolymers has the benefit of their widespread availability and biodegradability. Biopolymers can be used to provide coatings with antibacterial, UV resistance, insect repellent, and flame retardancy.

12.6.2.7 Ultrasonic Finishing

Ultrasonic Finishing uses ultrasonic which is the oscillations of frequencies more than the upper limit of hearable range for humans which is more than 20 kilohertz. This finishing process uses zinc oxide nanoparticles (ZnO-NPs). The textile finished with ultrasonic, exhibits optimal sustainability, ease of maintenance, and cost-effectiveness. It provides long-lasting antimicrobial device characteristics.

12.6.2.8 Anti-static Finish

The primary role of anti-static agents is to prevent the generation of static electricity within the textile fabrics. Oxygen plasma treatment at low temperatures can enhance the anti-static qualities of polyester fabrics. Fiber alterations brought on by low-temperature plasma treatment result in oxidation and surface deterioration of the fibers. Degradation alters the surface morphology of fibers, whereas oxidation raises surface energy. This results in a reduction of the fiber half-life. The fabric becomes more anti-static as a result of the increased electrostatic charge dissipation, as seen by the decrease in half-life [24].

12.6.2.9 Bleaching

In this application, the chemicals are treated in the fabrics. Although the main benefit is that it eliminates the natural color from the fabric, the main drawback is that it diminishes the strength of the fabric, which ultimately makes it weaker.

12.6.2.10 Absorbent and Soil Release Finishes

In this process, the textile materials modify their surface tension and other characteristics to improve water absorption or enable the discharge of dirt. If the textile's surface tension increases, it will likely retain water droplets. Conversely, lowering it will probably cause the droplets to naturally fall off the material's surfaces, carrying dust and other particles.

12.6.2.11 Water Repellent

In industries like sports, apparel, upholstery, automotive textile, bed linens, carpets, curtains, and workwear—where the technical ability to withstand wetting or penetration by water, oil, or soil particles is required—water and oil repellence are among the most frequently required functional properties for protective clothing [35].

12.6.2.12 Shrink Resistance

Agro-textiles are treated with a chemical treatment based on chlorine to stop shrinking and make them easily washable. For example, polyethylene (PE) fabric absorbs no water and is as strong when wet as dry. It doesn't decay or diminish with aging in time duration. This type of fabric is resistant to all chemicals and microorganisms. Thus, polyethylene fabrics are normally used for the agriculture sector, like ropes (float) and trawl nets, where low specific gravity is useful as the net tends to float clear of the bottom. Generally, polyethylene is dyed in the melting form during twists. High-density polyethylene (HDPE) is the most widely used for the manufacturing of fabrics or net-like barriers. Its excellent mechanical behavior, such as its high modulus, high tensile strength, and tear resistance with low specific density, makes it a good choice.

12.6.2.13 Waterproofing or Water Resistance

The water resistance properties of an agro-textile change as it receives a waterproofing treatment. The feasibility of employing nanoparticles such as SiO_2, ZnO, and TiO_2 has found good water resistance properties. A silicon chemical is sprayed on the fabric as part of the treatment. This process is usually applied to fabrics that are going to be used in agro-textile such as outdoor shade fabric. Water-resistant textiles are frequently high-density woven fabrics consisting of very fine threads or conventional materials treated with a hydrophobic finish. Waterproofing is the characteristic of a substance that prevents fluids from penetrating it. Waterproofness may be assessed using one of the two methods: one that simulates rain and the other that submits the cloth to hydrostatic pressure. Specific surface finishing treatments are used to create water-resistant fabrics. Waterproof materials are often achieved by surface finishing procedures such as coating. Waterproof and water-repellent fabrics have been found extensive functional applications across diverse sectors like agriculture, biomedical, food as well as packaging industries [36]. Hydrophobic chemicals, such as paraffin waxes, silicones, silanes, and fluorinated polymers, are traditionally utilized to impart water-resistant functionalities. A standard parameter for a superhydrophobic water-repellent coating on fabric is a contact angle of a drop greater than 150°, known as the "Lotus Effect". Hydrophobic/superhydrophobic surfaces on agro-textiles are developed by producing surface structures with low surface tension [35]. When water-repellent nano finishes are applied on the agro-textile, the fabric surface becomes hydrophobic, and drops of water readily fall off. However, this type of fabric significantly retains the fabric's air permeability and breathability.

12.6.2.14 Flame Retardant

Fabrics that are flame resistant are essential given the risk that fire poses to people's lives and property. The capacity of an agro-textile to withstand rapid fire defines a flame-retardant textile. In cotton fabric, atmospheric-pressure dielectric barrier discharge is mostly used to apply plasma treatment. The pad-dry-cure process is used to apply the flame-retardant finish. Alkaline phosphonate structured flame-retardant compounds are padded into polyester fabrics after undergoing low-frequency oxygen plasma treatment. Through this process, polyester fabric's hydrophilic qualities are further enhanced.

12.6.2.15 Anti-microbial Finishes

One of the main issues with natural fabrics is microbial deterioration. Bacteria can deteriorate textiles, mostly cotton, linen, or jute. Antimicrobial finishes are therefore crucial in this situation. When selecting antimicrobial finishes for agro-textiles, it is essential to consider environmental impact, durability under outdoor conditions, and potential effects on soil and crops. Natural antimicrobial agents, such as chitosan or plant extracts, may offer eco-friendly alternatives suitable for agricultural applications. While direct research on antimicrobial finishes specifically for agro-textiles is limited, the principles and agents used in antimicrobial textile treatments can be adapted to meet the unique requirements of agricultural textiles. Various antimicrobial agents are used in textiles, including synthetic agents derived through chemical processes. These agents are widely utilized for their fast and effective microbial action, easy preparation, accessibility, and long-term stability. Common types include metal salts (silver, zinc, copper, titanium oxide, gallium), quaternary ammonium compounds (QAC), polybiguanides (PHMB), halogenated phenols (triclosan), and N-halamines [37]. These antimicrobial agents can also be applied in agro textiles to enhance protection against microbial contamination and improve durability.

12.7 The Application of Agro-Textile

Textile fabrics provide a vast range of woven, non-woven, and knitted fabrics applied for different uses in agriculture, landscaping, horticulture, aquaculture, and floriculture. For instance, the land is used to clear unwanted grassy areas, block solar radiation, prevent excess water during rainfall, package agricultural equipment, make fishing nets, stop soil erosion, build roads for afforestation, create low tunnel rows, and protect crops from bird attacks. It also increases the cultivated land's capacity to hold water. Agro-textile products can be used in several applications and fields, as illustrated in Fig. 12.6.

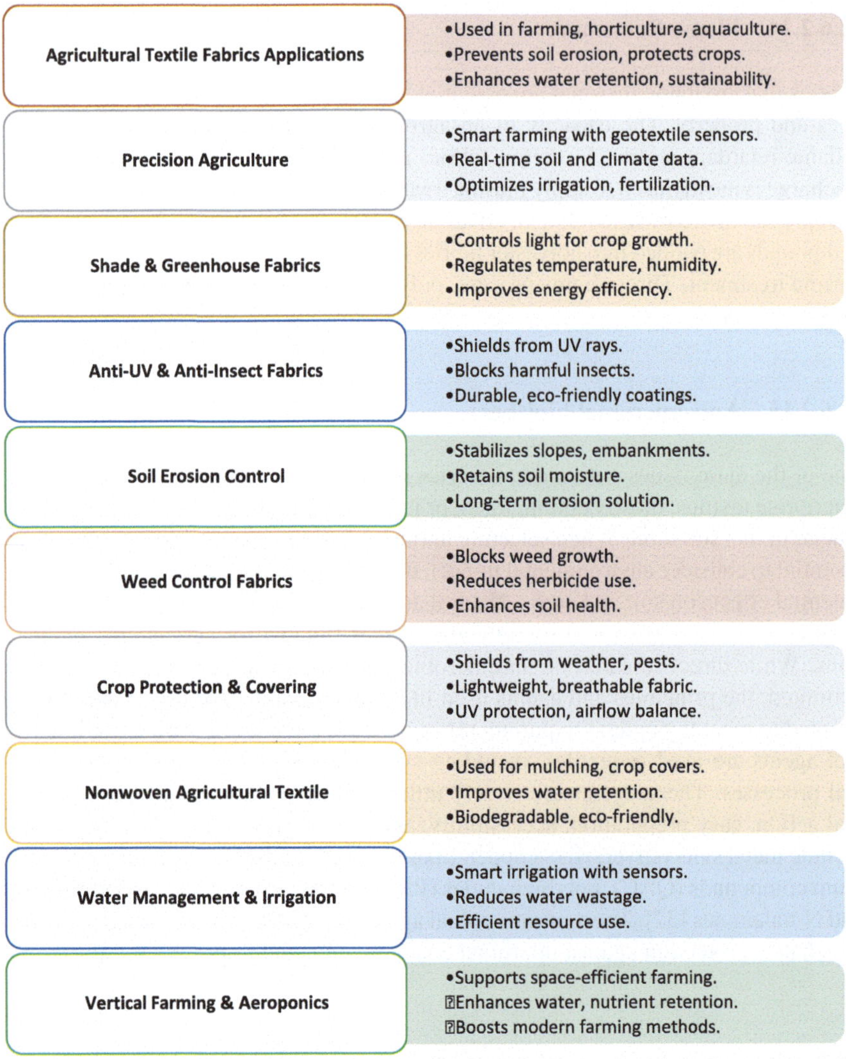

Fig. 12.6 Agro-textile applications [2, 35, 36, 38–42]

12.8 Conclusion

Agriculture is experiencing a significant shift due to technological advancements and a growing focus on sustainability. A key step toward more environmentally responsible practices in the textile industry is using eco-friendly textile finishes. By adopting biodegradable, non-toxic, and energy-efficient finishes, we can reduce environmental impact, conserve water, and minimize harm to human health. Agro-textile fabrics are leading the way in this change, with innovations that include precision agriculture solutions and biodegradable materials. These textiles promote resource efficiency and sustainable farming practices and enhance crop yield and quality. It is important for technical textile professionals to stay informed about these developments and applications in agricultural textiles. The textile industry can support the growing demand for sustainable and ethical agro-textiles by focusing on eco-friendly finishes and balancing environmental responsibility and practicality. Additionally, plasma technology plays a crucial role in advancing textile finishing, offering an environmentally friendly option that boosts the sustainability of the agro-textile industry. Agro-textiles are at the forefront of shaping the future of farming, providing materials beyond functionality to support sustainable agricultural practices.

Declaration

The authors declare that they have no known competing financial interests or personal relationships that could have appeared to influence the work reported in this chapter. Generative AI and AI-assisted technologies have been used in the writing process to improve the readability and language of the work only. After using this tool, the authors carefully reviewed and revised the content to ensure accuracy and quality. The authors take full responsibility for the final version of this publication.

References

1. Al-Madani AA, Al-Helal IM, Alsadon AA (2024) Assessing the effectiveness of reflective and diffusive polyethylene films as greenhouse covers in arid environments. Agronomy 14. https://doi.org/10.3390/agronomy14051082
2. Sharma N, Allardyce B, Rajkhowa R et al (2022) A substantial role of agro-textiles in agricultural applications. Front Plant Sci 13. https://doi.org/10.3389/fpls.2022.895740
3. Bhavani K, Mallikarjun N, Sunilkumar NM (2017) Agro textiles—their applications in agriculture and scope for utilizing natural fibers in agro tech sector. Int J Appl Home Sci 4:653–662
4. Restrepo-Osorio A, Álvarez-López C, Jaramillo-Quiceno N, Fernández-Morales P (2019) Agrotextiles and crop protection textiles. In: High performance technical textiles. Wiley, pp 279–318
5. Marasović P, Kopitar D (2019) Overview and perspective of nonwoven agrotextile. Textile Leather Rev 2:32–45. https://doi.org/10.31881/TLR.2019.23
6. Saberi Riseh R (2024) Advancing agriculture through bioresource technology: the role of cellulose-based biodegradable mulches. Int J Biol Macromol 255:128006. https://doi.org/10.1016/j.ijbiomac.2023.128006

7. Kopitar D, Marasovic P, Jugov N, Schwarz I (2022) Biodegradable nonwoven agrotextile and films—a review. Polymers (Basel) 14:2272. https://doi.org/10.3390/polym14112272
8. Dorogan A, Carpus E, Burnichi F, C-tin P (2018) Agrotextiles sustainable strategic development strategy of the converging economic sectors. In: ISB-INMA TEH international symposium, pp 171–176
9. Ministry of Textile (2021) Technical textiles : emerging opportunities and investments
10. Chowdhury MdJ, Nasrin S, Faruque MdA Al (2017) Significance of agro-textiles and future prospects in Bangladesh. Europ Scientific J ESJ 13:139. https://doi.org/10.19044/esj.2017.v13n21p139
11. Priyanka R, Sangeetha K (2016) Recent developments in agro-textiles. In: National level Seminar on "recent developments in innovative textiles" At: Jamal Mohamed College, Trichy, pp 1–6
12. Cărpuş E, Dorogan A, Burnichi F (2018) Agrotextile systems-strategic elements for sustainable development of the agriculture. In: ICAMS Proceedings of the international conference on advanced materials and systems, pp 307–311
13. Dorugade V, Taye M, Qureshi SA et al (2023) Agrotextiles: important characteristics of fibres and their applications—a review. J Natural Fibers 20. https://doi.org/10.1080/15440478.2023.2211290
14. Adriana Restrepo-Osorio, Catalina Álvarez-López NJ, Fernández-Morales and P (2019) Agrotextile and crop protection textiles. In: High performance technical textiles, pp 279–318
15. Desai A (2018) Textiles Used in Modern Agriculture to Smart Farming. 6
16. Azam F, Ahmad S (2020) Fibers for agro textiles. In: Topics in mining, metallurgy and materials engineering, pp 151–168
17. CARPUS Eftalea, DOROGAN1Angela, GROSU Cristina, STROE Cristina, SANDU Andreea BF (2019) Aspects regarding the eco-awareness of the role of agrotextile systems in the sustainable development of romania. Annals of the university of oradea fascicle of textiles, leather work 20:23–28
18. Folina A, Roussis I, Kouneli V, et al (2020) Evaluation of woven agrotextiles in the development of hemp (Cannabis sativa L.) in Greenhouse. BulletinUASVM Horticulture 77(1) 77:53–62. https://doi.org/10.15835/buasvmcn-hort
19. Nayak L, Baite H (2018) Emerging innovations in product diversification of Jute & Allied Fibres
20. Nagy N (2021) Selecting textile fibers to match the design & final product functional use to meet the challenges of the local & global market. Int Design J 11:265–278. https://doi.org/10.21608/idj.2021.162524
21. Raihan A (2024) A review of recent advances, challenges, and potential future directions of climate-smart agriculture. In: International conference on climate-smart agriculture, pp 1–22
22. Vikash SSM (2022) Agro-textile market, significance of their products and future prospect in India : a review. ANVESAK UGC Care Group 1 Journal 52:127–145
23. Kapila P, Dhillon BS (2020) Significance of agro textiles in crop production. J Agric Extension Manag XXI:84–91. https://doi.org/10.1177/2321024919883101.Why
24. Uddin F (2019) Introductory chapter: textile manufacturing processes. Textile Manuf Process, 1–11.https://doi.org/10.5772/intechopen.87968
25. Tarafder P (Dr) N (2019) Prospects of agrotextiles as technical textiles in the world market. Mantech Publ 4:26–38
26. Mihail N. DUDIN; Natalia AVEEFJAAPAK (2017) Strategic development of the textile industry in the context of the use of green manufacturing and logistics technologies 38:33
27. Massella D, Giraud S, Guan J et al (2019) Textiles for health: a review of textile fabrics treated with chitosan microcapsules. Environ Chem Lett 17:1787–1800. https://doi.org/10.1007/s10311-019-00913-w
28. Mandle SA (2019) Agro textile –a new innovation for crop management. Int J Adv Innov Res 6:137–141
29. Danihelová A, Němec M, Gergel' T et al (2023) The effect of flame retardants on the fire technical characteristics of recycled textiles. J Ind Text 53. https://doi.org/10.1177/15280837231202526

30. Anja Verbi˘c MG and BS (2019) Zinc oxide for functional textile coatings : coatings 550:17–23
31. Pratibha, Mona Verma PS and S (2024) Advance textile finishing used in textile industry. Vigyan Varta 5:62–67
32. U K Gangopadhyay, M R Mathur, P R Surwase RSS (2019) Energy efficient agro- textiles for agricultural applications. Asian Textile J 33–38
33. Mortaza Hajyzadeh YE (2023) Innovative and groundbreaking technologies for sustainable agriculture. In: SÜRDÜRÜLEBİLİR TARIM VE GIDA GÜVENLİĞİ, pp 259–277
34. Manasmita M, Kumar S, Mehra D (2024) Recent advancements in textile finishing focused on eco-friendly and plasma technology approach. J Univ Shanghai Sci Technol 26:310–321
35. Javaid S, Saleem N, Ur Rehman S (2024) Polymeric nano-emulsion in functional textile finishing. Nanoemulsions—design and applications [Working Title], 1–21. https://doi.org/10.5772/intechopen.1004397
36. Basuk M, Yadav A (2024) Eco-friendly natural fibers based sustanable agro-textile for packaging of agro-products. In: Shahid M, Aranya Mallick SD (ed) Advances in renewable natural materials for textile sustanability, 1st Editio. CRC Press, pp 1–14
37. Bibi A, Afza G, Afzal Z et al (2024) Synthetic versus natural antimicrobial agents for safer textiles: a comparative review. RSC Adv 14:30688–30706. https://doi.org/10.1039/D4RA04519J
38. Kayastha S, Behera A, Sahoo JP, Mahapatra M (2024) Growing green: sustainable agriculture meets precision farming: a review. Bhartiya Krishi Anusandhan Patrika 38:349–355. https://doi.org/10.18805/bkap697
39. Schwarz I, Kovačević S (2017) Textile application: from need to imagination. Textiles Adv Appl, 3–28. https://doi.org/10.5772/intechopen.68376
40. Grosu C, Stroe C, Sarbu T, Burnichi F, Adriana Muscalu CT (2020) Design and creativity in the field of agro-textiles. J Int Scientific Publ 8:107–119
41. Liu M, Chen X, Jiao Y (2024) Sustainable agriculture: theories, methods, practices and policies. Agriculture (Switzerland) 14:10–13. https://doi.org/10.3390/agriculture14030473
42. Manna K, Kundu MC (2018) Effect of nonwoven jute agrotextile mulch on soil health and productivity of broccoli (Brassica oleracea L.) in lateritic soil. Environ Monit Assess 190:1–10

Chapter 13
Assessment of Sustainable Textile Finishes

Sanat Kumar Sahoo and Somadatta Sharma

Abstract In today's world, there is a growing need for functional products that provide comfort and meet specific requirements. This need can be addressed using newly developed technologies in the industrial sector. In the textile industry, finishing is an essential phase in the production process, as it improves the functionality of the final products. The application of finishes involves various process parameters, which are influenced by multiple factors. These parameters must be carefully adjusted to ensure the product meets the required quality standards for its intended use. After the finishing process is complete, it is important to evaluate the final product to confirm that the desired finish has been achieved according to the established quality standards. This assessment is necessary to ensure consumer satisfaction. Additionally, the finishing processes should be sustainable to align with the expectations of the modern textile industry. This chapter discusses the functional finishing processes and the methods used to evaluate them based on established norms and standards.

Keywords Textile finishing · Processes involved · Standards followed · Assessment · Sustainability

13.1 Introduction

A crucial step in the textile manufacturing process, textile finishing involves applying various treatments to fabrics to improve their properties and functionality. Finishing procedures are essential in producing textiles that adhere to industry standards and satisfy consumer expectations, as they enhance durability and functionality while imparting color and patterns [1]. Nevertheless, it is critical to evaluate the efficacy and productivity of these finishing procedures to guarantee the quality, functionality, and environmental friendliness of the textile supply chain as a whole. Evaluation of textile finishing procedures entails assessing the results of every treatment stage,

S. K. Sahoo · S. Sharma (✉)
Textile Engineering Department, Odisha University of Technology and Research, Bhubaneswar, Odisha, India
e-mail: som.iihtb@gmail.com

S. Biranje et al. (eds.), *Sustainable Finishing Techniques in Textiles*,
https://doi.org/10.1007/978-981-96-4860-3_13

scrutinizing the influence on the properties of the fabric, and enhancing the efficiency of the production process as a whole. The extensive evaluation comprises various critical components, such as performance testing, cost-effectiveness, environmental impact assessment, and performance monitoring. Through the methodical assessment of these variables, textile manufacturers have the ability to optimize their finishing methodologies, reduce wastage, and bolster the environmental viability of their activities [2]. The textile industry is currently confronted with a critical juncture as it attempts to simultaneously satisfy consumer demand and reduce its ecological impact. In light of the growing significance of sustainability in numerous industries, the textile sector, due to its complex supply chains and substantial environmental footprint, is currently experiencing unprecedented scrutiny [3]. In the present context, textile treatments are of paramount importance. These textile finishing, which are utilized to improve the characteristics of the material, consist of an extensive variety of treatments, including stain and water repellents, flame retardants, and antimicrobial agents. This chapter aims to conduct an in-depth analysis of the evaluation of sustainable textile finishes, exploring the intricacies, progressions, and obstacles in sustainable textile finishing procedures. As consumer preferences are increasingly influenced by environmental consciousness and ethical considerations, the textile industry must adopt sustainable practices. Sustainability in textile finishes, on the other hand, necessitates the management of an intricate web of elements, such as resource usage, environmental impact, adherence to regulations, and consumer opinion [4]. Before proceeding, it is critical to comprehend the environmental ramifications associated with traditional textile finishes. Throughout history, numerous processes for finishing textiles have utilized hazardous compounds that adversely affect human health and the environment. The persistent presence of these chemicals, including heavy metals and perfluorinated compounds (PFCs), in the environment long after manufacturing the fabric poses a threat to ecosystems and fauna [5]. Additionally, certain finishing procedures that are energy-intensive contribute to the emission of greenhouse gases, thereby intensifying the issue of climate change. Therefore, there is an immediate need for sustainable alternatives.

Technological advancements and changing consumer demands are propelling a positive transformation in sustainable textile treatments. An auspicious strategy entails the advancement of biodegradable and biobased coatings sourced from renewable materials, including microbial cultures, enzymes, and plant extracts [6]. These environmentally sustainable alternatives provide functionality similar to their traditional counterparts while substantially diminishing their ecological footprint. Moreover, progress in nanotechnology has given rise to nano-scale finishes characterized by improved durability and reduced resource consumption, thereby creating novel opportunities for sustainable textile finishing. Nevertheless, the evaluation of textile coatings' sustainability extends beyond the simple replacement of components; it necessitates a comprehensive examination of the product's complete life cycle. With this in mind, life cycle assessment (LCA) is a useful instrument that enables producers to assess the ecological consequences of finishes throughout their entire life cycle, from the extraction of raw materials to their eventual disposal [7]. LCAs enable optimizing processes and reducing environmental burdens by quantifying variables such as energy consumption, water usage, and emissions.

In addition, industry standards and regulatory frameworks are pivotal in promoting sustainability initiatives in the textile industry. Certifications and guidelines for sustainable production practices, such as the Higg Index of the Sustainable Apparel Coalition and Bluesign®, promote accountability and transparency throughout the supply chain [8]. Adherence to these standards strengthens the brand's reputation and ensures that textile finishes meet strict environmental and social requirements. Notwithstanding these developments, obstacles continue to endure in pursuing sustainable textile finishes. A significant challenge is the intricate nature of supply chains, which frequently traverse multiple nations and entail a multitude of participants. Establishing and maintaining traceability and transparency across these supply chains is critical to detecting and alleviating environmental and social hazards. Furthermore, the extensive implementation of sustainable practices is hindered by financial constraints, as environmentally friendly alternatives may require greater upfront investments or ongoing operational expenditures.

13.2 The Objective of Assessment

The objective of assessing textile finishes is to ensure that the applied finishes meet the required quality, functionality, and compliance standards. This involves evaluating whether the finishes achieve the intended properties, such as enhanced durability, softness, or specific functional attributes like water resistance, antimicrobial protection, or flame retardancy. The assessment also ensures that the finished product complies with industry norms, environmental regulations, and social standards, reflecting the growing importance of sustainability in the textile sector. Additionally, the process aims to verify that the final product fulfills consumer expectations for performance, comfort, and appearance. Ultimately, thorough assessment supports the production of high-quality textiles, reduces waste, and reinforces the brand's reputation in the market.

13.2.1 Quality Assurance

The evaluation of textile finishing processes is incomplete without quality control, which guarantees that completed fabrics conform to predetermined criteria regarding color, appearance, tactile experience, and functionality [9]. Quality control measures commence during the raw material stage, wherein the consistency and uniformity of fabric characteristics are evaluated. During the refining process, samples undergo regular assessments of pertinent characteristics such as colorfastness, dimensional stability, and tensile strength [10]. Preventative measures are taken to resolve any identified deviations from quality standards to ensure the integrity of the product and maintain customer satisfaction.

13.2.2 *Performance Evaluation*

Performance testing aims to ascertain whether the functional properties of completed textiles satisfy the demands of particular end-use applications. As determined by their intended application, fabrics may be subjected to the following performance evaluations:

- Wear resistance and durability are evaluated via abrasion resistance testing.
- Moisture management and water resistance testing are utilized to assess the permeability and comfort of a fabric.
- Flame retardancy testing aims to ascertain resistance to flame propagation and ignition.
- Evaluation of the effectiveness of antimicrobial coatings in impeding microbial growth through antimicrobial efficacy testing.
- UV protection testing determines the capacity of a fabric to obstruct damaging UV radiation.

Performance testing is a critical tool in evaluating the appropriateness of completed textiles for various uses, thereby offering direction for product development and marketing tactics [11].

13.2.3 *Analysis of Environmental Impact*

It is essential to evaluate the environmental consequences of textile finishing procedures to promote sustainability and reduce pollution and resource consumption. An environmental impact analysis assesses various aspects of finishing operations, including energy consumption, water usage, chemical usage, greenhouse gas emissions, and refuse generation. Methods such as life cycle assessment (LCA) offer an all-encompassing structure for quantifying the ecological impact of textile products, spanning from the exploitation of raw materials to their eventual disposal at the end of their useful lives [12]. To mitigate their ecological footprint, textile manufacturers may adopt a range of tactics, which comprise:

- Implementing energy-efficient and water-conserving technologies, including heat recovery and low-water dyeing methods.
- One may implement closed-loop chemical management systems or substitute hazardous chemicals with safer alternatives to reduce chemical usage and waste.
- Integrating biodegradable coatings, recycled fibers, and renewable materials can foster circularity and decrease dependence on primary resources.

- Enhancing resource efficiency and minimizing waste generation through the optimization of production processes.

Textile manufacturers can foster a more environmentally conscious sector and reduce their ecological footprint through the adoption of sustainable practices and the execution of comprehensive environmental impact assessments [13].

13.2.4 Cost-Effectiveness

To determine whether textile finishing procedures are cost-effective, the balance between production expenses and the quality and functionality of the final products must be evaluated. Efficient finishing methods reduce production interruptions, maximize output, and optimize resource utilization while maintaining product quality standards. In a cost-effectiveness analysis, various elements are considered, including expenditures on basic materials, labor, equipment acquisition, and overhead.

13.2.5 Increase Efficiency

Invest in cutting-edge technology and apparatus to reduce labor costs and streamline production processes. Cultivate advantageous contractual agreements with suppliers to obtain raw materials and substances at competitive prices. Lean manufacturing principles should be implemented to reduce lead times, eradicate waste, and increase efficiency. Maintain a constant vigil over and optimize production workflows to detect opportunities for cost reduction and enhance profitability. Textile manufacturers can augment their competitiveness and profitability while ensuring the provision of superior-quality finished products to the market through adopting cost-effective practices [14].

13.3 Textile Finishing Process

Fabric is subjected to various treatments and techniques during the textile-finishing process to improve its properties and functionality. Although particular finishing processes may differ contingent on the fabric type, intended properties, and intended uses, the subsequent enumeration provides a broad synopsis of prevalent textile finishing procedures:

13.3.1 Pre-Treatment

Pre-treatment procedures eliminate impurities, lubricants, and natural waxes to prepare the fabric for subsequent finishing processes. Desizing (removing sizing agents utilized during the textile process), scouring (removing greases and natural impurities), and bleaching (lightening or whitening the fabric) are all possible pretreatments [15].

13.3.2 Dyeing

Various dyeing techniques impart color to the fabric during the dying process. Frequent dyeing techniques encompass:

- In direct dying, dyes are dropped into a receptacle and applied to the fabric.
- Disperse dyeing: Suitable for synthetic fibers; these pigments are applied to the fabric via heat after being dispersed in water [16].
- Reactive dyeing: Establishing covalent bonds between reactive dyes and fabric fibres imparts superior colour-fastness.
- Vat dyeing: To become soluble and adhere to the fabric, vat dyes, which are insoluble in water, require a reducing agent.

13.3.3 Printing

Printing involves the application of patterns and colors to the surface of fabric through diverse printing techniques, including heat transfer printing, screen printing, roller printing, and digital printing [17]. The printing process enables the application of elaborate designs and patterns onto the fabric.

13.3.4 Finishing Treatments

- Mechanical finishes: Mechanical finishing procedures modify the surface texture and visual appeal of the fabric by employing mechanical methods, including calendaring, sanding, brushing, or pressure and heat-induced smoothing.
- Chemical finishes: Chemical finishes are utilized to imbue fabrics with distinct characteristics, including but not limited to suppleness, water resistance, stain resistance, flame retardancy, and wrinkle resistance. Common chemical finishes include

  - o Lubricants and softeners that enhance the sensation of the hand.
  - o Water-repellent coatings comprised of silicone or fluorocarbon-based compounds.
  - o Finishes that are antimicrobial to impede microbial growth.
  - o Finishes that inhibit flames utilizing chemical additives.

- Biological finishes: Enzymes or microorganisms are utilized in biological finishing treatments to alter the fabric's properties. Enzymatic desizing, bio-polishing to eliminate surface fuzz, and biodegradable coatings are some examples [18].
- Nano-finishing treatments: These treatments use nanoparticles to imbue particular characteristics onto the surface of fabrics, including but not limited to stain resistance, UV protection, and antimicrobial activity.
- Curing and Drying: Following finishing treatments, printing, dying, or printing, the fabric is dried by a variety of means, including infrared drying, steam drying, or hot air drying [19]. Certain coatings may necessitate curing at higher temperatures to guarantee appropriate adhesion and longevity.
- Quality Control: Quality control measures are consistently applied during the finishing process to ascertain that the final fabric satisfies predetermined criteria regarding colour, appearance, tactile experience, functionality, and safety.

13.4 Requirement

Fabrics' value is increased through textile finishing, an integral part of the textile manufacturing process, which improves their appearance, feel, and functionality [20]. The following are essential criteria and goals of textile finishing:

- Textile finishing procedures are frequently utilized to augment the aesthetic appeal of fabrics. This may involve imbuing the object with color via printing or dying, establishing ornamental effects, or enhancing the surface texture.
- To enhance comfort and sensation, applying finishing treatments can modify the tactile qualities of the fabric, rendering it more opulent, silky, or velvety. This aspect holds specific significance in the case of textiles designed for home furnishings and apparel, where comfort is the foremost consideration.
- Enhancing Strength, Abrasion Resistance, Tear Resistance, and Dimensional Stability: Finishing treatments can potentially augment the durability and performance of textiles by improving pertinent properties [21]. This ensures that the fabric retains its integrity and appearance despite repeated laundering and use.
- By imparting particular functional properties to fabrics, textile finishing enables them to satisfy the needs of various end purposes. This encompasses remedies that provide UV protection, water repellency, stain resistance, flame retardancy, wrinkle resistance, and antimicrobial properties.

Fig. 13.1 Finished fabric to be assessed

- Certain finishing treatments can facilitate the management and processing of textiles throughout their lifecycle, including production, transportation, and use. This encompasses treatments that effectively mitigate shrinkage, creasing, and static electricity, simplifying the fabric's manipulation and upkeep.
- To ensure adherence to regulatory requirements, textile finishing processes must follow pertinent regulatory standards and certifications that regulate worker health and safety, environmental impact, and chemical safety [22]. By adhering to these standards, the safety and sustainability of finished textiles can be guaranteed to meet both legal obligations and consumer expectations.
- Textile finishing is essential in aligning with the ever-changing demands of the market and consumers' preferences. This encompasses the capacity to provide an extensive assortment of hues, designs, textures, and functional attributes to accommodate customers' varied requirements and preferences.
- Striving for equilibrium Cost and Efficiency: To balance the resources and time needed for production and the intended finishing effects, textile finishing processes must be both cost-effective and efficient. This entails the optimization of processes, the reduction of waste, the maximization of productivity, and the preservation of quality standards (Fig. 13.1).

13.5 Classification

Textile Finishing refers to a step-by-step sequence of operations carried out on the surface of grey fabric to enhance its overall functionalities and surface characteristics as required during specific end uses. The term finishing encompasses all the pretreatment processes bestowed upon the grey fabric, such as scouring, bleaching, dyeing, printing, etc., whereas the term "Functional finishing" refers to mechanical and chemical operations performed on the fabric to impart some special aesthetic and other end-use specific properties [23]. An overview of classification of functional finishes is shown in Fig. 13.2.

13.6 Assessment of Finishing Process

13.6.1 Mechanical Finishes

Mechanical finishes encompass a wide range of methods that alter the characteristics of textile materials via physical processes as opposed to chemical procedures. These treatments are of utmost importance in improving textiles' visual attractiveness, functionality, and performance in a wide range of sectors, including home furnishings, technical textiles, apparel, and industrial applications. In contrast to chemical finishes, mechanical finishes provide distinct benefits, including enhanced permeability, durability, and tactile qualities [23]. Furthermore, they promote sustainability by reducing dependence on potentially hazardous substances.

Mechanical finishes are an adaptable and environmentally friendly method for altering the characteristics of textile materials, providing an extensive array of techniques to cater to the varied demands of different industries. Nevertheless, ongoing collaboration and innovation must persist to tackle sustainability challenges and optimize mechanical finishing processes. By collectively striving to implement environmentally conscious methods and optimize resource utilization, the textile sector has the potential to utilize mechanical finishes as a catalyst for a more sustainable future [4].

13.6.1.1 Calendering

The purpose of calendering is to alter the texture of the fabric by either softening or solidifying it [24]. To bestow a silken sheen upon the fabric. Through the application of heat and pressure in an extruder, the molten polymer is transformed into a sheet or film using calendering rollers.

A multitude of finishes can be attained via the calendering procedure by manipulating distinct facets of the procedure. Beetling, watered, embossing, and Schreiner are the primary varieties. In comparison to extruded films, calendered material generally exhibits superior tensile properties, product uniformity, and gauge control that is remarkably precise [25]. Melt degassing is an additional benefit of the process mixing, which makes it particularly well-suited for materials and applications that are densely loaded. PVC is the predominant plastic material calendered. The assortment of products consists of upholstery fabrics and wall coverings, reservoir linings, and agricultural mulching materials.

Fig. 13.2 Classification of finishing process

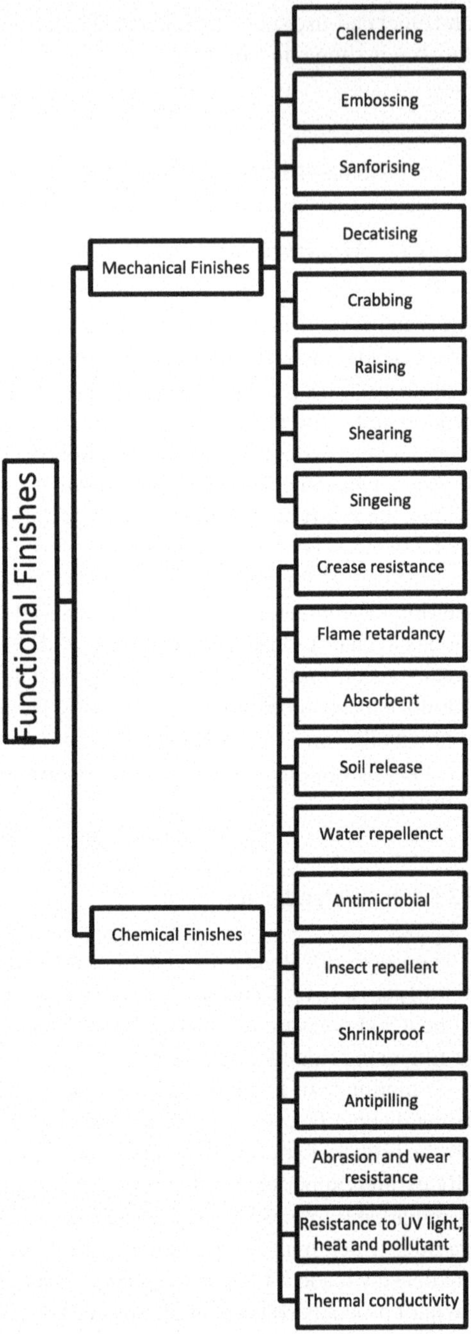

13.6.1.2 Embossing

Embossing is a printing method that creates a contrast in texture by applying one or more colors of ink to the surface of the paper and then another colour to the raised areas [26]. Passing the fabric through heated rollers bearing an engraved design results in the application of a raised design that is visually distinct from the backdrop. Even though embossing was previously a transient effect, processes have since been developed to render it permanent. An aesthetic effect is achieved with the use of embossed designs on fabrics. Embossing is accomplished with specialized calendering rollers (calendering refers to the process of altering the surface of a fabric by pressing between two rollers; this is what is meant by the term "embossing") [27]. While its primary function is decorative, embossing can also serve a functional purpose in certain circumstances. Most embossing is utilized to imbue materials and products with aesthetically pleasing designs. Embossing is a craft that improves quality and imparts a distinctive detail.

13.6.1.3 Sanforising

Sanforising is a mechanical finishing technique utilized on woven fabrics, with a specific focus on cotton. During this process, the fabric is stretched, fixed, and stabilized to prevent substantial shrinkage after laundering [28]. This method is also known as anti-shrinking finishing. Sanforization is the pre-shrinking of fabric before its incorporation into a garment. An item of fabric that has undergone the sanforization process is labelled "pre-shrunk" or "sanforized." Without sanforisation, denim is called "unsanforized" or "shrink-to-fit." Sanforizing and shrinking machines are utilized to stabilize and pre-shrink cotton and blended cotton fabrics, including denim. It consists of two components: the shrinking device, which regulates compressive shrinkage with steam and mechanical pressure, and a thick Sanfor rubber belt [29].

13.6.1.4 Decatizing

Decatising, also referred to as decatizing, is a textile finishing procedure that imparts a permanent finish to fabric, preventing it from shrinking during the garment-making process [30]. The term originates from the French décatir, denoting the process of eliminating the cati or finish from wool. Fabric is coiled into a roll using a cotton or cotton/synthetic wrapper, and steam is circulated through the roll for a maximum duration of 10 min. The fabric and wrapper are chilled by drawing ambient-temperature air through the roll after the steaming process. In the woollen fabric processing stage, a discontinuous decatizing machine is employed to impart handle (or surface smoothness) and stabilization to worsted and woollen-blended fabrics [31].

13.6.1.5 Crabbing

During the finishing process, controlled tension is applied to the fabric to attain a fabric structure that is both symmetrical and square. The term "crabbing" originates from the lateral displacement that occurs when tension is applied to the fabric, causing it to resemble the motion of a crab. To ensure the fabric's squareness and alignment, rubbing rectifies any warping or distortion that may have occurred during preceding processes like dying, printing, or weaving [32]. Bypassing the fabric through a crabbing machine, which causes the fabric to distort or move laterally by applying lateral tension and steam, the process is accomplished. By performing this lateral motion, any inherent bias in the fabric strands is mitigated. Two primary categories of crabbing procedures exist:

- *Wet Crabbing*—Before passing through the crabbing machine, the fabric is dampened with water or a steam-dampening system in wet crabbing [33]. Moisture facilitates the relaxation of fabric fibres, thereby enhancing the ease of manipulating and rectifying any distortions. It is common practice to wet crab natural fibres such as cotton and linen.
- *Dry Crabbing*—It is an upholstery process in which the fabric is passed through a crabbing machine devoid of any supplementary moisture. Conversely, the fabric is exposed to regulated heat and tension, frequently utilizing steam or heated air [33]. Dry crabbing is a method commonly utilized to clean synthetic fibres such as nylon or polyester.

13.6.1.6 Raising

Raising is another finishing procedure. As in the case of flannelette, the fabric surface is treated with sharp teeth during the raising process to elevate the surface fibres, which imparts warmth, softness, and hairiness. The purpose of a raising machine is to impart a hairy texture to a fabric [34]. This is accomplished using a series of needles that, following a mechanism that will be elaborated upon subsequently, penetrate, break, and elevate the fibres.

13.6.1.7 Shearing

Shearing is a type of mechanical finishing that improves the aesthetics of the fabric by precisely and uniformly slicing the raised surface or loops to a consistent height [35]. The apparatus potentially possesses a spiral blade akin to that of a grass-cutting machine. Shearing machines are capable of reducing the mass or loop to the desired height. A shearing machine is designed to cut the pile (or yarn) at a constant height that is evident on the surface of a fabric. This operation applies to a vast array of fabrics and is suitable for an extensive variety of purposes. By eliminating superfluous fibres and producing a sleek finish, shearing enhances the fabric's quality and value

[36]. A variety of textures and patterns can be produced on the fabric through the process of shearing, thereby augmenting its aesthetic appeal and texture.

13.6.1.8 Singeing

Singeing is a mechanical finishing or treatment utilized in textiles to achieve a sleek surface or filament with fewer hairs. The protruding fibres of threads or fabrics are burned in a singeing machine by exposing them to direct flames or heated plates [37]. The term for this is "gassing." The primary purpose of the singeing process is to eliminate protruding fibres from the fabric surface to reduce the tendency for fabric pilling and produce a clean surface. The singeing process is utilized to enhance the sheen and smoothness of textile or fabric surfaces. The purpose of singeing a fabric is to acquire a pristine fabric surface that permits clear visibility of the fabric's structure. Singed fabrics are more resistant to soiling compared to unsinged fabrics [38]. Especially with synthetics and their blends, the danger of pilling is diminished when fabrics are singed.

13.6.2 Chemical Finishes

The processing of every form of fabric utilized in the apparel and home fashion industries. It consists of fabric pretreatments, dying, printing, and finishing. To cultivate an understanding of fabric value addition, "Textile Chemical." The application of chemical coatings typically involves padding, curing, and drying. Foam application, coating, exhaustion, cushioning, and exhaustion are the primary methods of application, which vary according to the chemical finish, textile material, and equipment availability [39]. Using chemical products utilized in finishing formulas, fabrics acquire an array of technical properties. In particular, they can impart softness, fire resistance, and impermeable qualities to fabrics, as well as prevent creasing, felting, and static generation in materials [40]. As report says that a misdistribution and lack of management of the approximately 8000 compounds utilized in the textile industry's various manufacturing processes can contaminate natural water bodies.

13.6.2.1 Crease Resistance

Descriptive of a treated fabric or garment that, when laundered, retains its sleek appearance, form, creases, and/or pleats. These types of garments require minimal to no ironing, especially when dried in a tumble dryer. The application of crease-resistant treatments to cellulose fibres (cotton, linen, and rayon) that are susceptible to wrinkling [41]. Fabrics labelled "Permanent Press" feature crease-resistant finishes that not only impede creasing but also aid in preserving pleats and creases during laundering and use. Consequently, the finish is labelled anti-wrinkle, crease-resistant, or anti-crease. After wearing, garments treated with these finishes tend to become smoother when hung correctly. This process is alternatively referred to as resin finishing due to the utilization of resins. This texture is permanent and chemically based.

Important parameters that aid in determining maximal conversion and packaging speeds, board stiffness, and crease resistance testing may also be associated with the final dimensional stability of the product. The most prevalent method involves impregnating the fabric with a reagent solution that permeates the fibres. Following the drying and curing process, the fibre structure is cross-linked through the action of a catalyst and heat. The effect of resistance to creasing is washable and resistant to normal use.

Increase recovery refers to the capacity of a fabric to revert back to its initial state. This recuperation is quantified in the form of the crease recovery angle (CRA), which serves as an indicator of crease resistance [42]. This is determined with the aid of the Shirley crease recovery instrument. A fabric is given a crease by separating it between two glass plates that are subjected to a 500 g weight for one minute. Utilizing Three-Dimensional Crease Retention Replicas, fabrics that have undergone testing following AATCC TM88C are evaluated.

13.6.2.2 Flame Retardancy

Flame retardants are substances that are applied to combustible substances, including textiles, to prevent their ignition or spread. Fabrics treated with flame retardants can reduce fire damage and potentially save lives. Curtains, carpets, car seats, sofas, firefighting attire, and so forth. Using chemical flame retardants, fabrics are treated to impede or suppress the combustion process. They disrupt combustion at multiple phases, including heating, decomposition, ignition, and flame propagation. A gaseous phase reaction produces fire. In essence, flame-retardant coatings on fabrics are indispensable for safeguarding against fire incidents and promoting safety in a variety of industries [43]. In addition to impeding the propagation of flames, they diminish the emission of hazardous vapours and enhance the longevity of textile materials. The majority of polymers that possess inherent fire resistance are produced through the incorporation of aromatic cycles or heterocycles, which provide the polymers with rigidity and stability. From aromatic heterocycles, polymers such as polyimides, polybenzoxazoles (PBOs), polybenzimidazoles, and polybenzthiazoles (PBTs) are formed.

In order to ensure adherence to fire safety regulations, flame-retardant textiles are frequently utilized in hospitality establishments such as hotels, restaurants, theatres, and other public spaces. Curtains, draperies, bedding, upholstery, and carpets incorporate these materials to mitigate the potential for fire incidents and guarantee the well-being of individuals within. The TF319 Multipurpose Textile Flammability Tester is capable of assessing the flammability resistance of an extensive range of textiles, including but not limited to domestic items, safety protection nets, toys, and decorative fabrics. It is predominantly utilized for vertical combustion testing. As defined by ASTM D6413, this standard employs the vertical smouldering method to test flame retardants [44].

13.6.2.3 Absorbent

Absorbency refers to a fabric's capacity to absorb moisture; this is a critical property that influences a variety of other attributes, including skin comfort, resistance to static buildup, contraction, water resistance, and wrinkle recovery. With hydrophilic finishes, synthetic fibres that come into contact with the epidermis are uncomfortable due to their low moisture absorption. Water is absorbed by hydrophilic fibres due to the bonding of their molecules with water molecules. These surfaces retain moisture. Fabrics that absorb moisture, including cotton, are exceptionally hydrophilic. To impart absorbency, hydrophilic finishes utilize chemicals including ammonium compounds, humectants, and wetting agents [45]. Due to the fact that their molecular structure prevents water molecules from readily bonding with them, hydrophobic fibres repel water. Cotton remains the prevailing absorbent fibre utilized, although contemporary textile innovations have introduced more absorbent alternatives, including modal, micro-modal, Tencel®, and other viscose-based fibres. All of these are composed of the water-loving plant cellulose that serves as their foundation. A concentrated organo-hydrophilic finishing agent, Moisture-Guard is designed for use in cellulosic, polyamide, and their mixtures. It facilitates enhanced wicking, rapid moisture absorption, and exhalation, resulting in a cooling sensation for the wearer. The American Association of Textile Chemists and Colourists (AATCC) published the AATCC 79 test, which measures the water absorbency of apparel, fabrics, and fibres [46]. Embroidered, knitted, and non-woven textiles of any fibre content or structure may incorporate it.

13.6.2.4 Soil Release

A soil-release finish can aid in preventing the infiltration of various particles and liquids into the fabric [47]. The garment is significantly simpler to clean if the stain remains on the surface. An additional finish that enhances the fabric's absorbency facilitates the removal of stains at a quicker rate, improves the fabric's wicking action for increased wearer comfort, enables dry-cleaning without substantial depletion of soil release properties, and preserves the fabric's brightness even after repeated laundering. This is achieved through the augmentation of surface tension on the garment. The capacity of a textile to eliminate stains while undergoing laundering is improved. Although liquids may not bead up for a release finish, they typically absorb into the fabric. For evaluation, the Soil Release test (AATCC 130) is utilized [48]. In recent times, stain-release finishing has become increasingly prevalent in garment fabrics. This innovation effectively combats stubborn stains and soil to ensure the fabric retains its attractiveness and durability, as the stains are effortlessly eliminated during the laundering process. Nicosist-SRH is a soil-release finish characterized by its antistatic properties and supple handle. The treated products acquire a hydrophilic quality and enhanced moisture absorption as a result; thus, they possess the soil release property. Additionally, it has a silky texture and an antistatic effect.

13.6.2.5 Water Repellent

Water-repellent coatings are fabric surface treatments that provide a certain level of water resistance while maintaining the open pores of the fabric, thereby enhancing comfort during wear [49]. Wax and resin compounds, aluminium salts, silicones, and fluorochemicals are examples of such finishes. Water-resistant coatings are succeeded by water-repellent coatings. Products bearing the water-repellent label possess hydrophobic properties, which effectively dissuade water from entering. Water-repellent materials are more resilient in the face of aqueous conditions. A coating substance on water-repellent fabrics causes water to collect and roll away. However, this will help keep you dry during light rain, but it will not provide much protection during severe rain. Additionally, they are frequently more affordable and breathable than materials that are impermeable or resistant to water. Water-repellent surface protective coatings offer several advantages, including enhanced resistance to corrosion, protection against UV radiation, and improved aesthetic appeal. In addition to reducing energy expenses, these coatings can increase the longevity of structures. AATCC TM22 is frequently used to screen the water-repellency of fabric in a fast and inexpensive manner [50]. This testing procedure is universally pertinent to textile fabrics, irrespective of whether or not they have been treated with a water-repellent coating. It determines the degree to which fabrics resist water soaking.

13.6.2.6 Anti-Microbial

Fabrics with anti-microbial finishes prevent the growth of microorganisms like mould, mildew, algae, bacteria, viruses, and fungi [51]. The methods used to apply antimicrobials to fabrics differ. Certain antimicrobials function by transferring or evaporating off the surface they are applied. Either to shield the fibre from microbial attack or to shield the user from the spread of harmful germs is the dual function of the antimicrobial finishes. When compared to cotton, silk may have better mildew resistance, but both its mildew resistance and storage stability are lacking. Antimicrobial technology strengthens fabric's defences against mould, mildew, and bacteria that cause odours. This prolongs the freshness of products made of antimicrobial fabric, enabling you to wear them with assurance. To assess its potential as a long-lasting antimicrobial finish, cotton garments were treated with Bio shield AM 500 3-(trimethoxy silyl)-propyl-dimethyl octadecyl ammonium chloride using a cold pad-batch process in the presence of an alkaline catalyst [52]. For textile finishing, a wide range of antimicrobial agents is available, including triclosan, silver nanoparticles, quaternary ammonium compounds, etc. But because they are synthetic, they have an impact on the environment. The effectiveness of antibacterial treatments applied to textile fabrics is assessed quantitatively using the AATCC 100 test method. Antibacterial treatments applied to fabrics protect consumers from dangerous microorganisms and lower the chance of illness transmission.

13.6.2.7 Insect Repellent

Textiles that are finished with insect-repellent chemicals are treated to deter insects like ticks, mosquitoes, and ants [53]. These finishes can be used as a post-treatment for the fabric or as part of the production process. This is a summary of the procedure:

- Choosing Repellent Chemicals: Permethrin, DEET, picaridin, and IR3535 are a few of the substances that work well as insect repellents [54]. Regulations, the target insects, and durability requirements all have a role in the chemical choice.
- Techniques of Application: There are several ways to apply insect repellents on fabrics, such as clothes are dipped or submerged in a repellent chemical solution, and any leftover solution is then wiped off.
- Padding: A solution containing the chemical repellant is used to cushion fabrics.
- Spraying: With the right tools, repellent compounds can be sprayed onto textiles.
- Encapsulation: Chemicals that repel people are contained in minuscule capsules that stick to the surface of the cloth.

Diethyl Benzamide N–N Cotton materials were treated with a mosquito-repellent finish by employing a straightforward padding technique. The N-NDiethyl Benzamide treated cotton fabrics that repel mosquitoes were assessed using the Cage Test. 100% mosquito repellency was achieved with the fabric treated with 12% N–N Diethyl Benzamide [55]. It is frequently used to ward off biting pests like ticks and mosquitoes.

13.6.2.8 Shrink Proof

Sanforizing is a mechanical finishing technique used on woven textiles, especially cotton, in which the fabric is stabilized, stretched, and fastened to prevent excessive shrinkage after washing. The anti-shrinkage finishing procedure is another name for it. Finishing clothing and similar items to avoid shrinking is known as anti-shrinking finishing [56]. The mechanical method, which makes use of potential shrinkage already present in the fabric and pre-retracts it before it becomes a finished product, is an efficient way to address warp direction shrinkage. A sanforizing or zero-zero machine is used in the anti-shrink finish process to apply controlled shrinkage to fabrics. The fabric is dampened and placed onto the stretched surface of a rubber blanket by the machine, which employs roller contraction and steam heating to shrink the fabric uniformly as the blanket contracts. Additionally, the ISO created ISO 3759, a standard that gauges how a fabric's dimensions change after washing and drying. While there is a tumble-drying phase added, this test is comparable to the AATCC test. AATCC 135 and AATCC 158 are two commonly used shrinkage test procedures that are internationally recognized [57]. ISO 3759 is another. The ISO guidelines also state that a fabric's shrinkage may not exceed three to five percent. A few methods of fabric processing can be applied to lessen the rate at which fabric shrinks. Examples include pre-shrunk and merchandising. Cotton cloth is the primary material for these techniques. Heat frequently causes synthetic fibres, such as nylon, to shrink. However, if materials are heat-set during production, shrinkage can be minimized.

13.6.2.9 Anti-Pilling

The fabric's resistance to abrasion during wear causes tiny balls to develop on the surface. This might be a built-in characteristic of the fabric or the result of a specific treatment applied to stop pilling. Using finishing auxiliaries in resin form is the greatest way to minimize the creation of pilling since they function as binders and stabilize the fibres, avoiding migration to the surface of yarns and fabrics. The foundation of anti-pilling finishes is the use of chemical treatments designed to lessen synthetic fibres' mechanical resistance and inhibit their capacity to slacken [58]. One example that is suggested for the bio-polishing procedure is AVCOSON DNL. By breaking down tiny cellulose fibres on the cloth's surface, AVCOSON DNL produces a softer, smoother fabric that has less pilling during future washing cycles. Using the standard test procedure, ASTM D 4970 (500 rpm) on Martindale, the treated fabric's anti-pilling efficacy was assessed [59].

13.6.2.10 Abrasion and Wear Resistance

Abrasion-resistant fabric protects the wearer and prolongs the life of a product by withstanding surface wear from rubbing. Companies whose products depend on consistent safety and wearability look for textiles that maintain their construction after prolonged use. According to Abdullah et al., abrasion is the physical breakdown of fibres, yarns, and fabrics caused by rubbing one textile surface against another [60]. Abrasion is one of the main causes of textile fabrics becoming unusable, while there are other reasons as well. The ability of a cloth to withstand surface wear brought on by flat rubbing against another material is known as abrasion resistance. The textile industry typically uses the Wyzenbeek and Martindale test methods to evaluate abrasion resistance. In actuality, the abrasion test procedure (ISO 4649 / DIN 53,516) is rather straightforward [61]. A test rubber piece is moved across the surface of an abrasive sheet that is fixed to a rotating drum to determine the abrasion resistance. It is stated as a loss of volume in cubic millimetres, such as 150 mm^3. Polyethylene is by far the most important finishing agent discovered for enhancing the abrasion resistance of non-durable press cotton textiles.

13.6.2.11 Resistant to UV Light and Pollutant

Ultraviolet (UV) protective finishes are specialized treatments applied to fabrics to shield them from UV radiation, protecting human skin from potential damage. Common chemical finishes used for UV protection include UV absorbers, fluorescent whitening agents (FWAs), and titanium dioxide (TiO_2) treatments [62]. A cloth becomes less permeable to UV light when it contains TiO_2 particles, which scatter UV rays. Certain plastics have built-in UV stability, including acrylic, Ultem®, PVDF, and PTFE. However, when exposed to UV radiation, most plastic polymers show color changes and elongation loss. Tightly woven materials with a high thread count and bright or dark hues are the best defense since they absorb light. Canvas, nylon, polyester, and lightweight wool are a few examples. Fabrics coated with acrylic are renowned for their superior UV resistance and color preservation. These textiles are created by covering a base material, like polyester or cotton, with acrylic. The acrylic treatment improves the fabric's resilience to fading and durability.

The spectrophotometric analysis technique uses a spectrophotometer to measure the amount of UV light transmitted through a fabric [63]. This method evaluates both the UV radiation absorbed and transmitted by the fabric sample after exposure to light.

The European standard EN 13,758–1, the American Association of Textile Chemists and Colourists Test Method (AATCC-183), the Australian/New Zealand standard AS/NZS 4399:1996), and the UV standard 801 [64]. These test standards define different conditions for the substance under test. The UV resistance of the fabric is examined using two different test types: in vitro (UPF) and in vivo (SPF). In vivo examinations are performed on volunteers' backs who have healthy skin. A spectrophotometer operating in the 280–400 nm range is used in in vitro testing to determine the material's UPF.

13.6.2.12 Thermal Conductivity

Thermal conductivity is a significant thermal property of a fabric that is closely linked to the fabric's breathability. A fabric with a high thermal conductivity allows heat to be transferred from a hot surface, such as the human body, to a cooler surface, such as the air on the other side of clothing. Thermal conductivity refers to the capacity of a substance to transfer heat from one side to the opposite side [65]. The thermal conductivity coefficient λ represents it. A smaller value of λ implies that the material possesses enhanced heat insulation and preservation properties. One notable characteristic is that fabrics do not naturally have high heat conductivity. Indeed, textiles are employed for thermal insulation rather than heat dissipation. Conductive fillers can enhance the thermal conductivity of both polymeric and conventional textile materials. Methods for Measuring Thermal Conductivity. There are two techniques available for evaluating the thermal conductivity (TC) of materials: steady state and transient methods. These approaches vary in their operational mechanisms. Before commencing measurements, steady-state procedures necessitate the sample and reference components to reach thermal equilibrium. The prescribed technique is ASTM D7984 [66]. Apparel that offers optimal thermal comfort can offset the absorption or release of heat, maintaining a stable body temperature and allowing the user to feel content with the thermal conditions, without needing to rely on their thermoregulatory mechanisms.

13.7 Conclusion

Sustainable approaches to textile finishing involve processes that enhance textiles' visual appeal, tactile properties, and functional attributes while minimizing environmental impact and promoting responsible practices across the supply chain. One such approach is using natural dyes derived from plants, minerals, or other organic sources, which reduces reliance on synthetic chemicals and ensures easier biodegradability. Water conservation methods, such as low-water or waterless dyeing techniques, along with recycling and treating wastewater, are vital for reducing the ecological footprint of textile production [68]. Energy efficiency can be achieved by optimizing production processes, adopting energy-efficient equipment, and incorporating renewable energy sources like solar or wind power. Biodegradable finishes ensure textiles decompose naturally at the end of their lifecycle, reducing waste and pollution [69]. Using recycled materials, including post-consumer or post-industrial fabrics and fibers, enhances circularity and reduces dependency on virgin resources. Employing non-toxic chemicals for wrinkle resistance, flame retardancy, and stain resistance minimizes harm to consumers, workers, and the environment [70]. Closed-loop systems that recycle resources such as water and chemicals further reduce waste and environmental impact. Biological treatments, including enzymatic and microbial agents, provide eco-friendly alternatives to conventional chemical treatments [71]. Adherence to globally recognized certifications and standards, such as OEKO-TEX, Bluesign, and the Global Organic Textile Standard (GOTS), ensures sustainable practices throughout production [72]. Implementing transparency and traceability measures in the supply chain empowers consumers to make informed choices and encourages accountability among manufacturers and suppliers. These sustainable methods collectively contribute to mitigating the ecological impact of textile finishing while fostering environmentally responsible practices. It is crucial to evaluate textile finishing processes to guarantee product quality, functionality, and environmental friendliness across the entire textile supply chain. By integrating quality control protocols, performance evaluation, environmental impact assessment, and cost-effectiveness evaluation, manufacturers can optimize finishing methodologies, reduce their ecological imprint, and improve overall operational effectiveness. In the ever-changing textile sector, continuous evaluation and enhancement of finishing procedures will be indispensable for satisfying consumer expectations, complying with regulatory mandates, and achieving sustainability objectives.

Declaration The authors declare that they have no known competing financial interests or personal relationships that could have appeared to influence the work reported in this chapter. Generative AI and AI-assisted technologies have been used in the writing process to improve the readability and language of the work only. After using this tool, the authors carefully reviewed and revised the content to ensure accuracy and quality. The authors take full responsibility for the final version of this publication.

References

1. Becerra, Liliana (2016) CMF design: The fundamental principles of colour, material and finish design. Frame Publishers
2. Kılıç C, Atilla G (2024) Industry 4.0 and sustainable business models: An intercontinental sample. Bus Strateg Environ 33(4):3142–3166
3. Choudhury R, Kumar A (2013) Green chemistry and the textile industry. Text Prog 45(1):3–143
4. Curteza, Antonela (2011) Sustainable textiles. Radar. 2(1)
5. Brigden, Kevin, et al (2012) Organic chemical and heavy metal contaminants from communal wastewater treatment plants with links to textile manufacturing, and in river water impacted by wastewater from a textile dye manufacturing facility. China Greenpeace Res Lab Tech Rep 7
6. Selvaraj, Chandrabose, Sanjeev Kumar Singh (2021) Eco-friendly microbial biopolymers: recent development, biodegradation, and applications. Microb Polym: Appl Ecol Perspect. 547–577
7. Ciroth, Andreas, et al (2011) Towards a life cycle sustainability assessment: making informed choices on products
8. Cederfeldt, Sofia (2018) Achieving a Sustainable Fashion Industry through Global collaborations and standardized measurement tools: an exploratory case study on the sustainable apparel coalition
9. Niinimäki, Kirsi (2013) Sustainable fashion: New approaches. Aalto University
10. Vigo, Tyrone L (2013) Textile processing and properties: Preparation, dyeing, finishing and performance. Elsevier
11. Vezzetti E, Alemanni M, Macheda J (2015) Supporting product development in the textile industry through the use of a product lifecycle management approach: a preliminary set of guidelines. Int J Adv Manuf Technol 79:1493–1504
12. Zbicinski, Ireneusz (2006) Product design and life cycle assessment. 3. Baltic University Press
13. Islam, Md Mazedul, Patsy Perry, Simeon Gill (2021) Mapping environmentally sustainable practices in textiles, apparel and fashion industries: a systematic literature review. J Fash Mark Manag: Int J 25.2 (2021): 331–353
14. Pal R, Harper S, Vellesalu A (2018) Competitive manufacturing for reshoring textile and clothing supply chains to high-cost environment: A Delphi study. Int J Logist Manag 29(4):1147–1170
15. Al-Sayed, Wafaa, Shimaa H Abdelrahman (2021) Sustainable chemistry in textile processes (pretreatment, coloration and chemical finishing). Green Chemistry for Sustainable Textiles. Woodhead Publishing. 93–111
16. Islam, Md Touhidul, et al (2022) Synthetic dyes for textile colouration: process, factors and environmental impact. TextE Leather Rev. 5: 327–373
17. Ragab MM, Othman H, Hassabo A (2022) An overview of printing textile techniques. Egypt J Chem 65(8):749–761
18. Dash, Ashwini Kumar, Sanat Kumar Sahoo (2021) Role of enzymes in textile processing. Bioprospecting Enzym Ind, Healthc Sustain Environ. 395–410
19. Holme, Ian (2016) Coloration of technical textiles. Handb Tech Textes. Woodhead Publishing, 231–284

20. Bullon, Juan, et al (2017) Manufacturing processes in the textile industry. Expert Syst Fabr Prod
21. Holme I (2007) Innovative technologies for high performance textiles. Color Technol 123(2) 59–73
22. Amutha K (2017) Sustainable practices in textile industry: standards and certificates. Sustain Text Ind. 79–107
23. Paul R (2015) Functional finishes for textiles: an overview. Funct Finish TextEs, Improv Comf, Perform Prot 1–14
24. Stukenbrock K-H (2002) Mechanical finishing. Nonwoven Fabr: Raw Mater, Manuf, Appl, Charact, Test Process 411–420
25. Scheirs, John (2009) A guide to polymeric geomembranes: a practical approach. John Wiley & Sons
26. Morlok, Franziska, Till Beckmann (2013) Extra: encyclopaedia of experimental print finishing. Walter de Gruyter
27. Peel JD (1992) Calendering and embossing. Dictionary of Converting. Dordrecht: Springer Netherlands, 1–27
28. Chowdhury, Muksit Ahamed, Pintu Pandit (2022) Chemical processing of knitted fabrics. Adv Knitt Technol. Woodhead Publishing, 503–536
29. Kangle PJ (1956) Mechanical processes of shrinkage control of cotton textiles. Bombay Technol 6:52–60
30. Shaw, Trevor, et al (2018) The chemical technology of wool finishing. Handb Fiber Sci Technol 2. Routledge, 317–442
31. Rovero, Giorgio, Claudio Tonin, Silvio Sicardi (2019) Study and scientific rationalization of the last finishing stages for high quality wool fabrics
32. Karthik T, Ganesan P (2016) And D. CRC Press, Gopalakrishnan. Apparel manufacturing technology
33. Read AI (1951) Finishing of face cloths, pile fabrics, etc. Journal of the textile institute proceedings 42(5):P205–P208
34. Kayseri, Gonca Özçelik, Nilgün Özdil, Gamze Süpüren Mengüç (2012) Sensorial comfort of textile materials. Woven Fabr. 235–266
35. Choudhury, Asim Kumar Roy (2017) Principles of textile finishing. Woodhead Publishing
36. Johnston, Amanda, Clive Hallett (2014) Fabric for fashion: the complete guide: natural and Man-made Fibres. Hachette UK
37. Achaw, Osei-Wusu, Eric Danso-Boateng (2021) Textile and fabric manufacture. Chemical and process industries: with examples of industries in Ghana. Cham: Springer International Publishing. 67–105
38. Shamey, Renzo, T. Hussein (2005) Critical solutions in the dyeing of cotton textile materials. TextE Prog 37, 1–2 1–84
39. Basyigit, Zeynep Omerogullari (2021) Application technologies for functional finishing of textile materials. Texts Funct Appl 101
40. Venkatraman, Praburaj (2015) Fabric properties and their characteristics. Mater Technol Sport Perform Appar 53–86
41. Nuessle AC, Fineman MN, Heiges EOJ (1955) Some controversial aspects of crease resistance of cellulosic fabrics. Text Res J 25(1):24–40
42. Pervez, Md Nahid, et al (2018) Multi-response optimization of resin finishing by using a taguchi-based grey relational analysis. Mater 11.3, 426
43. Tsapko, Yuriy, et al (2022) Application of coating for fire protection of textile structures. Key Eng Mater 927 115–121
44. Nguyen, Thach-Mien, et al (2014) Fire self-extinguishing cotton fabric: development of piperazine derivatives containing phosphorous-sulfur-nitrogen and their flame retardant and thermal behaviors. Mater Sci Appl 5.11, 789
45. Rajendran, R (2010) Biotechnological application in textile Industry—Antimicrobial textiles. Ind Exploit Microorg 129

46. Tang K-P, Kan C-W, Fan J-t (2017) Comparison of test methods for measuring water absorption and transport test methods of fabrics. Meas 97:126–137
47. Özek, Hikmet Ziya (2018) Silicone-based water repellents. Waterproof Water Repel TextEs Cloth. Woodhead Publishing, 153–189
48. Gururajan A, Hequet EF, Sari-Sarraf H (2008) Objective evaluation of soil release in fabrics. Text Res J 78(9), 782–795
49. Loghin, Carmen, et al (2018) Introduction to waterproof and water repellent textiles. Waterproof Water Repel TextEs Cloth. Woodhead Publishing, 3–24
50. Schindler, Wolfgang D, Peter J Hauser (2004) Chemical finishing of textiles. Elsevier
51. Kamel, Mariam Yasser, Ahmed G Hassabo (2021) Anti-microbial finishing for natural textile fabrics. J TextEs, Color Polym Sci 18.2, 83–95
52. Krishnaveni V (2013) Bio shield AM 500 finish on cotton. Melliand Int 19.2
53. Xin JH, Wang XW (2018) Insect-repellent textiles. Engineering of High-Performance Textiles. Woodhead Publishing, 335–348
54. Diaz JH (2016) Chemical and plant-based insect repellents: efficacy, safety, and toxicity. Wilderness & Environ Med 27(1):153–163
55. Mittal PK et al (2011) Efficacy of Advanced Odomos repellent cream (N, N-diethyl-benzamide) against mosquito vectors. Indian J Med Res 133(4):426–430
56. Islam, Mohammad Tajul, Syed Asaduzzaman (2019) Environmentally-friendly textile finishing. Texts Cloth, 101–129
57. Mohsin, Muhammad, et al (2017) Textile dyed and finished fabric. Adv TextE Test Tech. CRC Press, 189–207
58. Luo, Hangyu et al (2022) Research progress on lignocellulosic biomass degradation catalyzed by enzymatic nanomaterials. Chemistry–An Asian J 17.18, e202200566
59. Sanches, Regina Aparecida et al (2014) Wearbility Analysis of knited fabrics produced with colored organic cotton. Bamboo rayon, corn, recycled pet/cotton and recycled pet/polyester. Am Int J Contemp Res 4, 28–37
60. Abdullah, Iram, et al (2006) Abrasion phenomena in twill tencel fabric. J Appl Polym Sci 102.2, 1391–1398
61. Hakami, Ferial, Alokesh Pramanik, Animesh Kumar Basak (2022) Materials and methodology. Tribol Elastomers. Singapore: Springer Nature Singapore, 53–60
62. Zhang, Mingwen, et al (2017) Synthesis of TiO2&SiO2 nanoparticles as efficient UV absorbers and their application on wool. TextE Res J 87.14, 1784–1792
63. Becheri, Alessio, et al (2008) Synthesis and characterization of zinc oxide nanoparticles: application to textiles as UV-absorbers. J Nanoparticle Res 10, 679–689
64. Hatch, Kathryn L (2002) American standards for UV-protective textiles. Cancers of the Skin: proceedings of the 8th world congress. Berlin, Heidelberg: Springer Berlin Heidelberg
65. Fan L, Khodadadi JM (2011) Thermal conductivity enhancement of phase change materials for thermal energy storage: a review. Renew Sustain Energy Rev 15(1), 24–46
66. Islam S, Mominul SM, Alam, (2018) Investigation of the acoustic properties of needle punched nonwoven produced of blend with sustainable fibers. Int J Cloth Sci Technol 30(3), 444–458
67. Rahaman, Md Tareque, et al (2024) Green production and consumption of textiles and apparel: importance, fabrication, challenges and future prospects. J Open Innov: Technol, Mark, Complex, 100280
68. Hussain T, Wahab A (2018) A critical review of the current water conservation practices in textile wet processing. J Clean Prod 198, 806–819
69. Radhakrishnan, Shanthi (2022) Sustainable development goal: sustainable management and use of natural resources in textile and apparel industry. Sustainable approaches in textiles and fashion: circular economy and microplastic pollution. Singapore: Springer Singapore, 165–205
70. Sfameni, Silvia, et al (2023) Inorganic finishing for textile fabrics: Recent advances in wear-resistant, UV protection and antimicrobial treatments. Inorganics 11.1, 19
71. Sarayu K, Sandhya S (2012) Current technologies for biological treatment of textile wastewater–a review. Appl Biochem Biotechnol 167, 645–661
72. Standard, Global Organic Textile. Global Organic Textile Standard. Recuperado el 27 (2008)

Chapter 14
Sustainability in Textile Finishing: An Introduction to Current Challenges and Innovations

Mohammad Shahid and Mohd Yusuf

Abstract The textile industry is among the most environmentally damaging due to its intensive water, energy, and chemical usage, especially during the finishing stages. Traditional textile finishing processes rely on hazardous chemicals and energy-intensive methods, contributing to significant pollution, resource depletion, and health risks. This paper explores the critical sustainability issues in textile finishing, including water consumption, chemical pollution, and energy intensity, and highlights recent advancements in sustainable alternatives. Innovative technologies such as enzymatic treatments, nanotechnology, irradiation-based methods, and bio-based materials offer promising solutions for reducing environmental impacts while maintaining product quality. Techniques like bioscouring, enzymatic desizing, biobleaching, and advanced water-recycling systems such as zero-liquid discharge (ZLD) demonstrate practical and scalable approaches for a greener industry. The adoption of renewable energy sources and efficient machinery further supports sustainability goals. By integrating these advancements, the textile industry can mitigate its ecological footprint, align with regulatory demands, and meet growing consumer expectations for sustainable products.

Keywords Sustainability · Enzymatic treatments · Textile finishing · Nanotechnology in textiles · Environmental impact · Resource efficiency

M. Shahid
Department of Chemistry, Koneru Lakshmaiah Education Foundation, Guntur, Andhra Pradesh, India
e-mail: mshahid96@gmail.com

M. Yusuf (✉)
Department of Natural and Applied Sciences, School of Science and Technology, The Glocal University, Saharanpur, Uttar Pradesh, India
e-mail: yusuf1020@gmail.com

S. Biranje et al. (eds.), *Sustainable Finishing Techniques in Textiles*,
https://doi.org/10.1007/978-981-96-4860-3_14

14.1 Introduction

The global population is projected to reach 8.6 billion by 2030 and 9.8 billion by 2050 [1]. The growing global population drives higher demand and consumption, leading to greater resource depletion and environmental stress. Sustainable development refers to meeting our present needs in a way that does not compromise the ability of future generations to meet their own. A holistic concept of sustainability interweaves three key dimensions: economic, environmental, and social sustainability [2]. However, the environmental dimension of sustainable development has received particular attention in the textile industry [3]. Environmental sustainability, a key consideration in product decisions for textiles and apparel companies, balances economic productivity with environmental impact through the preservation and responsible management of natural resources, pollution reduction, and climate change mitigation—factors with significant implications for both current and future generations [4]. Textile manufacturing is one of the world's largest industrial sectors, with enormous environmental and socio-economic consequences [5]. Sustainability has long been a pressing concern in the textile industry, which operates globally with diverse production methods, raw materials, and environmental regulations. Due to these variations, there is no universal solution to achieving sustainability. However, textile manufacturers can make significant progress by adopting best practices, leveraging existing sustainable solutions, making informed material and process choices, and integrating emerging environmentally friendly technologies.

The textile and apparel industry is a key sector in consumer goods and one of the most environmentally harmful industries. It consumes much energy and water, particularly in fabric and garment manufacturing. Untreated wastewater from wet processing contains toxic chemicals that can cause serious environmental harm, impacting air quality, water sources, and the fabric itself [6]. The textile industry encompasses the production of fibers, yarns, fabrics, clothing, and goods for various purposes, spanning raw materials to finished products across its diverse and complex manufacturing sub-sectors. Textile products generate waste and environmental impacts throughout their life cycle, from fiber and yarn production to dyeing, finishing, manufacturing, distribution, use, and disposal [7]. Each production stage contributes to environmental degradation by consuming water, materials, chemicals, and energy [8]. Textile finishing, the final step in the production chain of textile products, involves various chemical and mechanical processes to enhance textiles' appearance, texture, and performance. These finishing treatments are essential to the textile value chain and play a critical role in textile products' end-use quality and durability. Traditional textile finishing processes heavily depend on synthetic chemicals, many of which have adverse environmental and health effects [9]. These chemicals often end up in wastewater streams, contaminating ecosystems and posing risks to aquatic life and human health. In addition, many traditional finishes require large amounts of water and energy, contributing to the overall environmental burden of textile manufacturing. The growing environmental footprint of the textile industry, combined with increasing awareness and regulatory requirements around the globe, has necessitated the search for sustainable alternatives to the highly polluting traditional industrial practices. Continuous efforts are being made to find substitutes at both material and technological levels to reduce harmful chemicals, minimize water and energy consumption, and decrease waste generation without compromising the functionality and quality of the finished textile products [10]. The industry is researching and implementing innovative finishing technologies with lower ecological footprints to address this issue. While the textile industry encounters considerable sustainability challenges and opportunities, research gaps remain in understanding managerial approaches and perceptions regarding adopting sustainable practices [11]. Addressing these gaps presents valuable opportunities to advance the field. By continuously innovating and embracing sustainable solutions, the textile industry can move toward more responsible production practices, reducing its environmental impact while meeting the growing consumer demand for eco-friendly products.

14.2 Sustainability Issues in Textile Finishing: Water, Chemicals, Energy

The textile industry is among the most environmentally damaging sectors, with its significant ecological footprint stemming from high energy and water consumption, extensive chemical usage, textile waste generation, and the release of microfibres into the environment during production and use [12]. The industry annually accounts for 8–10% of global emissions and 20% of wastewater pollution. It is the second-largest industrial water polluter, using high loads of toxic chemicals that generate large amounts of sludge and wastewater [13].

14.2.1 Water Usage

Textile manufacturing practices, particularly in poorer and emerging countries with the highest global production, have seen little evolution despite technological advancements, leading to severe environmental impacts such as pollution and deforestation [13]. The textile industry's significant water footprint is a major sustainability concern, driven by its water-intensive processes and the large volumes of polluted wastewater it produces, prompting a need to reassess and reduce water use and wastewater impact due to increasing consumer awareness, stricter environmental regulations, and global water scarcity [14, 15]. Textile industry wastewater typically exhibits a high chemical load influenced by the number of processing steps, application methods, and features such as varying pH levels, Biochemical Oxygen Demand (BOD), Chemical Oxygen Demand (COD), Total Organic Content (TOC), Total

Dissolved Solids (TDS), heavy metals, and elevated electrical conductivity due to the presence of electrolytes [16]. Despite significant water consumption and pollution in the textile finishing sector, the implementation of water reuse remains limited, with challenges including high treatment costs, complex wastewater characteristics, and resource constraints [17].

14.2.2 Chemical Pollution

While dyeing is frequently emphasized as a major source of chemical pollution in textile manufacturing, finishing processes also involve substantial chemical usage, leading to significant pollution, especially through wastewater discharge and its chemical load. If not properly managed, the specialty chemicals and auxiliaries used in finishing processes can have considerable environmental and health impacts. The wastewater produced during textile finishing contains a variety of pollutants, including surfactants, solvents, heavy metals, and other toxic chemicals that can severely impact air and water quality. Toxic chemicals are released during various stages of the textile finishing chain, posing great health and environmental risks. Mechanical, wet, and thermal pre-treatment are required to eliminate natural and added impurities from textile raw materials before downstream processing. Mechanical pre-treatment has minimal environmental impact compared to wet and thermal processes. Airborne emissions from textile finishing encompass both captured emissions released through stacks and uncaptured fugitive emissions, with the primary sources of captured waste gases originating from processes such as singeing, heat setting (particularly in the finishing of man-made fibers), drying, printing, chemical and mechanical finishing (including dust generation in shearing and raising), coating, laminating, and energy production in boiler houses [18]. The primary environmental issues arising from wet textile pre-treatment processes include high water consumption, significant chemical pollution from chemicals (ammonium persulfate, hydrogen peroxide, sodium hydroxide, chlorine-based compounds, and peracetic acid, etc.) used in desizing, scouring, and bleaching, potential air emissions from chemical reactions, and generation of the hazardous waste which can contaminate water sources if not properly treated [19–21]. Perfluoroalkyl or polyfluoroalkyl substances (PFAS) are widely used in durable water-repellent treatments in the textile industry due to their water- and oil-repellent properties. Still, their persistence in the environment and potential health risks, including hormone disruption and cancer, have raised significant concerns, leading to efforts to phase out their use [22]. Halogenated flame retardants release persistent organic pollutants, VOCs from flame-retardant finishes contribute to indoor air pollution, and formaldehyde emissions from cross-linking agents cause respiratory irritation and carcinogenic risks [23]. The release of formaldehyde, a known human carcinogen, from traditional easy-care and wrinkle-resistant finishes can occur during manufacturing, wear, and washing, posing health risks [24]. Anti-bacterial and anti-fungal agents are widely used to treat textiles, especially in sportswear, medical textiles, and home textiles. Many agents, such as triclosan and silver nanoparticles, are persistent in the environment and can disrupt ecosystems [25].

14.2.3 Energy Intensity

Energy is crucial throughout every stage of the textile production chain, making it one of the most significant energy consumers among industries [26]. Energy sources for the textile industry are primarily electricity and thermal energy, largely derived from non-renewable resources such as petroleum, natural gas, coal, and hydrocarbon gas liquids or supplied by government entities [27]. Energy is a significant cost factor in the textile industry, with energy costs accounting for 5% of manufacturing expenses in Brazil and 12% in India and potentially even higher in countries with elevated energy costs, making energy efficiency a critical priority, particularly during periods of energy price volatility. The textile sector has limited scientific studies addressing its energy challenges, though notable research highlights energy-efficiency measures, savings, and policies in countries like Turkey, Germany, Colombia, Taiwan, and India. Additionally, various reports, such as those from the Carbon Trust, CIPEC, and Japan's Energy Conservation Center, provide comprehensive guidance on energy-saving technologies and best practices for the industry [28].

14.3 Recent Developments in Sustainable Finishing—An Overview

To address these challenges, sustainability in textile finishing has gained significant traction in recent years. Sustainable textile finishing aims to minimize the negative environmental and social impacts associated with traditional textile finishing processes while maintaining or improving the quality and performance of the final product. This involves adopting cleaner technologies, renewable materials, and innovative processes that use fewer resources and generate less pollution. Recent research has introduced several innovative technologies for textile pre-treatment processes, focusing on improving efficiency, reducing environmental impact, and enhancing sustainability. These advancements address traditional issues such as excessive water, energy, and chemical usage while maintaining or enhancing fabric quality. The textile industry has adopted many of these sustainable methods, which are included in the best available techniques reference document [18].

Enzymatic pre-treatment processes such as bioscouring [29], enzymatic desizing [30], biopolishing [31], biobleaching [32], enzymatic degumming [33], etc., have been proposed as alternatives to conventional chemical pre-treatments in textile wet processing. State-of-the-art industrial enzymes in textile processing include amylases for cotton desizing and cellulases for denim washing and biopolishing, while emerging enzymatic processes, such as cotton scouring and bleaching, bast fiber retting/degumming, wool scouring and shrink-proofing, silk degumming, bio-dye production, and enzyme-assisted dyeing and finishing, are at various stages of development [34, 35].

Nanotechnology in textiles, driven by the growing demand for enhanced appearance, functionality, and connectivity in fashion, incorporates various nanomaterials such as nanoparticles, carbon nanotubes, and nanoelectronics through methods like particle impregnation, spray coating, and direct weaving, enabling high-performance applications across industries like healthcare, military, sports, fashion, and wearable electronics, with features like antibacterial properties, odor control, UV protection, water repellence, wrinkle resistance, strength enhancement, and integration of sensors, drug release systems, and fiber optics, all while maintaining lightweight, flexible, and seamless designs [36, 37]. Nanotechnology has significantly advanced textile finishing processes, offering sustainable alternatives to traditional methods. These nanofabricated finishes reduce the need for chemical treatments and frequent laundering, thereby conserving water and energy. For instance, applying silver nanoparticles imparts antimicrobial properties to textiles, reducing the necessity for chemical disinfectants and extending the lifespan of garments [38].

Different irradiation-based technologies [39] such as ultrasound [40], plasma [41], gamma [42], ultraviolet [43], and microwave technology [44] are advanced methods used in textile processing to improve surface modification, reduce chemical and water consumption, enhance cleaning, scouring, and bleaching, and shorten processing times for finishing operations and lowering energy requirements.

Various low-liquor ratio finishing processes are tested and used with the benefits of water savings, energy savings, and minimal liquid discharge [45]. Low-liquor ratio machines, such as jet dyeing and continuous washing machines, are designed to process fabrics efficiently by minimizing water and chemical usage [46].

Intensive efforts are underway to find eco-sustainable alternatives to traditionally used textile finishing agents, addressing each step of the finishing process. Table 14.1 gives some common examples of textile finishing steps and sustainable alternatives for traditionally used chemicals.

Efforts in textile finishing waste management and wastewater remediation are also becoming essential in advancing sustainability within the textile industry. Innovative methods for treating textile wastewater have been developed to address environmental impacts, including integrating advanced oxidation processes with membrane reactors and bioreactors for various wastewater treatment applications [60]. Membrane technologies, including reverse osmosis and nanofiltration, are increasingly employed for water recovery and reuse in textile finishing plants [61]. Bioremediation using microbial enzymes and bio-adsorbents has also shown promise in breaking down harmful substances and removing heavy metals from textile effluents [62, 63]. Zero-liquid discharge (ZLD) technology is gaining widespread attention in the textile industry for its ability to optimize water recycling, minimize wastewater volumes, and ensure complete treatment, recycling, and reuse of water, effectively eliminating the release of contaminants into the environment [64]. These advancements improve

Table 14.1 Some common examples of eco-friendly substitutes for the chemicals traditionally used in these finishing steps

Finishing step	Traditional chemicals/methods	Sustainable alternatives	Refs.
Desizing	Synthetic detergents, oxidizing agents (e.g., hydrogen peroxide)	Enzymatic desizing (e.g., amylases)	[47]
Scouring	Sodium hydroxide (caustic soda)	Bioscouring with pectinase enzymes	[29]
Bleaching	Chlorine-based bleaches, hydrogen peroxide	Biobleaching with laccase or peroxidase enzymes	[48]
Wrinkle resistance	Formaldehyde-based crosslinkers	Non-formaldehyde resins, citric acid, or succinic acid esters	[49]
Water-repellent finishes	Perfluoroalkyl substances (PFAS)	Bio-based repellents (e.g., beeswax, castor oil, PLA)	[50]
Flame retardant finishes	Halogenated retardants, organophosphorus compounds	Ammonium polyphosphate, lignin-based retardants	[51]
Antimicrobial finishes	Triclosan, heavy-metal-based agents (e.g., silver, copper)	Natural antimicrobials (e.g., essential oils, bio-synthesized nanoparticles)	[52]
UV protection	Synthetic UV absorbers	Zinc oxide or titanium dioxide nanoparticles	[53, 54]
Softening	Silicone- and petroleum-based softeners	Bio-based softeners (e.g., soybean or palm oil derivatives)	[55]
Odor control	Synthetic odor absorbers	Activated charcoal, finishes	[56]
Anti-shrinkage treatment	Chlorine-based treatments, synthetic polymers	Enzymatic shrink-proofing (proteases), bio-based polymer coatings	[57]
Stonewashing	Pumice stones, synthetic chemicals	Enzymatic stonewashing (cellulases), laser treatments	[58]
Insect repellent finishes	N,N-Diethyl-meta-toluamide, picaridin	Citriodiol, essential oils	[59]

the sustainability of textile finishing processes and align with global efforts to protect water resources and reduce pollution.

14.4 Conclusion

The textile industry faces significant sustainability challenges, particularly in the finishing stages that rely heavily on water, energy, and hazardous chemicals. Traditional processes, such as desizing, scouring, bleaching, fixation, and antimicrobial treatments, often involve synthetic agents and energy-intensive methods, contributing to pollution and resource depletion. Innovative, sustainable alternatives are transforming these practices, offering environmentally friendly solutions without compromising quality or performance. Enzymatic treatments like bioscouring, enzymatic desizing, and biobleaching provide effective, biodegradable replacements for harmful chemicals. Similarly, plant-derived and bio-based materials, including chitosan, natural waxes, and cellulose derivatives, are employed for water repellency, softening, and stiffening processes. Advanced technologies such as nanotechnology, irradiation techniques, and laser treatments enable more efficient finishing with reduced chemical and water usage. Sustainable approaches to wrinkle resistance, flame retardancy, and anti-shrinkage treatments now use bio-based or non-toxic alternatives, addressing both environmental and health concerns. Using renewable energy sources and low-energy machinery further supports energy efficiency in textile finishing. These advancements demonstrate the industry's commitment to adopting greener solutions. By transitioning to sustainable materials and innovative technologies, the textile sector can significantly reduce its ecological footprint, improve resource efficiency, and respond to increasing consumer demand for environmentally responsible products.

Declaration AI tools were used in the writing process only to improve the readability and language of the work. After using this tool, the authors carefully reviewed and revised the content to ensure accuracy and quality. The authors take full responsibility for the content in the final version of this publication.

References

1. https://www.un.org/en/desa/world-population-projected-reach-98-billion-2050-and-112-billion-2100
2. Purvis B, Mao Y, Robinson D (2019) Three pillars of sustainability: in search of conceptual origins. Sustain Sci 14:681–695
3. Shamsuzzaman M, Islam MM, Hasan HMRU, Khan AM, Sayem ASM (2023) Mapping environmental sustainability of knitted textile production facilities. J Clean Prod 405:136900
4. Luo Y, Song K, Ding X, Wu X (2021) Environmental sustainability of textiles and apparel: a review of evaluation methods. Environ Impact Assess Rev 86:106497
5. Farhana K, Kadirgama K, Mahamude ASF, Mica MT (2022) Energy consumption, environmental impact, and implementation of renewable energy resources in global textile industries: an overview towards circularity and sustainability. Mater Circ Econ 4:15
6. Jia F, Yin S, Chen L, Chen X (2020) The circular economy in the textile and apparel industry: a systematic literature review. J Clean Prod 259:120728
7. Schmutz M, Som C (2022) Identifying the potential for circularity of industrial textile waste generated within Swiss companies. Resour Conserv Recycl 182:106132

8. Niinimäki K et al (2020) The environmental price of fast fashion. Nat Rev Earth Environ 1:189–200
9. Madhav S, Ahamad A, Singh P, Mishra PK (2018) A review of textile industry: wet processing, environmental impacts, and effluent treatment methods. Environ Qual Manag 27:31–41
10. Hasanbeigi A, Price L (2015) A technical review of emerging technologies for energy and water efficiency and pollution reduction in the textile industry. J Clean Prod 95:30–44
11. Desore A, Narula SA (2018) An overview on corporate response towards sustainability issues in textile industry. Environ Dev Sustain 20:1439–1459
12. Leal Filho W et al (2022) An overview of the contribution of the textiles sector to climate change. Front Environ Sci 10:973102
13. De Oliveira CRS, Da Silva Júnior AH, Mulinari J, Immich APS (2021) Textile re-engineering: eco-responsible solutions for a more sustainable industry. Sustain Prod Consum 28:1232–1248
14. Mikucioniene D et al (2024) Understanding and addressing the water footprint in the textile sector: a review. AUTEX Res J 24:20240004
15. Haque MdS, Nahar N, Sayem SM (2021) Industrial water management and sustainability: development of SIWP tool for textile industries of Bangladesh. Water Resour. Ind. 25:100145
16. Uddin MA et al (2023) Water and chemical consumption in the textile processing industry of Bangladesh. PLOS Sustain Transform 2:e0000072
17. Vajnhandl S, Valh JV (2014) The status of water reuse in European textile sector. J Environ Manag 141:29–35
18. European Commission, Joint Research Centre (2023) Best available techniques (BAT) reference document for the textiles industry: industrial emissions directive 2010/75/EU (integrated pollution prevention and control). Publications Office, LU
19. Panda SKBC, Sen K, Mukhopadhyay S (2021) Sustainable pretreatments in textile wet processing. J Clean Prod 329:129725
20. Dos Santos TF, Da Silva Santos CM, Do Nascimento JHO, Paul R (2024) Innovations in textile pretreatments. In: Sustainable innovations in the textile industry. Elsevier, pp 315–343. https://doi.org/10.1016/B978-0-323-90392-9.00010-0
21. Han F, Lang C, Qiu Y (2022) Cleaning behavior based on waste textile before resource conversion and its challenges in China—a comprehensive review. J Clean Prod 370:133509
22. Holmquist H et al (2016) Properties, performance and associated hazards of state-of-the-art durable water repellent (DWR) chemistry for textile finishing. Environ Int 91:251–264
23. Horrocks AR (2011) Flame retardant challenges for textiles and fibres: new chemistry versus innovatory solutions. Polym Degrad Stab 96:377–392
24. Herrero M et al (2022) Early-life exposure to formaldehyde through clothing. Toxics 10:361
25. Windler L, Height M, Nowack B (2013) Comparative evaluation of antimicrobials for textile applications. Environ Int 53:62–73
26. Ozturk E, Cinperi NC, Kitis M (2020) Improving energy efficiency using the most appropriate techniques in an integrated woolen textile facility. J Clean Prod 254:120145
27. Kousar S et al (2022) Sustainable energy consumption model for textile industry using fully intuitionistic fuzzy optimization approach. Comput Intell Neurosci 2022:5724825
28. Hasanbeigi A, Hasanabadi A, Abdorrazaghi M (2012) Comparison analysis of energy intensity for five major sub-sectors of the textile industry in Iran. J Clean Prod 23:186–194
29. Niaz A, Malik QJ, Muhammad S, Shamim T, Asghar S (2011) Bioscouring of cellulosic textiles. Color Technol 127:211–216
30. Zhang X et al (2022) Effects of exogenous proteins on enzyme desizing of starch and its mechanism. Int J Biol Macromol 218:375–383
31. Prathaban M et al (2025) Efficient synthesis, purification and characterization of novel carboxymethyl cellulase from newly isolated Streptomyces sp. CSMPJR101, with potential use in denim biopolishing. Bioresour Technol Rep 29:102035
32. Unuofin JO, Moloantoa KM, Khetsha ZP (2022) The biobleaching potential of laccase produced from mandarin peelings: impetus for a circular bio-based economy in textile biofinishing. Arab J Chem 15:104305

33. Subash MC, Muthiah P (2021) Eco-friendly degumming of natural fibers for textile applications: a comprehensive review. Clean Eng Technol 5:100304
34. Shahid M, Mohammad F, Chen G, Tang R-C, Xing T (2016) Enzymatic processing of natural fibres: white biotechnology for sustainable development. Green Chem 18:2256–2281
35. Tochetto GA, Aragão AMI, De Oliveira D, Immich APS (2022) Can enzymatic processes transform textile processes? A critical analysis of the industrial application. Process Biochem 123:27–35
36. Yetisen AK et al (2016) Nanotechnology in textiles. ACS Nano 10:3042–3068
37. Syduzzaman M, Hassan A, Anik HR, Akter M, Islam MdR (2023) Nanotechnology for high-performance textiles: a promising frontier for innovation. ChemNanoMat 9:e202300205
38. Saleem H, Zaidi S (2020) Sustainable use of nanomaterials in textiles and their environmental impact. Materials 13:5134
39. Shahid-ul-Islam, Mohammad F (2015) High-energy radiation induced sustainable coloration and functional finishing of textile materials. Ind Eng Chem Res 54:3727–3745
40. Vouters M, Rumeau P, Tierce P, Costes S (2004) Ultrasounds: an industrial solution to optimise costs, environmental requests and quality for textile finishing. Ultrason Sonochem 11:33–38
41. Zille A, Oliveira FR, Souto AP (2015) Plasma treatment in textile industry: plasma treatment in textile industry. Plasma Process Polym 12:98–131
42. Khafaga MR, Ali HE, El-Naggar AWM (2016) Antimicrobial finishing of cotton fabrics based on gamma irradiated carboxymethyl cellulose/poly(vinyl alcohol)/TiO_2 nanocomposites. J Text Inst 107:766–773
43. Unal F, Avinc O, Yavas A, Eren HA, Eren S (2020) Contribution of UV technology to sustainable textile production and design. In: Muthu SS, Gardetti MA (eds) Sustainability in the textile and apparel industries. Springer International Publishing, Cham, pp 163–187. https://doi.org/10.1007/978-3-030-37929-2_8
44. Yu S et al (2021) Facile microwave assisted flame retardant treatment for cotton fabric using a biobased industrial byproduct: phytic acid. Cellulose 28:10655–10674
45. Mohsin M et al (2022) Water efficient, eco-friendly and effluent free nano bubble finishing of cotton fabric. J Nat Fibers 19:12586–12595
46. Juraev Z, Sattarov M, Paul R (2024) Innovations in textile wet processing machinery. In: Sustainable innovations in the textile industry. Elsevier, pp 285–314. https://doi.org/10.1016/B978-0-323-90392-9.00018-5
47. Ibrahim NA, El-Hossamy M, Morsy MS, Eid BM (2004) Optimization and modification of enzymatic desizing of starch-size. Polym-Plast Technol Eng 43:519–538
48. Sawada K, Tokino S, Ueda M, Wang XY (1998) Bioscouring of cotton with pectinase enzyme. J Soc Dye Colour 114:333–336
49. Ji B, Sun G (2020) Advances in durable press finishing of fabrics with formaldehyde-free agents. In: Advances in functional and protective textiles. Elsevier, pp 81–116. https://doi.org/10.1016/B978-0-12-820257-9.00005-9
50. Shahid M, Maiti S, Adivarekar RV, Liu S (2022) Biomaterial based fabrication of superhydrophobic textiles—a review. Mater Today Chem 24:100940
51. Liu J et al (2024) Durable flame-retardant, anti-UV, photothermal, and antibacterial cotton fabrics based on lignin and ammonium polyphosphate. ACS Sustain Chem Eng 12:12670–12682
52. Bibi A et al (2024) Synthetic *vs.* natural antimicrobial agents for safer textiles: a comparative review. RSC Adv 14:30688–30706
53. Montazer M, Pakdel E (2011) Functionality of nano titanium dioxide on textiles with future aspects: focus on wool. J Photochem Photobiol C Photochem Rev 12:293–303
54. Verbič A, Gorjanc M, Simončič B (2019) Zinc oxide for functional textile coatings: recent advances. Coatings 9:550
55. Hayes DG, Smith GA (2019) Biobased surfactants: overview and industrial state of the art. In: Biobased surfactants. Elsevier, pp 3–38. https://doi.org/10.1016/B978-0-12-812705-6.00001-0

56. McQueen RH, Vaezafshar S (2020) Odor in textiles: a review of evaluation methods, fabric characteristics, and odor control technologies. Text Res J 90:1157–1173
57. Hassan MM, Carr CM (2019) A review of the sustainable methods in imparting shrink resistance to wool fabrics. J Adv Res 18:39–60
58. Shahid M, Zhou Y, Tang R-C, Chen G (2017) Enzymatic washing of Denim: greener route for modern fashion. In: Muthu SS (ed) Textiles and clothing sustainability. Springer Singapore, Singapore, pp 67–83. https://doi.org/10.1007/978-981-10-2474-0_3
59. Chatha SAS et al (2019) Environmentally responsive and anti-bugs textile finishes—recent trends, challenges, and future perspectives. Sci Total Environ 690:667–682
60. Sathya U, Keerthi, Nithya M, Balasubramanian N (2019) Evaluation of advanced oxidation processes (AOPs) integrated membrane bioreactor (MBR) for the real textile wastewater treatment. J Environ Manag 246:768–775
61. Amar NB, Kechaou N, Palmeri J, Deratani A, Sghaier A (2009) Comparison of tertiary treatment by nanofiltration and reverse osmosis for water reuse in denim textile industry. J Hazard Mater 170:111–117
62. Sudarshan S et al (2023) Impact of textile dyes on human health and bioremediation of textile industry effluent using microorganisms: current status and future prospects. J Appl Microbiol 134:lxac064
63. Reddy S, Osborne J (2022) An insight on the advancements of biological technologies in the bioremediation of textile effluents. Urban Water J. 19:468–480
64. Pundir A et al (2024) Innovations in textile wastewater management: a review of zero liquid discharge technology. Environ Sci Pollut Res 31:12597–12616

The manufacturer's authorised representative in the EU is Springer
Nature Customer Service Centre GmbH, Europaplatz 3, 69115 Heidelberg,
Germany. If you have any concerns regarding our products, please
contact ProductSafety@springernature.com

Printed and bound by CPI Group (UK) Ltd, Croydon, CR0 4YY
01/10/2026
02225850-0010